Towards a Theory of Rationality in Science

A Plea for Reasonableness

Towards a Theory of Rationality in Science

A Plea for Reasonableness

S. Stephen Jayard

GLOBAL VISION PUBLISHING HOUSE
20, Ansari Road, Daryaganj, New Delhi-110002 (INDIA)

GLOBAL VISION PUBLISHING HOUSE
F-4, 1st Floor, 'Hari Sadan' 20, Ansari Road,
Daryaganj, New Delhi-110002 (INDIA)
Tel.: 011-23261581, 23276291, Mob: 9810644769
Email: nsingh_2004@vsnl.net, info@globalvision.pub.com
Website: www.globalvisionpub.com

Towards a Theory of Rationality in Science

First Edition 2012

ISBN: 978-81-8220-453-9

PRINTED IN INDIA

Published by Dr. N.K. Singh for Global Vision Publishing House, New Delhi-2 and *Printed at* G.S. Offset, Naveen Shahdara, Delhi-32.

They will get it straight one day at the Sorborne.
We shall return at twilight from the lecture,
Pleased that the *irrational is rational*

(Wallace Stevens, *Notes toward a Supreme Fiction*, X, II. 16-18),
as quoted by Toulmin, 2001, p. xii

Dedication

A thought-provoking Advertisement by the *Child Relief and You* (CRY) Foundation on the Television…

> **If you are not able to read this because**
>
> **it is written upside down, don't feel bad.**
>
> **You are more for fortunate than millions**
>
> **in the world; those are the illiterate children**
>
> **and they will not even know that something**
>
> **is written in a language; to them it will look**
>
> **like only a picture of lines and curves!**

My heart goes out to think of such unfortunate children in millions,

Who never get a chance of stepping into a school in their lifetime.

They are the beautiful flowers that go unnoticed, and their

Great potentialities sadly go unrealized!

Let us, who are fortunate to be educated,

Put our heads and hearts together

To enable these beautiful flowers to see their full bloom!

Acknowledgement

It is a pleasure to recognize the debt I owe to *Prof. Prajit K. Basu*, Research Guide of my doctoral programme at the Hyderabad Central University, Hyderabad. I am thankful for his professional and patient guidance, his constant encouragement and friendly suggestions. I thank *all the Faculties* at the School of Humanities and the Department of Philosophy of the University.

I express my heartfelt thanks to *Prof. Dudley Shapere,* Professor Emeritus at the Wake Forest University, North Carolina, U.S.A., who was kind enough to spend time with me for discussions and clarifications, when I visited him at his place. I need to thankfully acknowledge *Prof. S. V. Dhurandhar* of the Inter-University Centre for Astrophysics and Astronomy (IUCAA), Pune, for throwing light on the Gravitational Wave Experiments. The personal interviews that I had with *Prof. Howard Sankey,* of the University of Melbourne, Australia, have helped me a great deal in shaping my ideas in this study. I thank him as well.

Rt. Rev. Anthony Devotta, my Bishop, all *the priest-friends* of the Diocese of Tiruchirapalli, Tamil Nadu, *the Staff and the Students of Papal Seminary and Jnana-deepa Vidyapeeth* (JDV), Pune, *the innumerable people of the various places of my priestly ministry, especially in Pune and Hyderabad* – all have really contributed to the success of this study in some way or the other. My hearty thanks to all of them.

No words would be adequate to express my gratitude and affectionate appreciation that I owe to *my parents*, *Mr. A. Susainathan* and *Mrs. Jeya Christy,* whose love and blessings constantly accompany me. I do enjoy the care and the concern of *my sister, Jessintha & Leo,* as well as, *my brother, Britto & Roseline, and their dear families,* all *other relatives and friends*, who have offered their support and prayerful greetings – to all of them I offer tons of grateful smiles.

Dr. S. Stephen Jayard

Preface

This journey towards reasonableness, I trust, will be pleasant and useful not only for those who are familiar with philosophy of science, but also those who are really interested in science. All those who like to have a constructively critical look at science and all its activities and all those who dearly expect science to be human and holistic in all its endeavours, I hope, will find this journey inspiring and informative. In this Introduction, after I set out the objectives of the journey, I discuss a few ideas, with the hope that it would set the stage ready for the further discussions in the ensuing Chapters. I close this section with a brief outline of the journey to reasonableness.

Objectives of the Journey towards Reasonableness

As I begin the journey towards Reasonableness and to spell out some of its implications for modern science and society, I like to present my objectives, which are simple and clear. I make attempts:

i) To show that science is neither *the* paradigm of absolute knowledge, rational and objective to the core, nor an enterprise where 'anything goes', with no rationality, no stability, no methodology and no objectivity;

ii) To show that both the rational and the non-rational models of science don't give the actual picture of science, as it is revealed and substantiated by the recent historians of science;

iii) To show the need to go beyond the traditional accounts of rationality in order to have an adequate of understanding of what happens in the realm of science, including the issues of rationality in science;

iv) To explicate some features of the *notion of reasonableness* and show how these factors, especially the notion of *Consensus in the Scientific Community*, which are usually neglected in the traditional accounts of science, need to be earnestly considered to arrive at *an integrated understanding of rationality in science* in terms of reasonableness; and

v) To bring out some of the implications of taking reasonableness seriously in the Cotemporary Science and Society.

The Need for a Critical Look at Science

The gap between the popular understanding of science and the actual practice of science seems to be widening in the 21st century. The highly elitist image of science is being constantly and consistently questioned with better awareness of the actual proceedings of science down the history. Gone are the days when science was regarded as the highest pinnacle of knowledge, and believed to give solutions to any problem. Ironically along with every solution that science and technology bring in, many more complicated problems, hitherto unknown to the humanity, pop up. The claim that science is totally objective, giving the real state of affairs of reality, seems to have long been mitigated by new discoveries in the deeper realms of the physical and biological worlds. An element of belief is very essential in the world of science too. Even to accept that science gives us the true picture of the world we need to believe in science. One of the elements of the popular image of science is that science is absolute; it has answers for everything; once science has given its verdict there is no further appeal whatsoever. Be it in the matters of legal system or the world of medicines, they all look up to science for the final word. This is, I think, due to two reasons: a) science is thought to have sufficient knowledge about all the possible fields; it has powerful tools and method(s) to probe into everything under the sun and above the sun; science has no limits, intrinsic or extrinsic. It has no boundary in its exploration, and in fact it has to break the boundary, if any, in its investigations; and b) scientists, worthy of their names, are to be always unbiased, objective and unaffected by their personal likes and dislikes, untouched by their regional or religious affiliations. But sadly enough, very often, we fail to realize that just as *irrationalism* is anti- scientific and is to be rejected, so also is that *scientism*, which tries to solve the most basic problems of reality, only by means of scientific worldview, making metaphysics and religion superfluous, *is also to be rejected.* I do believe that the misgiving about 'unlimitledness' of science needs to be corrected.

That science is in crisis is not a mere allegation by the anti-science people. Even serious philosophers like Husserl (1954) or Horkheimer (1941) and scientists, like Oppenheimer and Weissacker, bring out the deficiencies in science, which tend to sweep away the 'humanness' from science. No one is ready, after the constructive criticisms by neopositivists, like Richard Rorty, the later Quine and the later Wittgenstein, to see science as the absolute enterprise, as *all-in-all.* In spite of all the success stories of science, it is painful to see science losing human touch in its approaches and achievements. As science is seen as the epitome for reason, crisis in science is seen as the crisis of reason, or even vice versa. In the postmodern society humanity looks for wholeness in everything. Fragmented approaches to life and reality have always resulted in unsatisfactory conclusions. Now humanity seems to realize that a healthy and holistic living requires something more than mere science. That is why, several disciplines like Philosophy, Sociology, Psychology, Political Science, Economics, Religion and so on have started looking at science critically, forcing science to act constructively.

Clarifying the terms 'Science' and 'Rationality'

Since the whole journey towards reasonableness is undertaken with a view to have an adequate understanding of rationality in science, the notions of 'science' and 'rationality' are very important. A few introductory remarks clarifying these two important notions are in order, and I believe, this will help the readers in the further discussions.

Understanding Science(s)

Etymologically, the English word 'science' comes from the Latin verb *scire*, meaning "to know". Knowing is done through senses and the faculty of intellect. Many languages have two different verbs to know through senses and to know through intellect. (Tamil: *arithal* and *Uithunarthal*; Sanskrit: *laukika* and *alaukika pratyaksha,* Greek: *epitamai* and *gnôskô*; Latin: *scire* and *novisse*; French: *savoir* and *connaître*; German: *wissen* and *kennen*; Dutch: *weten* and *kennen*; Italian: *sapere* and *conoscere*; Spanish: *saber* and *conocer*). *Sensitive Knowledge* is obtained through the faculties of five external senses (sight, hearing, taste, smell and touch) and four internal senses (common sense, imagination, memory and the estimative sense) of animals and human beings. It is concerned with things in their individual concreteness, like, this table or that pen. *Intellectual Knowledge* is obtained through the faculty of intellect, which is believed to be unavailable for the non-human creatures. Intellect enables humans to rise above the concrete individuality to the level of *abstraction* or *universality* and to form an *idea* or *concept*. By manipulating these two concepts, intellect helps humans to arrive at a *judgment* (e.g. Cotton is soft). Through the act of *reasoning,* intellect derives a new proposition from other sets of propositions. *Scientific Knowledge* involves both sensitive and intellectual knowledge and still goes further; it must satisfy (at least) three conditions: it must be at the *intellectual* level (to ensure universality, not merely individuality of sense-perception); it must have at least some element of *certainty* and it must have an *insight into the causes* of the object under study. So scientific knowledge is not merely stating of facts but also knowing the causes and their working among themselves.

Speaking about the meaning of the term 'science' we see that there are layers of meaning. In a *subjective sense*, it implies the actual operation of the human intellect, by which one comes to have a scientific insight, for instance, into a mathematical theorem, a physical or historical problem. Also it indicates one's *habitual ability* or an internal capacity to tackle scientific problems with confidence. In this sense we may call one 'a person of science', 'a woman with a scientific mind'. In an *objective sense,* it denotes the object of 'science in the subjective sense'. This is what is grasped by the intellectual operations. Science also means *the whole of cognitive activities*, both intellectual and sensitive, by which humanity acquires knowledge about itself and the surroundings. In this sense it can be differentiated from other disciplines of human activities; e.g. faith and science, science and art, science and politics. Science also denotes *a particular realm of human knowledge*, with proper characteristics. Finally, 'Science' is used as a collective noun to mean *the whole body of people devoted to science*. E.g. Science needs to be responsible in its researches. Of course, all these meanings are closely related to one another. The common element in all these meanings is that 'science' is always somehow related to humans and their specific activities of intellectual cognition. Science has always elements of both subjectivity and objectivity; it is never purely subjective or objective. For, though it is a result of human thinking, it can be spoken or written down for the others to work on it. Science is never static but dynamic, having birth, growth, modifications and it changes in its scope and depth.

Generally speaking, *sciences have particular characteristics*; and when they are fulfilled we can have different sciences, like Life Sciences, Physical Sciences, Experimental Sciences, Cultural or Social Sciences, Astronomy, Economics, Ethnology, Philology, Psychology etc. *Science is a System* – with coherence, interconnectedness and order among the parts. So, an enlisting of unrelated facts or data does not amount to a science. *Science is concerned with a definite field of knowledge.* Demarcation of definite fields is necessary for the development of that particular science. *Scientific knowledge is generally expressed in universal statements.* A mere description of a subject matter does not make it science. It must be able to arrive at universal laws or rules, giving us the intellectual understanding of 'so it is' or 'so it ought to be'. *The Statements of science must be true or probably true.* To be true is to be objectively in conformity with reality and to be certain is subjective conviction without any fear of being in an error. *Science is concerned with the essences, foundation, causes and definite properties of its objects* It provides, *not just a description, but an account of what is going on* behind the phenomena we experience. *Science must involve abstraction and necessity.* Along with the universality and the essences, it must show an element of necessity present in the connection between the subject and the predicate of the propositions.

In short, a *descriptive definition* of science can be given as follows: Science is a logically ordered system of true, or at least probably true, and universal statements concerning the essences, the causes and the nature of objects and processes in a definite field of knowledge, with reference to the investigations, arguments, explanations and demonstrations upon which the conclusions are based.[1]

A General Description of Rationality

Rationality is understood in very many ways. It has come to stand, from the times of Aristotle that humans are taken to be rational beings, meaning that they have the ability of thinking, deliberating and inferring conclusions and so on. However, there does not seem to be any one particular notion of rationality, acceptable to all domains of knowledge. There is no consensus among philosophers in giving features of rationality. The question of human rationality is one of the important concerns for philosophers and scientists (cognitive psychologists). While the philosophers generally attempt to arrive at *a normative account*, the scientists attempt to provide *a descriptive account* of rationality. The normative version tells us, how we *ought* to reason out, while the other how we *actually* do it. Philosophers regard humans as essentially rational beings and hence by their very nature humans cannot but be rational. The cognitive psychologists attempt to discover the actual mechanisms and processes that underlie the pattern of reasoning.

No one, it seems, can deny the fact of rationality because even the very denial has to assume its existence! To be rational seems to be such a basic existential experience of humans that rationality can't be denied without assuming its existence. To deny rationality I need to give reasons; this whole process definitely involves rationality, which I aim to deny. However a serious investigation into the functions and nature of reasoning (rationality) helps us in tackling the issues of epistemology and in appreciating our human innate capacity and the abilities to learn more new things in life. The discussions on rationality become more crucial in the context of universality or relativity of rationality: Are there standards of rationality valid at all times, at all places, across all groups of peoples?... Or are those standards culturally, as well as temporally, variable?... The traditional notion of universal, *a historical* and *a temporal* notion of rationality is being questioned by different trends in modern times for different reasons.

According to the traditional understanding (known as the 'received view'), *rationality is essentially based on relevant evidence.* A rational decision must be based on evaluation of relevant evidence by the application of appropriate rules. Rationality is often equated to *logical consistency*; so it is irrational to affirm both p and ~p at the same time. By expanding this notion, one can understand that it is irrational to try to achieve a goal (g) with improper means (m), and have a conscious knowledge about the improperness of the means. Rationality is also seen as *the fundamental assumption for human existence* and for their relationship with others, and it is the 'gift' of the long evolutionary process. Philosophers like Quine (1960) put forth, what is known as, 'the argument from charity', by which they argue that unless we assume that the other person is rational, we cannot communicate, nor can we attempt to understand or translate the other's language. *Rationality is equated to justification;* a justified belief will necessitate (entail) a rational belief; and when a belief is unjustified it is irrational.

However, *the new awareness and reflection upon rationality* seem to suggest that not all our activities can be brought under strict scanner of such traditional understanding of rationality. Philosophers generally make the distinction between *rationality of belief* (which is also referred to as theoretical / evidential / epistemic reasoning) and *rationality of action* (which is known as practical / pragmatic / means-end reasoning). The former deals with how well a person's network of beliefs represents the external world, while the latter gives us the account how well a person's actions maximize the satisfaction of their desires, given their beliefs. Both are inter-related. Charles Taylor argues that the concept of rationality is richer than mere inconsistency. It involves more than avoiding inconsistency; it depends on different judgments we make when we compare incommensurable cultures and activities. Not just formal criteria of rationality, but new understanding of rationality must involve "human activities of articulation which give the value of rationality its sense".[2] More critical reflections on the various traditional accounts of rationality are given in the Chapter III.

Rationality as Understood in Science

Contemporary debates in Philosophy of Science, as stated earlier, don't fail to focus on the issue of rationality in science. By the turn of the 20th century it was taken for granted that science was the repository of rationality. Science, however, was soon brought under severe scrutiny by various disciplines. Kuhn's stunning statement, "History, if viewed as a repository for more than anecdote or chronology, could produce a decisive transformation in the image of science by which we are now possessed",[3] predicted the drubbing of the received image of science. More knowledge in history of science revealed that many non-rational elements too shaped the course of science. That is why Ian Hacking also, like Doan, is convinced that there is a crisis of rationality.[4] Soon there emerged two powerful paradigms of science, *rational* and *non-rational*, underscoring or underestimating the role of rationality respectively. The proponents of both these paradigms stood at the extremes, and, in fact, *that triggered my interest* to undertake this journey to study this issue a bit deeper to find out what the actual status of affair is. The recent decade seems to see lots of research in two broad areas of *rationality* and *realism*: the former includes the epistemological issues of what reason it, what evidence and method are, what the rationale behind giving special importance (if any!) to observation language over theoretical language, while observation itself theory laden; and the latter involves ontological questions such as what reality is (if ever we will be able to know it!), what is in the world, what is true about the world and so on. *I am interested more in the former issues,* which are related to rationality, because *what* we know about the world largely depends on *how* we know it.

Science is usually seen as the exemplar of rationality; and because it is a rational enterprise there is always progress within science. It always accumulates more and more knowledge, moving towards the true picture of the world. Therefore Popper is convinced that "[Science] is one of the very few activities – perhaps the only one – in which errors are systematically criticized and fairly often, in time, corrected... in other fields there is change but rarely progress".[5] Rationality in science consists in the unbiased applications of the scientific method, where each step is taken towards the goal. All the proponents of the rational model of science agree on the following basic things: a) science always pursues to attain a particular goal, likes to know more about the facts of the world, or at least somewhat closer to the true picture of the world, or to produce theories with more truth content. The theory that comes later, being richer in content, always moves towards truth; b) there is a specific method to arrive at the goal; and c) theory choice is to be explained by *internal* factors, that is, it depends on the relation between the theories and available evidence.

This rational model of science has been severely criticized during the second half of 20th century. It was earlier considered to be scientific, is rational; and its corollary would be unscientific, is irrational; but Putnam defends the view that 'scientific' is not co-extensive with rational. He discusses three important areas of our human lives, where science has nothing to say, not even it can affirm that those facts exist: a) the domain of objective values; b) the domain of freedom; and c) the domain of rationality itself.[6] The traditional picture of science is now shown to be highly wanting with more knowledge about actual historical episodes that the highly rational picture of science has to be changed. It is claimed that the scientific changes and the theory choice can't be rationally explained as there are many social, psychological and non-rational elements at work. *A rationalist model has to tackle the following hurdles:*[7] a) the issue of incommensurability, which claims that with major theory changes, the meanings of the terms in those theories radically change; b) the goal of science has to be explicated; c) they have to show that the principles of comparison are in fact a means to arrive at that goal; d) they have to show that adhering to these principles ensures progress in future and also in the past; and e) they have to show that the actual history fits with this model and that the social and psychological factors have only a minimal role in the course of science. All these hurdles are not going to be easily overcome as we shall see later. Thus we are forced to look for a more adequate notion of rationality.

Philosophy of Science in the 20th Century

The 20th century has seen a lot of changes and improvements in all the fields, including the theories of manners and morals. Especially the progress-rate in the areas of science and technology is barely conceivable to most people lived before this century. Perhaps, in Philosophy this age can be described as 'Age of Analysis'. Though from Plato, philosophers have employed analytic tools as a philosophical technique, now the analytic technique has come to dominate the scene and is considered to be the most fruitful approach to philosophical questions. The main contention of analytical philosophy is to do away with metaphysics, as it is meaningless. In the ancient philosophy, metaphysics was the main concern; in the medieval times philosophy was involved with Religion; in Modern times Rationalism, Empiricism and the Epistemological questions dominated the scenes. But in the 20th century, Reality is studied by natural and human Sciences (like Physics, Chemistry, Biology, Psychology, Sociology, Political Science etc.) in a much detailed manner. Though all these are 'children' of Philosophy, since they have become individual disciplines in themselves, now Philosophy is left with nothing but to analyze the language, meaning, methods of sciences etc.

The domain in Philosophy that critically studies (natural) Sciences is known as "Philosophy of Science". During the times of the ancient Greek Philosophy there was no clear distinction between philosophy and science as such. Science was known as 'natural philosophy'. This continued up until the modern times. In fact, Newton (1643-1727) was known as a 'natural philosopher'. The very terms 'science' and 'scientists' (also 'physicist' and 'hypothetico-dedcutive') are of a recent origin, as William Whewell (1794 –1866), an English Anglican priest and philosopher, theologian and historian of science, introduced them in 1833. So philosophy and science have a long and natural relationship down the centuries. However, 'Philosophy of Science' as an individual discipline is popular only by the turn of the 20th century.

One way of looking at the trends of philosophy of science in the 20th century is to roughly divide them into three schools: School of *Logical Positivism*, popular till 1920s, School of *Historicism*, predominant from 1930s till 1960s, and School of *Historical Realism*, powerful in the later decades of the 20th century. Another way of looking at the schools of philosophy of science would be to divide them into *Positivist* and *Post-Positivist Schools*. Logical Positivism glorified science and put it on a high pedestal, making it the custodian of rationality. Logical Positivists (members of the Vienna Circle) and Logical Empiricists (likeminded group from Berlin), insisted that Philosophy of Science should be precise and rigorous giving importance to the logical and linguistic analyses of the scientific concepts and theories. Positivists like Carnap, Hempel, Reichenbach, Schlick, Popper etc. treated scientists as *little philosophers*, incorruptible investigators who are maximally rational and worried about epistemological and ontological issues".[8] They remove all that is not rational as meaningless, including metaphysics and religion. On the other hand, Historicism went to the other extreme of relativizing science, rendering it almost irrational. Historicism, a popular school of thought in Germany by the 19th century, looks at everything in terms of historical developments. Reacting against the normative view of knowledge and cognition of the Enlightenment period, it argues that *life and reality are nothing but history*. This made a great impact on philosophy of science as well. Historicists, like Hanson, Feyerabend, Kuhn, Lakatos etc., tried to demythologise the logical positivists' view of science in order to arrive at a picture that is more faithful to the actual historical developments and sociological factors of science. In short, one can say: "Hume thought science was inductive and irrational, Popper thought it was non-inductive and rational, and Carnap thought it was inductive and rational", and in contrast to them, as we shall see later, "Kuhn seems to argue that science is both non-inductive and non-rational".[9]

However, Historical Realism attempts to envisage a picture of science and philosophy of science, which would be more comprehensive, by avoiding the extremes of both Logical Positivism and Historicism. Historical Realists like, Imre Lakatos, Larry Laudan, Dudley Shapere and others, pay due importance to the historical, social and non-rational aspects of science, while maintaining the rational character of science intact. Opposing Logical Positivism, they claim that everything in science is changeable, as nothing is final and absolute; however, against historicists, they show that these changes take place in a rational and responsible manner, and therefore 'anything cannot go' in science. They understand rationality in science in a non-traditional way and new criteria for rationality emerge from the scientific community. Thus it can be said to be a mid-way position between Historicism and Logical Positivism. The issue of rationality in science is one of the important contentions for all the schools of philosophy of science. Before we focus on the issue of rationality as such, let us see how Historicism challenges Logical Positivism.

Some of the Claims of Logical Positivism (LP) Challenged by Historicism

a) *Rationality of Science* – LP claims that science is rational, because it strictly follows scientific method (of verification or Popper 's falsification), and that too, one method ('methodological monism'). But Historicism argues that science is non-rational and there can be many methods in science;

b) *Cumulative growth* – Science progresses cumulatively, where the latter theories are the betterment of the previous ones with more empirical content; such progress is not available in any other enterprise, where knowledge grows only in a rough sense. But for Historicism growth is transformative and latter theories are not necessarily better than the previous ones;

c) *Objectivity in Science* - Science is objective, while Historicism shows that science has got both subjective and objective elements;

d) *Monopoly of Truth* - Science is the storehouse of truth-claims. Historicism challenged this and showed that other resources are also reliable.

e) *Reductionism of science* – All sciences can be reduced to physics, as there is a single set of fundamental methods for all of them. Historicism does not agree with this as there can be different methods for science and reductionism becomes more controversial with more growth in various domains of sciences;

f) *Superiority of Scientific theories* - Scientific theories are superior to any other belief systems, for scientific theories are immutable. But Historicism argues that scientific theories do change and there is nothing special about scientific enterprise and its theories;

g) *Dichotomy between the context of discovery and the context of justification* – This distinction is very important because a hypothesis may arise out of any sort of personal or social consideration, but it has to be justified on the basis of observational evidence; the context out of which it emerges is not relevant to philosophy of science as only its justification matters. But Historicism denies any such dichotomy, for, as we shall see later, the social, psychological, non-rational factors and belief systems play a big role in science, and therefore justification is a complex manner;

h) *Clear Distinction between observational terms and theoretical terms:* LP insists upon the distinction of observational terms and theoretical terms (and thereby observational and theoretical statements also). The observational terms can be traced to some sort of observational experiences, while there is no such observational basis for theoretical terms. However Historicism denies this distinction because there is no 'pure' observation and all observations are theory-laden;

i) *Weltanschauung (World view) of the scientists:* According to LP, the value-system, the beliefs, the passion, prejudices and the whole world-view of scientists have no role in science, as the scientists remain beyond all these elements.

Historicism clearly shows that the *Weltanschauung* of the scientists has a crucial role to play in their science, by colouring and controlling the process of science. (*Weltanschauung* can be understood as the totality of background, upbringing, personal preferences, prejudices, and expectations etc. of an individual or a community);

j) *Role of History in Philosophy of Science:* For LP history of science is not necessary to do philosophy of science, while for Historicism, history of science is important to do philosophy of science, because historical factors do play an important role in having an adequate picture of science. (Most of these arguments will be elaborated in Chapter III).

Reason, Rationality and Reasonableness

In the ancient times reason and rationality have been understood as two different faculties, with a clear-cut hierarchy. Reason, as it provides first principles, is said to be superior to rationality. Greek philosophers made a distinction between reason (*nous*) and rationality (*dianoia* or *logos*), and in Latin they are *intellectus* and *ratio* respectively. Reason is the faculty providing the first principles; rationality is expected to operate upon those principles. The core of rationality is in giving argumentation but the argumentation has its limits; that is, the conclusion depends on the certainty of the premises. Tran van Doan, for instance, classifies reason under three dimensions: reason, rationality and reasonableness. Reason can be seen as the ultimate causes, logos, or the *principium principiorum*; Rationality is seen as the effective instrument to deal with technological knowledge, and advances, to dominate nature and human destiny, and reasonableness is seen as the principle of human practical life, which embraces different forms of human life such as morality, laws, arts etc. Major part of the problem is due to the undue stress on one form of reason in the scientific enterprises. As Doan explains, "as long as human beings are integral and total, and as long as they are complex and developmental, then all forms of human life-world may generate different forms of reason".[10]

However, in the modern times and more so in the contemporary times, reason and rationality are not seen as two different faculties, and even they are treated as two layers or functional modes of our reflective activity. They are seen as the two sides of the same coin, because in all our rational activities we always use reason; reason is the capacity which allows us to come to grips with the conditions we find ourselves in.[11] Different functions of our reflective activity, rooted in all forms of human life-world, and must be taken seriously. Unfortunately rationalists and empiricists attempt to build human life only on the edifice of rationality, ignoring the other forms of reason. This has drastically distorted the wellbeing of humanity. The crisis of reason is described by Husserl as "the loss of the life-meaningfulness".[12] Martin Luther labelled reason, may be more for theological considerations than for logical ones, as a beast and a whore. After the Enlightenment reason was kept as a high pedestal and many schools of thought, positivism, materialism, behaviourism, glorified reason so much so it sadly resulted in the loss of the meaning in life.

Of course, reason is a great strength for humans and it is, as Voltaire says, "God's incomprehensible gift to mankind". It gives the distinguishing identity to humans, "Like freedom, (reason) is an undeniable fact. But this does not mean that we uncritically share the optimism of Origen, Aristotle, Leibnitz, Kant and especially of the Enlighteners".[13] Criticizing reason does in no way mean glorification of irrationalism. With no proper understanding of reason, humanity will easily be led to lunacy, disorder, pandemonium, nihilism or anarchism, with no respectable place left for morality or law. "Without an authentic and true understanding of reason, reason would be nothing but a means, an instrument of self-destruction",[14] which is obviously a destruction of humanity itself. Doan strongly recommends for a natural balance among all the three aspects of reason, rationality and reasonableness; and "any attempt to separate the trio of reason, rationality, and reasonableness and any effort to use only any one of the threes principles, it would cause: a) A crisis of knowledge and human life (as in the case of the crisis of science), b) A stagnation, or backwardness in terms of intellectual development (as in the case of the Confucian society); c) Mysterious, utopian, and a-*human* features (as seen in some religion and ideologies)".[15]

Several philosophers try to respond to this crisis in various ways: Peter Winch (1972) searches for *social rationality*; Stephen Nathenson (1994) looks for an *ideal type of rationality in terms of reasonableness*; Heidegger looks for *an authentic reason*, as reason is the most authentic form of Being, and for him the misunderstanding of science and technology is basically the result of our misunderstanding of Reason; Habermas (1971) insists on constructing *practical rationality*, and Tran Van Doan (2001) invites to us *return to Confucius' notion of reasonableness.* Similarly, Feyerabend warns us to be cautious of science, so much so, as we will see later, he demands that *society be liberated from science*. We shall return to the issue of revising and reconstructing the notion of rationality in science in Chapter III.

The Outline of the Journey towards Reasonableness

The First Chapter offers a brief presentation of the 20th century theories of scientific rationality. *Section 1* deals with the views of two important *proponents of rational models of science*, namely, Hempel and Popper. *Hempel* insists upon, among other things, the goal-means relation to explicate the rationality in science. As in any other human activity the rationality of the scientific activity is also determined by the goal-means relation. An action is rational which maximizes the possibility of achieving the goal. *Popper*, who has been popular among various quarters for several reasons, has made a great contribution in philosophy of science as well. For him science is highly rational; it is basically a critical enterprise. So he is even known as 'critical rationalist'. Rationality depends on method;

science is rational because it has a unique method and that is the method of Falsificationism. In the *Second Section,* I take up the staunch critiques of the rational model, namely, Kuhn and Feyerabend: *Kuhn* can be rightly given the credit of awakening philosophers of science from their "rationalistic slumber", to see the historical and social elements, which have been sadly ignored, that play a very vital role in science. Interestingly even Kuhn, who is the most prominent critiques of rational model of science, agrees that science is a rational enterprise, though for different reasons (i.e. the principle of rationality emerges from the scientific community itself). *Feyerabend* is the most vigourous and influential critique of the received view of science. He vehemently attacks the high pedestal-status given to science in modern times and shows that there is no unique method in science. He even appeals to save the modern society from science, as it once needed liberation from religions.

In second Chapter, I present some critiques of both the proponents of the rational and non-rational models of science. *Section1* critiques the proponents of the rational model. I spend a considerable time in seeing how a few prominent critiques look at *Hempel* and then I try to spell out how rationality in science can't be confined only to goal-means relation. Similarly, while acknowledging *Popper's* immense contributions to philosophy of science, I do attempt to bring out a few elements where philosophy of science has moved much forward since he published his *Logic of Scientific Discovery.* In *Section 2* I try to show how certain precautionary steps need to be taken to stop *Kuhn* from slipping into relativism. While assessing his notion of incommensurability, I refer to Shapere's notion of 'chain-of-reasoning' connections as one possible way to face the difficulty of incommensurability. I rely on Kuhn's idea of the role of the community and the notion of theory choice, among other factors, in developing my understanding in rationality in science. Turning to *Feyerabend* next, I critically look at his famous claim, 'anything goes'. Comparing Kuhn's and Feyerabend's notions of incommensurability I argue that Kuhn is in a better position to overcome the dire consequences of the thesis of incommensurability, with his linguistic overtones in his latter period. However I don't fail to point out that incommensurability, though it is a matter to be considered with attention, yet, if it is taken literally, it creates a lot more serious problems. I do, finally, appreciate Feyerabend's efforts to make science with a human face, by showing how harmful scientism is to human society and to the environment.

The Third Chapter is primarily dedicated to accomplish four things: a) to analyze some of the traditional accounts of rationality, showing the need for a revision of the notion of rationality in general, and in science in particular; b) to put forward my understanding of rationality in science in terms of reasonableness; c) to elaborate the notion of 'Consensus among the Scientific Community', one of the important features of making science a reasonable enterprise; and d) to bring out some of the implications of the notion of reasonableness for the Contemporary Science and the Society. Our analyses so far have helped us to realize that rationality is very important for our lives and science as well. But it is not everything even for science: "Rationality helps", as Henry Harris has it, "but it is not a prescription for making discoveries".[16] In *Section 1*, I briefly analyse some of the traditional accounts of rationality to show their limitations. For instance, rationality is often identified with truth or very popularly, with scientificity. But a bit deeper analysis will make us realize that rationality/irrationality cannot be merely identified with true/false beliefs, nor that it can be reduced to scientificity. I focus more on the inadequacies of equating rationality with deducibility. As I criticize the traditional accounts of rationality, I am not under the illusion that every aspect of the received view has to be rejected or reformulated. I point out the need to appreciate and celebrate inductive reasoning, a sort of positive legacy of the traditional accounts of rationality. However the notorious problem of inductive reasoning is quite alarming. Having briefly examined some of the solutions and dissolutions of the problem, I offer a short discussion on *Bayesian theory of reasoning,* which seems, at least hitherto, to be rather a satisfactory solution to the problem of induction. Induction may be seen as a part of our neural wiring. No logical or deductive evidence is needed to support its working. 'Inter-subjective' testing enables us to remove the subjective elements and biases to the extent of making science *(reasonably!) rational.* Basing on such analyses I try to show that science is rational in spite of inductive reasoning.

Section 2 takes up the views of some philosophers, who, being convinced of the short comings of the traditional views of rationality, look for an alternative account of rationality. Philosophers like, *Hilary Putnam* ('Integrated View of Rationality'), *Stephen Toulmin* (Avoiding 'Over-Reliance on Rationality'), *Christopher McMohan* ('Collective Rationality') and Habermas ("Consensus as a Form of Rationality'), propose alternatives to the traditional accounts of rationality, which seem to capture the notion of rationality in a more adequate manner. In *Section 3* I present my understanding of rationality in science in terms of *Reasonableness*. Reasonableness plays a very important role, as in the common day-to-day life and in the world of science as well. Though it is not easy to define the notion of reasonableness, the features of reasonableness, which I discuss, I am sure, will enable us to understand it better. I see the role of reasonableness, in science, at three levels, though they cannot be categorized in water-tight compartments, nor are they exhaustive: *Reasonableness at the Personal Level of the Scientists* – Imagination, the Agent, Judgment and Intuition; *Reasonableness in Scientific Methodology* - Rejection of Zero-tolerance, An Enriched Notion of Objectivity, The Need of Skeptical Attitude, and The Need for Common Sense; and *Reasonableness in Scientific Practice* – Embracing Pragmatism and the Consensus of the Scientific Community. These features, though overlapping sometimes, as I attempt to show, constitute the notion of reasonableness in science.

Section 4 is entirely devoted for the discussion of the *Consensus in the Scientific Community*. For, of all the features of reasonableness, consensus, as I am trying to show, plays an important role especially in the context of research frontiers in contemporary science, where the experiments or empirical observations often don't seem to arrive at unambiguous conclusions. Certain experiments need the consensus of the community to reach some meaningful and useful conclusions, without which those experiments are incomplete and unproductive. Analysing some aspects of gravitational wave experiments, *I try to explicate how the role of scientific community begins to have a very prominent role in deciding upon important concepts in science, for instance, observation.* So we are forced to include the aspect of consensus in the holistic understanding of rationality (reasonableness) in science. When one considers the various features of Reasonableness seriously, science obviously becomes a very reasonable enterprise. Perhaps, more awareness of the social elements and the actual practice of science seem to suggest that *as long as these features of reasonableness are part and parcel of science, it seems not to be of much use to speak of Rationality* (in the traditional sense) *in science, but only of Reasonableness in science.*

In the last Chapter, I bring out some of the implications of the notion of reasonableness for the contemporary science and society. Taking the notion of reasonableness seriously, *Section 1*, makes an attempt to show that scientific knowledge is relative and mutable, not absolute and unchangeable; we realize that social elements and value-considerations, including gender issues, play an important role in science; the traditional notion of observation as 'pure' and 'secure' is shown to be untenable; in spite of all great success and achievements, we need to go beyond them to achieve 'human fullness', though we don't have easy and measureable parameters; we need to learn 'how to learn', transcending the traditional rationalism and empiricism; and finally, reasonableness invites us to settle with 'reasonable' realism. In *Section 2*, we discuss how society needs to take the notion of reasonableness seriously, as the notion of mere rationality is insufficient in many ways. Stressing the need to be free from the clutches of scientism, we show how the contemporary society needs a holistic thrust in education. If science, which is supposed to be the most rational enterprise, has to have a relaxed notion of rationality in terms of reasonableness, then our day-to-day lives need to be comfortable with uncertainties and mysteries of reality, limits and limitations of life. Finally, society has to learn that spirituality and mysticism are not meaningless no matter how advanced we get in the modern scientific world. For, contemporary science may solve many mysteries of the past, but it creates many other new mysteries. So only the number and nature of mysteries may change, but mystery always remains. This fact provides us with a platform for deeper and closer interactions between science and other disciplines, including philosophy, religion and mysticism. *From these interactions whether science and/or religion gains or not, we human beings are going to gain a lot, as the society will become a well-integrated and holistic one.*

In the Conclusion I briefly show how my objectives with which we began our journey have been fulfilled, more or less at least, and leave some parting reflections.

Notes

1 For more elaborate discussion on these notions, please see sources, like *The New Encyclopaedia of Britannica,* 1997 and *Routledge Encyclopaedia of Philosophy*, 1988, from where I have drawn some ideas.

2 Charles Taylor, "Rationality", in *Rationality and Relativism*, Martin Hollis and Steven Lukes, eds., 1982, 87-105, 105.

3 Thomas Kuhn, 1970, 1.

4 Ian Hacking, 1983, 1.

5 K. R. Popper, *Conjectures and Refutations*, 1963, 216.

6 Hilary Putnam, "The Place of Facts in a World of Facts," in Douglas Huff and Omer Prewett, eds., 1979, 113-140.

7 Newton-Smith, *The Rationality of Science*, 1996, 267. First print 1981.

8 Klee, 1999, 199.

9 James Ladyman, 2002, 94.

10 Tran Van Doan, 2001, 4.

11 For a short exposition on this, see: http://www.uni-jena.de/welsch/reason.htmll. In my arguments also I too don't stress on the difference between reason and rationality too much.

12 Edmund Husserl, 1954.

13 Tran Van Doan, Reason, 2001, 18-19.

14 Ibid., 19.

15 Ibid., 267.

16 Henry Harris, 1981, 46.

Contents

ACKNOWLEDGEMENT (*vii*)

PREFACE (*ix*)

CHAPTER 1: 20TH CENTURY'S THEORIES OF RATIONALITY IN SCIENCE 1

- **Introduction** 1
- **Section 1: Proponents of Rational Model of Rationality in Science** 2
  - Carl Gustav Hempel: Goal Oriented Rationality! 3
  - Karl Popper: A Falsifying Rationalist 13
- **Section 2: Proponents of Non-rational Models of Rationality in Science** 27
  - Thomas Samuel Kuhn: A Paradigm for Paradigms 28
  - Paul Feyerabend: Methodology – Not a Corner Stone for Science 50
- **Notes & References** 66

CHAPTER 2: ASSESSMENT OF THE 20TH CENTURY'S THEORIES OF RATIONALITY IN SCIENCE 77

- **Introduction** 77
- **Section 1: Assessment of the Rational Models of Rationality in Science** 78

Hempel: Underestimating Goals in Science and Life 79

Karl Popper: A Fallible Falsificationist! 92

➤ **Section 2: Assessment of the Non-rational Models of Rationality in Science** **105**

Thomas Kuhn – A Rational Relativist! 106

Feyerabend: An Intellectual, but Human to the Core! 121

➤ **Notes & References** **135**

CHAPTER 3: REASONABLENESS – AN INTEGRATED VIEW OF RATIONALITY IN SCIENCE **143**

➤ **Introduction** **143**

➤ **Section 1: Traditional Accounts of Rationality and their Constraints** **144**

Various Accounts of Rationality and their Constraints 145

Inductive Reasoning – A Positive Legacy from the Traditional Accounts of Rationality 154

➤ **Section 2: The Need to Revise the Traditional Accounts of Science** **169**

Putnam's 'Integrated View of Rationality' 169

Stephen Toulmin's Criticism about 'Over-reliance on Rationality' 171

Stephen Nathanson's 'Reasonable Form of Rationalism' 172

McMohan's 'Collective Rationality' 174

Jurgen Habermas' Notion of Consensus as a form of Rationality 175

➤ **Section 3: Reasonableness: An Integrated Account of Rationality in Science** **179**

The Notion of Reasonableness 180

Confucian Notion of Reasonableness 184

Some Features of Reasonableness in Science 188

Reasonableness and Other Domains 198

➤ **Section 4: Consensus in the Scientific Community** **203**

The Need for Consensus in Science 203

Consensus in Cosmology 204

Consensus in Observation 206

Gravitational Wave (GW) Experiments – A Case Study 207

'Consensus' and the Experimenters' Regress 233

Consensus – Making Rationality Social, Yet not 'Irrational' 237

➢ **Notes & References** **240**

CHAPTER 4: THE IMPLICATIONS OF REASONABLENESS FOR CONTEMPORARY SCIENCE AND SOCIETY **253**

➢ **Introduction** **253**

➢ **Section 1: Implications of Reasonableness for Contemporary Science** **254**

Prominent Place for Values in Science 254

Gender Issues in Science 261

Indispensable Role for Social and Non-Cognitive Elements in Scientific Knowledge 270

Social and Moral Responsibilities of Scientists 272

Untenability of 'Pure' Observation 275

Nature of Scientific Knowledge 280

Some of the Major Axioms of Science 282

Total Comprehension of Reality – Ever Elusive! 284

Learning to Learn 287

Reasonable Realism 291

Transcending Rationality (Science) for Human Fullness 295

➢ **Section 2: Implications of Reasonableness for the Contemporary Society** **300**

The Need for a Holistic Thrust in Education 301

The Need for 'Intellution' 303

Freedom from the Clutches of Scientism 308

Enlightened to Live with Mysteries and Limits of Science 310
Meaningful Role for Spirituality and Mysticism 319
➤ **Notes & References** **342**

CONCLUSION **351**

Recap of the Journey towards Reasonableness 351
Parting Remarks 360
➤ **Notes & References** **366**

BIBLIOGRAPHY **369**

INDEX **373**

Chapter 1

20th Century's Theories of Rationality in Science

Introduction

This chapter seeks to make a brief presentation of some of the important theories of rationality in science that were dominant in the 20th century. One way of categorizing these theories would be to look at them under two broad categories: *Theories of Rational Model* and *Theories of Non-rational Model.* Under each category, only two leading figures have been chosen for our discussion, though there are many other prominent philosophers of science can be legitimately thought of. In Section one, Carl Gustav Hempel and Karl Popper are presented as the spokespersons of the rational models. Hempel, being a staunch proponent of rational character of science, pitches his argument on the goal-means relation to show that science is rational to the core, as science always has clear-cut goals to be achieved and it always chooses the best means available to achieve them. The commonsensical view of rationality, that is, an action is rational which maximizes the possibility of achieving the agent's goals, is very true in realm of science, making it highly rational. Karl Popper is deeply convinced of the rational face of science because of the very strong notion of falsification, which, according to him solves many problems in science. Holding onto falsification he claims to have solved the age-old problem of the inductive reasoning in science and the demarcation of science and non-science.

Section two elaborates the contributions of Thomas Kuhn and Paul Feyerabend, who can be comfortably said to be the champions of non-rational models of theories. Both of them, being committed critiques of the rational models, clearly bring out the fact how science involves many social factors in a very important manner. They show how these factors make science a social enterprise. Kuhn may not really object to THE claim that science is rational, provided that the notion of rationality emerges from the scientific community itself, rather than being imposed upon science from other sources. Feyerabend is determined to dethrone science from its rational pedestal and to make it clear that science is also equally a human and social enterprise. He vehemently criticizes the orientation towards unique methodology of science to show that there is no such all-binding methodology and in fact, science progresses only when the traditional methodological norms are violated. He laments over the harm done by the strict rational approaches of science and he wants the society to be liberated from science. This Chapter just seeks to familiarize the readers with the salient contributions of these authors and their ideas will be taken up for critical evaluation in the next Chapter.

SECTION 1
PROPONENTS OF RATIONAL MODEL OF RATIONALITY IN SCIENCE

In this Section, the proponents of rational models of science are discussed. C. G. Hempel argues for the rational character of science basing on the simple, but profound, notion of 'goal-oriented rationality'. No action can be said to be rational if it is not performed to attain the goal(s) of the agent, and the action must be done in such a way to maximize the chances of obtaining the intended goal(s). He analyzes the theories of analytic-empiricist (prescriptive) and the historic-sociological (descriptive) schools. His evaluation of Kuhn's notion of rationality reveals that there are several agreements and disagreements between him and Kuhn. He goes along with Kuhn in recognizing the socio-historical factors seriously in understanding science. However he takes objection to Kuhn on some grounds, like the problematic nature of the desiderata for theory-choice provided by Kuhn. Hempel shows the need to integrate the rationalistic and pragmatic construal of rationality and he proposes a fresh account of rationality as "Relaxed Rational Reconstruction of Scientific Rationality". Karl Popper, a Critical Rationalist, as he likes to call himself, is convinced of rationality in science, as it is the most critical and progressive enterprise. It is also objective which is found in its 'inter-subjective testability'. With the weapon of 'falsificationism' he solves two problems: the problem of inductive reasoning and the demarcation between science and non-science. Induction is no more seen as necessary for science, as hypothetico-deductive method suffices the needs of science. Since falsification is the strict and unambiguous criterion the demarcation between science and non-science is no more a problem. Both these stalwarts, thus, argue for the rational approach towards science, from different perspectives though. Their contributions will also be appraised in the next Chapter.

Carl Gustav Hempel: Goal Oriented Rationality!

C. G. Hempel (1905-1997) was born in Oranienburg, Germany. He studied mathematics, physics and philosophy at various places, including Vienna and Berlin. Working with Hans Reichenbach, he received his Ph.D. in 1934. He has made important contributions to the "covering law model" analysis of the structure of scientific theories. He has many important books to his credit, like: *Fundamentals of Concept Formation Empirical Science*; *Aspects of Scientific Explanation, and Philosophy of Nature*. Carnap got a research grant for Hempel and took him to Chicago in 1938. Later Hempel found teaching posts at City and Queen Colleges in New York and Yale. With many requests he agreed to teach in other universities, like those in Princeton (1955-75) and Pittsburgh (1977-85). He finally retired at Princeton. He has been awarded many honorary doctorates.

Hempel's contributions to philosophy in general and philosophy of science in particular have made a great impact in those fields of the 20th century. His philosophical thinking focuses mainly on three areas[1]: a) Confirmation, b) Explanation and scientific laws, and c) theoreticity and cognitive significance. Interestingly all these are intimately related to major issues of philosophy of science, like the rationality of science, the problem of confirmation and the structure of scientific theories, the most crucial issue of induction and so on. He wanted his solutions to the traditional philosophical problems to be vigorous and critical. That is why, as Fetzer puts it, "He [Hempel] cared more about finding the right solutions than whether his solutions were right".[2] Kuhn too expresses similar sentiments when he declares, "In him [Hempel] I learned to recognize the stance of a man who intends philosophical distinctions to advance truth rather than to win debates".[3] W. Salmon too admires Hempel's acumen in philosophy, especially in logical empiricism, by saying, "His [Hempel's] writings have uniformly exhibited the kind of clarity, depth, incisiveness, and fairness that ideally suit them to provide a solid basis for understanding the fundamental philosophical issues".[4] His students included very popular personalities such as Larry Laudan, Robert Nozick, Philip Kitcher and so on. Though Hempel's contributions to philosophy of science are multi-dimensional, I like to limit the discussion only to his ideas of rationality in science in this Section.

Two Major Paradigms of Scientific Rationality

Hempel, while discussing the issues of philosophy of science, spends a considerable time and energy on the question of scientific rationality. There is no second opinion in considering science to be the most rational enterprise for rationality, which produces the most reliable knowledge of all the human endeavours. But there is no such unanimity with regard to the conceptions of its methodology. One can easily note two major schools of thought regarding methodology: *Analytic-empiricist school (Prescriptive),* developed by logical positivism. For them the methodological principles serve as the criteria for rationality in formulating and in testing scientific claims and for theory choice: "The methodology of science, thus understood, is concerned solely with certain logical and systematic aspects of science which form the basis of its soundness and rationality – in abstraction from, and indeed to the exclusion of, the psychological and historical facets of science as a social enterprise".[5] One of such methodological criteria is the 'testability in principle' of the claims. Therefore, for instance, they reject neoviatlism (as entelechy is not testable), the doctrines of Marxism and psychoanalysis. On the other hand, according to the *historic-sociological school (Descriptive)*, methodological principles cannot be formulated merely on a philosophical or logical basis but the actual behaviour of the scientists is to be considered: "It insists that an adequate theory of scientific method must be based on a close study of the practice of scientific inquiry and should be able to explain at least some aspects of actual scientific theorizing, past as well as present".[6]

These two trends can also be seen as: *methodological rationalism* and *methodological naturalism*. The methodological rationalists demand *a priori* norms to which all scientific claims must conform. For example, Hempel claims that these "norms are established on *a priori* grounds, by logical analysis and reconstruction of the rationale of the scientific search for knowledge".[7] On the other hand, methodological naturalists, like Kuhn and Feyerabend insist that scientific inquiry has to be understood in terms of the actual practices of the scientists. The main focus of both the trends is 'theory choice'. Important questions like - 'Are there general principles, governing the choice between two competing theories?' and 'If so, can those principles be the criteria for rationality as envisaged by logical empiricism?' - preoccupy the debate. Logical empiricism looks for criteria to establish general rules to determine higher probability or rational credibility, while the historicist schools reject such an effort as doomed to failure. They agree that there are certain general considerations which do influence scientists in the preference of theories. But these considerations are the shared preferences or values of the given scientific community and never can act as absolute binding principles to decide upon the theory choice.

Hempel's Evaluation of Kuhn's Notion of Rationality

Hempel's evaluation of Kuhn's ideas throws some light on Hempel's own understanding of rationality. He gives much importance to Kuhn's notion of rationality in science. He agrees with Kuhn on many aspects, while he is quite critical of him on certain other fronts. Kuhn's account of the development of science is descriptive (explanatory) and yet it has certain normative or prescriptive characteristics too, so as to render it clearly rational. As we are going to see shortly, he maintains the rational character of science in spite of the fact that theory-choice in science cannot be undertaken purely on rational grounds. Hempel *agrees with Kuhn* on two accounts. First, normative account of science has to be compatible with actual practice of science. The aim and function of a normative account is to improve the particular domain. Therefore a normative account cannot lead to improvement unless it is compatible with practice. Second, scientific inquiry is always a social event, taking place in a socio-historical milieu. To comprehend science rather well only logic and other a priori considerations are not enough; rather it needs to be understood in terms of what the scientists do in actual practice of science.

However, Hempel *disagrees with Kuhn* on several grounds. First, to understand whether an action is rational or not, we need to have a clear notion of rationality. To analyze the notion of rationality, one needs to know what kinds of considerations support or deny attributions of rationality. But Hempel feels that Feyerabend, Kuhn, Popper and Lakatos don't seem to have a clear and common understanding of rationality. For instance, Lakatos[8] accuses Kuhn of irrationalism and mob psychology though Kuhn declares that his account is very rational.

Second, Hempel finds serious problems with the Kuhn's desiderata. For instance, the notion of simplicity: "none of the various efforts made by logicians and philosophers of science to explicate the notion of simplicity for theories has yielded a satisfactory generally applicable criterion for the comparison of theories in point of simplicity".[9] Further, one cannot arrive at one overall criterion of rational preference for competing theories. If, as it happens sometimes, each of the two competing theories fulfills various desiderata in different degrees, then it becomes difficult for one to prefer one theory over the other. According to Hempel, "to secure one general standard of comparison, the various desiderata would have to be rank-ordered in point of relative importance: and there is no plausible way in sight to achieve such an ordering".[10] Further, sometimes a well-established theory determines test findings as acceptable data. Significant deviations in experiments are overlooked in order to suit the theoretical predictions and expectations. For example, in the famous Millikan's oil drop experiment,[11] often the observed motion of the oil drops did not match the theoretical predictions, but they were overlooked and were not mentioned in the final publication of the results. All these were swept under a big carpet, called 'disturbing factors' of the experimental set up, upholding the intactness of the theoretical predictions. As it is quite natural that individual scientists would differ in their tastes and relative priorities these norms cannot be deciding factors in all the situations. Therefore, Hempel argues that in the case of theory choice "commitment to those norms does not ensure a uniform decision as to which of two theories outranks the other, by way of satisfying the entire set of desiderata".[12] Therefore, these desiderata can't be used for deciding unequivocally in theory choice; they can't be taken as providing a mathematical algorithm. Scientists' individual tastes always matter. More importantly there is a hitch, which is like the unresolved problem of induction: "the question whether there are any reasons to expect that a theory which, as judged by the desiderata, is preferable to its competitor, at a given time will continue to prove superior when faced with further, hitherto unexamined, occurrences in its domain".[13]

Third, there is no adequate explanation in Kuhn's account to consider current practices as a normative one. Hempel argues that "the assignment of prescriptive reading to a descriptive account of scientific research is not ... straightforward".[14] Purely descriptive account of scientific rationality is impossible. It is impossible for the same reason as pure observation is impossible. Any kind of explanation or understanding of descriptive realm is in some sense a kind of interpretation of it. And an interpretation must have a certain basis other than the descriptive realm. Therefore an account of methodology of science must have both the justificatory (goal-means relations) and descriptive aspects (context-related elements of the action).

Fourth, against Kuhn, Hempel maintains that the behaviour of scientific community can't be the only yardstick for the rationality of science. Kuhn tries to supplement the desiderata with other factors like learning from experience, demanding for consistency, considerations of language in which scientific terms are expressed and so on. In spite of all these Hempel asserts, "there cannot be, of course, be a definitive, unconditional justification".[15] *Kuhn attributes a prescriptive reading to a descriptive account of scientific research.* He accuses that Kuhn assumes science to be the best example of rational enterprise and this is a broad epistemological valuation. Kuhn gives much importance to the behaviour of scientists, but unfortunately often we find their behaviours not as contributing to scientific growth, but interfering with it – for instance, out of jealousy they may conceal their methods or results; or just for a short-cut victory they may create or manipulate evidences and experimental evidences or overlook falsifications. How can all these be included as prescriptive?

Fifth, according to Hempel, the desiderata are not sufficient conditions to determine the theory choice. When there is a conflict of desiderata, theory choice can however, be decided by larger considerations, which are traceable to the goal(s). However, Kuhn does not take into account the goal of the scientific community. Just because desiderata do not provide any strict, unwavering algorithmic procedure for theory choice, it does not follow that values are the only determining factors. Hempel claims that besides the desiderata, there are other severe constraints imposed by certain quite generally acknowledged norms, like "the standards of deductive logic, and the prohibition of logical inconsistencies: even though there is no general algorithmic test procedure for consistency, there is insistence in avoiding or in some way quarantining inconsistencies that may have been discovered…".[16] These norms represented by the desiderata are rather implicit and imprecise. However they don't permit any eccentric whims and fancies of a particular scientist as justification for theory choice. Thus, theory choice can be based upon the better fulfillment of the desiderata, when the scientific inquiry aims at more comprehensive and systematized understanding of the world.

Sixth, nevertheless Hempel does not accuse Kuhn of irrationalism. For, one can be irrational only if one goes against any well-established standards of rationality. Hempel claims that he is "not aware of any rule or standard that could be seriously held to be a binding requirement of scientific rationality that has been neglected or rejected by Kuhn".[17] In the same breath it should also be noted that though Hempel defends Kuhn from the attacks of irrationalism, Hempel does not seem to be at ease to call Kuhn's scientific understanding as rational either. It is because, *rationality consists in adopting appropriate and well-reasoned out means to obtain desired ends,* and Kuhn's desiderata include non-rational elements. Therefore, Kuhn's account of theory choice does not always seem to require application of well-reasoned out procedural rules. Moreover, Kuhn falls back upon the decision of the scientific community in case of a conflict in theory choice; but it can happen that the whole community may decide upon something which is obviously irrational. So Hempel prefers to call Kuhn's position "*adaptive*, but surely not *rational*: they are not adopted as a result of goal-directed *reasoning*".[18]

Features of Rationality in Science

For Hempel an account of rationality requires a reference to goal-means relationship. He further elaborates the notion of rationality with the notion of context-dependence. He also proposes conditions which make a claim rational.

Rationality in Terms of the Goal of an Action*:* An action by itself can't be judged to be rational or irrational. Rationality of an action is decided upon three factors: i) with reference to the goal the action is supposed to lead to, ii) the relevant information available to the agent prior to performing any action, and iii) the beliefs entertained by the agent regarding the pursuit of his goal. Therefore, for Hempel *the concept of rationality is decided by the goals and the means (methods) to achieve them.* Thus according to him a rational account of scientific inquiry must stipulate certain goals and some appropriate methodological principles to be pursued. It has also to show the instrumental rationality of the principles in relation to the goals and "only to the extent that this can be done does the conception of science as the exemplar of rationality appear to be viable".[19] In this understanding the methodological principles don't give us the rules in absolute categorical norms to dictate what is to be done, but "rather what way of proceeding is rational in the sense of offering the best chance of attaining a certain scientific objective".[20]

Context-dependence of Rationality: Rationality of an action does depend on its context; an action which is rational in one context may be irrational in another. However, *in all the contexts the goal of an action decides the rationality of the given action:* "A given action or a mode of procedure cannot be qualified as rational or as irrational just by itself, but only in considerations of the goal that it aimed at...". He gives an example: it is obviously rational when one jumps into a river to save a drowning child, though it is a big risk for one's own life. But will we see one's action rational if one jumps into the full flooded river to get to the other side as faster, say, instead of using the long-winding bridge?... Thus he makes it clear that "A mode of procedure is rational, relative to a certain goal and a given body of means-end information, if, judged by that information, the procedure offers an optimal chance of attaining the goal".[21] He acknowledges that the methodological principles are necessary but not sufficient to decide upon rationality. This is to an extent compatible with some aspects of Kuhn's position. He also points out that the logical positivists, by mistake, thought them to be sufficient. They expected the direct implementation or conformity of these methodological principles in the practices of sciences. But it is not the case. The adherence of scientists to the methodological rules is quite embedded in the socio-historical context. To sufficiently explain whether a particular scientific activity as rational or not, one has to analyze how the standards are elaborated or understood or interpreted in that particular context. Thus the analysis of the context as well as the conditions of goal is necessary to determine the rationality of an action.

Conditions for Rationality: Hempel delineates a few conditions to rationally accept a claim: i) if the claims are testable; ii) if there is a logical consistency with other sets of knowledge statements and iii) if every acceptable set is deductively closed (i.e. if p2 is deduced from p, and if q2 is logically deducible from p2, then q2 also must be included in p). These conditions are not for theory choice but for the more fundamental issue of ascertaining rationality of science; and these are "some modest necessary conditions of scientific rationality".[22] These conditions are applicable if the goal of science is to arrive at empirical belief of the world with more accuracy, comprehensiveness, simplicity and so on. In case, if we aim at the knowledge of the world with more emotional security and aesthetic satisfaction then different standards of rationality would have to be brought in. Perhaps, in that case, factual accuracy need not be insisted upon. If the goal of science is to obtain a reliable knowledge about the world, which enables correct predictions, a knowledge which helps us to escape or prevent harm, a knowledge that indicates means for achieving desired ends, then we need to check the hypothesis / theories with the relevant data derived from the world.

Values and Objectivity in Science

Scientific inquiry is said to be objective if the personal attitudes and ideological tendencies of the investigator and other extra scientific values don't influence the scientific inquiry. The question is, do the values not have any role at all in the scientific inquiry? The question of value-dependency can be looked at two levels: i) the actual behaviour of scientists and ii) the methodological standards for critical appraisal & justification of theories. The personal likes and dislikes and moral convictions of the scientists do affect their behaviour. Even the social and political values, strong orientation and interests of the society, availability of funds etc. too influence the scientists. Such motivational factors can only describe their behaviour but to justify scientists' accepting or rejecting a theory one needs, according to Hempel, "a critical appraisal of the theory in light of the available evidential and other systematic grounds that have a bearing on its acceptability".[23]

Hempel finds some role for values in the scientific inquiry, even in the Popperian and Carnapian schemata of science. For instance, Popper has a rational conception of scientific methodology, which is explicated by theories of falsification, corroboration etc. Also he has epistemic values (not aesthetic or moral), that is, a theory must have testability and a larger empirical content than the previous one. Further, the objectivity of science is ensured, in spite of the role played by the value-consideration, by the inter-subjective agreement among scientists. Similarly, Carnap, though he follows a normative methodology and though he tries to build a program for inductive logic in order to attribute some sort of deductive validity to it, yet he too employs some valuational component; namely he relies on intuition to explain inductive logic. In the words of Carnap, "the reasons to be given for accepting any axiom of inductive logic... are based upon our intuitive judgments concerning inductive validity".[24] But because it is left to the intuitions it is not up to the whims and fancies of the individuals; there is a wide body of shared intuitions and it is this body that makes the reasoning rational.

Hempel, thus, claims to show that both the naturalist and the analytic-explicatory construal of scientific rationality have a descriptive facet and a fact reflecting epistemological valuation for justifying the grounds for their methodologies. Neither of the methodologies is completely *a priori*, or totally descriptive. He also finds some strong affinities between the two extreme positions with regard to the rationality of science: rationalistic and pragmatic construals. An ideal rational account of scientific inquiry must involve both the *descriptive* (explaining) and *prescriptive* (justifying) accounts of scientific inquiry. An account may justify *why an agent followed certain rules* to attain a particular goal, but this need not explain *why the agent chose to act in that way.* So we need an explanation to that effect. "An account which does possess this double aspect of explanation and justification might be called an *ideal rational account*",[25] and he calls it 'explanatory-normative methodology' (E-N methodology).

Relaxed Rational Reconstruction of Scientific Rationality

Due to the status of the desiderata, as proposed by Kuhn, there is a danger of objectivity being lost. But though the desiderata are vague, and though they can't impose an algorithmic procedure in the choice, Hempel argues, they can impose certain considerations as to direct science towards certain goal; thus providing an objective orientation. External factors influence the theory choice but those factors are constrained by desiderata, and methodological principles; they are necessary (though may not be sufficient); thus the external influences need not violate the necessary conditions of rationality. He introduces what he calls *relaxed rational reconstruction*, where some methodological principles are explicit and precise but others are vague, though they may be important ones. In this relaxed but objectivist enterprise, scientific claims are appraised not by mere subjective standards, but also by objective norms, "like the demand for conformity with the standards of deductive logic, and the prohibition of logical inconsistencies".[26]

This notion of 'relaxed rational reconstruction' leaves open the possibility of having methodology of science with more precise standards; it sees the goal of science not as something fixed once and for all but changes as scientific inquiry advances. Though there may be some constraints on objectivity and they may be partly vague, it is not "a view of science as an enterprise in which 'anything goes'".[27] It is possible to maintain science as an exemplar of rationality, only when scientific inquiry is shown to have specific goal and methodological principles to achieve it; and the means "will have to exhibit the instrumental rationality of the principles in relation to the goals".[28] Finally, for Hempel, *goal-oriented action decides the rationality of the scientific inquiry*, and if it is not so, "scientific inquiry would have to be viewed neither as rational nor as irrational, but as a-rational".[29]

Karl Popper: A Falsifying Rationalist

Karl Popper (1902 – 1994) is yet another powerful proponent of a rational model of science in the 20th century. Popper, born in Vienna, an influential philosopher of science, can be seen as the bridge-figure between Logical Positivism and Historicism. He showed a lot of interests in a variety of subjects. He never accepted any theories or concepts without a critical examination, which is a philosophical activity for him. Knowledge can progress only thorough criticism. He was the first-one to explicitly and systematically react against Positivism, which was in full bloom in his days. He began with Logical Positivism but slowly moved away from it to Historicism. He began writing in 1930s and at that time the view of Logical Positivism was taken to be the official view. One can find, in his writings, elements both of Logical Positivism and of Historicism. He published his *Logik der Forschung* in 1934, which was translated into English, as *The Logic of Scientific Discovery* in 1959. Popper is rooted in the ideas of positivism and yet he moves on to reflect some of the insights of Historicism. Of course, he is critical of Historicism also, so much so that he treats Plato, Hegel and Marx as enemies of the open society, in his famous works, *The Poverty of Historicism* (1945) and The *Open Society and Its Enemies* (1957). His attack on totalitarianism in those books is well-appreciated in many quarters. Among his popular works, we find *Conjectures and Refutations, The poverty of Historicism, Objective Knowledge, Unended Quest* etc. He has enormously contributed to the concept of liberty and to the philosophy of science. He holds fourteen Honorary Doctorates from U.S., British, German, Austrian, New Zealand and Canadian universities. In 1978, the University of Vienna gave him the honorary doctorate in the natural sciences too. He is a member (or honorary member) of twelve academies. He passed away on 17 September 1994, which was indeed a great loss to humanity.

Here in this Section I would like to present some important features of his contributions to philosophy of science, with special focus on his theory of falsification. The problem of demarcation between science and non-science and the problem of induction are, according to Popper, inter-related. When the former is solved the latter gets dissolved.[30]

Critical Rationalism – Popper's Theory of Rationality

Karl Popper claims himself to be a *Critical Rationalist.* He proposed one method which is common for both philosophy and natural sciences as well. According to him, the method "is that of stating one's problem clearly and of examining its various proposed solutions *critically*".[31] Like Socrates, he demands that everything must be put through severe scrutiny. He is a rationalist in his approach to epistemology and metaphysics. He prefers to be *critical* because, he claims, the traditional empiricism and rationalism make a fundamental mistake of answering a wrong question: 'What is the source of knowledge?' It is a mistake because this question assumes i) that there is only one source of knowledge and ii) that once a piece of knowledge-claim is traced to that source then that claim automatically becomes correct. But more knowledge of history of epistemology and deeper reflections on history and philosophy of science show that the first assumption is presumptuous and the second one is dangerous, as it would land us in dogmatism. No matter what the source of knowledge is, what matters is how one can check the mistakes in it. Both the traditional empiricism and rationalism, therefore, are equally unacceptable to Popper. Nevertheless he prefers to call himself 'critical rationalist', in the sense, he is a rationalist only to the extent he is not an empiricist. For him mind is not a *clean slate*, which has to receive everything from the world outside through the sense experience. Mind plays an active role in acquiring experiences. So it is more proper to say, 'we *make* observations', rather than 'we *have* observations'.

Rationality in Science

In Popper's paper, "The Demarcation between Science and Metaphysics",[32] one finds his strong criticism against Logical Positivism. For him science is not an accumulation of observations, but it is the frequent overthrow or replacement of old theories, with better ones and the attempt to refute them, that can be called as proper science. He is convinced that every time we do an experiment we aim at refuting some theory or the other. So he says, "…from a logical point of view all experiments are attempted refutations".[33] He sees a sort of evolution in the growth of natural science; a tentative solution is presented to the problem-situation; with that solution errors are eliminated; then as a result we have a new problem-situation. He declares that there is only one type of rationality, entertained in all our efforts to get to know the world, and that is 'the critical examination of already existing theories'. Only this justification aspect of theories is rational and only that must be the concern of philosophy of science, whereas to arrive at a theory or a hypothesis, one does not need any particular logic or rationality.

Philosophy of science would do well if it confines itself only to the realm of the context of justification. The context of discovery can be relegated to the non-rational domains. Hence, Popper is committed to the distinction between the *Context of Discovery* and *Context of Justification.* The former is concerned with the process of creation of a theory, which may be explained by sociology, psychology and the like. Creative imagination, intuition, inspiration and even dreams can be an origin for a scientific theory. Context of justification is a rational process, using the principles of logic. Hypotheses are projected as sheer guess; these conjectures are then put to the test of falsification. So science does not proceed with "theories (that) are obtained from observation by means of *a priori* principles, as rationalist philosophers think, or by probability inferences, as empiricists hold".[34]

Progress in Science

Very often people think, as the Logical Positivists did, that cumulative growth is the right mode of growth for science. But Popper says that this is not what we see in the history of science. We find the progress as a *transformative process* or as an *evolutionary process* (he does not use the word evolutionary). There is a proposed theory that is helpful, productive and successful. But it does not mean that the proposed theory would meet all our expectations. It leads to a better theory transforming the former theory. The theory of gravitation is a good example. Aristotle explained gravity in terms of natural place. It was not an empty theory. It met the expectations of that time. But in course of time that theory had to face serious problems. So it had to be changed by Buridan and others. Then came along Descartes, Kepler, Galileo, Newton and others with their mathematical and mechanical ideas. Newton put forth a better theory of attractive force. He said that all material bodies attracted each other. But soon his theory also met with problems. Einstein was able to give a better theory. According to Einstein, gravitation is not a force at all; it is a geometrical property of the distortion of the space-time continuum. Right now people question Einstein's theory of gravitation too. There is a good chance that his theory will be superseded in future. Popper, therefore, does not use the phrase '*true theory*'. He would rather say *better theory.* If we go by correspondence theory of truth there can be only one truth. A better theory is closer to truth, though we will never attain truth. This is known as the 'theory of verisimilitude' that is, scientific theories always get us nearer to the truth but never at the truth. More and more rigorous tests are applied on scientific theories to get better theories. Thus the progress in science is not just accumulation of knowledge in course of time but there is a sort of evolution and transformation in the process of knowledge.

Scientific progress faces both economic as well as ideological obstructions. The manipulations by the economically, and otherwise, powerful do sometimes hinder even very genuine theories. So, "Big science may destroy great science, and the publication explosion may kill ideas. Ideas, which are not only too rare, may become submerged in the flood".[35] Similarly history has ample examples to show how ideological and religious intolerance can stifle the growth of science, though sometimes it turns out to be a blessing in disguise. For instance, when a movie is banned for some reasons or the other, people in general show more interests to view it! So Popper opines that "The martyrdom of Girodano Bruno and the trial of Galileo may have done more in the end for the progress of science than the Inquisition could do against it".[36]

Popper's Denial of Induction

Though Popper is known for his opposition to inductive reasoning in natural sciences, it is interesting to note that at the early stages of his philosophical career, that is, by 1920s, he was an inductivist. He clearly held inductivism as the method of science, as the distinguishing feature of empirical science. In his own observation, "The present work, although highly theoretical in its main parts, has arisen out of practical experiences [with pupils], and has eventually to serve practice again. Its method is therefore essentially inductive".[37] Only by 1929 Popper began to see induction as a problem, and connected with it the problem of demarcation of science from non-science. His thesis in 1929 shifted his focus from psychology to the philosophy of science. This thesis was the first step in his long journey into inquiry of philosophy of science.

In Popper's view, the problems of demarcation and those of induction are the main problems of philosophy of science. When we solve the former, it becomes easier to solve the latter, for the latter is only an aspect of the former. He rejects inductive logic in very strong terms: "As for inductive logic, I do not believe that it exists. There is, of course, logic of science, but that is part of an applied deductive logic: the logic of testing theories, or the logic of the growth of knowledge".[38] Inductive method largely relies on the process of evidence collection to support its conclusion. However, it is permanently away from any conclusive truth. Only deductive arguments are legitimate in guaranteeing certainty, and therefore for Popper, the method of science is basically 'proposing conjectures and refuting them'. He agrees that scientific theories are always beyond conclusive verification and also there is no meaningful solution for the problem of induction.

The most severe attack on induction, as we know, came from Hume. He put forth the problem of induction first in his *Treatise of Human Nature* (1739): "Let men be once fully persuaded of these two principles, that there is nothing in any object, considered in itself, which can afford us a reason for drawing a conclusion beyond it; and, that even after the observation of the frequent or constant conjunction of objects, we have no reason to draw any inference concerning any object beyond those of which we have had experience".[39] He gave further criticism of it in his *Enquiry concerning Human Understanding*. Nature consists of innumerable regularities. But from the observed regularities one cannot logically infer a conclusion about future. But as per *the principle of empiricism* scientific theory must be based upon the results of observational experiments. Due to the insolvable clash between these two demands he decided to give up rationalism in science and declared that our knowledge of laws relied on 'habitual association', rather than reason. Thus he shook the very foundation of al! cognitive exercises as he showed that inductive inference was not rational. But Popper, though he acknowledges the clash between the probable nature of inductive reasoning and the empirical demand for certainty in science, does not end up with anti-rationalism, as the scientific theories depend on *critical reasoning*. That is why, even scientific revolutions are rational, because "it is rationally decidable whether or not a new theory is better than its predecessor".[40]

Hypothetico-Deductive Method of Science

Going against the Inudctivist Method of Logical Positivism, Popper advocates Hypothetico-deductive method. Popper agrees with Hume about the problem with the inductive reasoning, that there is no logical justification to make general laws only from the known instances. However, he disagrees with Hume to maintain that science is still rational. For science can do away with inductive reasoning as it relies on, what he calls, hypothetico-deductive method. A conjecture hypothesis is proposed; testable consequences are deduced; these are put to the test of falsification; if it fails the test, (i.e. if falsified), the hypothesis is rejected; but if they withstand the test, then hypothesis becomes corroborated (not 'proved') and it may be falsified in future. Since science is actually a series of "conjectures and refutations", we get better, but not true, theories. Science can never attain the complete and absolute truth, rather it can only get "nearness to the truth" (*verisimilitude*). Since science uses only H-D method, there is no need for inductive reasoning.

Falsification – Popper's Solution to the Problem of Induction

Going against the current of the then-scientific community Popper strongly advocated the principle of falsification as the solution to the problem of demarcation and which in turn solves the problem of induction as well. He reformulated its basic tenets in his lecture at Peterhouse, Cambridge, in Summer 1953, as follows[41]:

a) It is easy to obtain confirmations, or verification, for nearly every theory – if we look for confirmations;

b) Confirmations should count only if they are the result of risky predictions...;

c) Every 'good' scientific theory is a prohibition: it forbids certain things to happen. The more a theory forbids, the better it is;

d) A theory which is not refutable by any conceivable event is non-scientific. Irrefutability is not a virtue of a theory (as people often think) but a vice;

e) Every genuine *test* of a theory is an attempt to falsify it, or to refute it. Testability is falsifiability; but there are degrees of testability: some theories are more testable, more exposed to refutation, than others; they take, as it were, greater risks;

f) Confirming evidence should not count *except when it is the result of a genuine test of the theory*; and this means that it can be presented as a serious but unsuccessful attempt to falsify the theory. (I now speak in such cases of 'corroborating evidence').

g) Some genuinely testable theories, when found to be false, are still upheld by their admirers – for example by introducing *ad hoc* some auxiliary assumption, or by re-interpreting the theory *ad hoc* in such a way that it escapes refutation. Such a procedure is always possible, but it rescues the theory from refutation only at the price of destroying, or at least lowering, its scientific status.

Though Popper does not recognize induction as a problem, yet he claims to have solved the problem of induction, by strongly refuting it. He in fact finds problem with the traditional formulation of the inductive problem, which for him, is incomplete and misleading: "What is the justification for the belief that the future will resemble the past? What is the justification of so-called *inductive inferences?*" – He argues that this is a bad formulation as "there is simply no such logical entity as an inductive inference".[42] Therefore he demands that "I think that I have solved a major philosophical problem: the problem of induction.... This solution has been extremely fruitful, and it has enabled me to solve a good number of other philosophical problems".[43] Of course one must not fail to note that he includes dissolution or rejection of a problem too as a solution of the problem. Therefore, he seems to dissolve, not to solve, Hume's problem of induction.

Popper agrees with Hume in affirming that induction is invalid and is never justified and therefore the demand for justification of inductive reasoning is invalid. He joins with Hume in stressing its denial that the experiences with the known instances cannot justify a conclusion about the unknown ones. He acclaims Hume for the same: "I found Hume's refutation of inductive inference clear and conclusive".[44] However unlike Hume he does not think that induction is essential ad necessary in our reasoning process. Popper argues that induction "is simply a mistake" and "It is a kind of optical illusion. What we do use is a method of trial and of the elimination of error; however misleadingly this method may look like induction…".[45] Only when one has the *justification* as the central problem one faces the problem that Hume faces. But for a critical rationalist, like Popper, the inability to justify a theory is not at all the main concern. Therefore he says, "(the inability to justify) never worries the critical rationalist".[46] Since all our human knowledge, including scientific knowledge, is only conjectural and hypothetical, we don't need any validity for that knowledge. The problem of justifying induction disappears. Therefore, "The place of the problem of induction is usurped by the problem of the comparative goodness or badness of the rival conjectures or theories that have been proposed".[47]

Popper identifies two aspects in Hume's description of the problem of induction: one is *the logical aspect*, i.e. the past observations never justify the unobserved future, not even the probability of future; and the second is *the psychological aspect*, i.e. induction, in spite of its rational invalidity, is a psychological fact, by which we are conditioned. Popper attempts a way out in order to avoid Hume's falling into a permanent skepticism, because, "According to Hume, all our knowledge, and especially all our scientific knowledge, is just irrational habit or custom, and it is rationally totally indefensible".[48] Popper faces the *logical problem* by 'preferences for more informative content and more ability to stand tests, and this solution can be applied by *the principle of transference* to meet the *psychological problem*: "… the solution of the logical problem of induction, far from clashing with those of the psychological or pragmatic problems, can, with some care, be directly transferred to them"[49] and thereby we can escape the irrational conclusions. Though the universal theories are perpetually conjectures and hypotheses, *we have a right for preference based on falsification.* This ability to withstand tests differentiates a scientist from a lunatic. It is not that the scientists always propose their theories on firm observational basis and the lunatics lack that; rather the only difference between them is that the scientists' theories are able to pass the tests (based on observation), whereas those of the lunatics easily fail the tests.[50]

Falsifiability gives us the clear demarcation between scientific and non-scientific theories or statements. For instance, Einstein's theory of relativity is falsifiable, and therefore it is scientific, whereas Marx's theory of history, Freud's psycho-analysis and Alfred Adler's so-called 'individual psychology' "...though posing as sciences, had in fact more in common with primitive myths than with science; ... they resembled astrology rather than astronomy".[51] The proponents of these three theories would turn any relevant phenomenon as verifying their claims and practically no instance is given which can falsify those theories. This constant verification was taken to be the source of their being scientific but, Popper realizes, "It began to dawn on me that this apparent strength was in fact their weakness".[52]

As scientific theories are unrestrictedly general, they can't be verified by any possible accumulation of observational evidence. No collection of particular observations will verify a general statement; at the most, it may make it probable. So the criterion to find out whether a statement is scientific or not depends on whether the statement is falsifiable or not. If a statement is falsifiable, then it is scientific, and if not, then it is non-scientific. He does not say that a non-scientific statement is meaningless or vacuous. So *falsifiability* (or refutability) *by observation* must be the criterion of the empirical and scientific character of a theory: "A theoretical system which is of such a kind that, whatever facts in its domain, there is a method of bringing the theory into agreement with the facts cannot be regarded scientific".[53] His criterion does not demand actual refutation, but it is enough that it is *refutable in principle*. He prefers this to verification, especially for natural sciences, as they are concerned with general laws (natural laws), which can never be proved by one instance, whereas can be refuted by one instance. To falsify means we can specify a situation or a set-up by which we can decide empirically whether a theory has been ruled out or not. A theory, as a universal statement can never be conclusively verified, whereas falsifying can disprove it conclusively. Thus, *the Principle of Falsificationism* (PF) is one of Popper's significant contributions to Philosophy of Science. Though it is contrasted against Positivists' *Verificationism*, it is proposed as the criterion of demarcation between science on the one hand and non-science or pseudo science on the other. It is a criterion for *scientificity*. It is not a criterion for *meaningfulness* as the Positivists would hold. For, "…if a theory is found to be non-scientific, or 'meta-physical' (as we might say), it is not thereby found to be unimportant, or insignificant, or 'meaningless', or 'nonsensical'",[54] though they cannot be given the status of science as it is not backed by empirical or observational evidence in the normal sense of scientific evidence. Scientificity consists in testability, which is identified with falsifiability. Testability cannot be identified with verifiability or probability.[55] Falsifiability has nothing to do with the content of the statement, rather it is an entirely logico-empirical characteristic of any statement to be scientific. Falsifiability is the basis for the criterion for being scientific or not; no other criterion must be incompatible with it; so Popper makes it clear that all the other rules in science must not question falsification.

Popper gives many examples[56] – from ancient times of Parmenides – Leucippus through the modern times of Galileo, Newton, and as late as 20th century's Quantum Physics – to show that science advances by falsifications, which lead to problem-situations. They in turn compel the scientists to come out with new innovative and ingenious hypotheses. Therefore, problems, not observations, are the starting point for science. When the scientific community encounters a problem, it proposes a solution. This hypothesis can originate from any source, including an intense intuition or a creative imagination. Then tests are carried on to falsify it. If it passes the test it is tentatively accepted (Corroborated) or else it is discarded. Therefore falsifiability plays a crucial role and therefore progress of science depends on highly falsifiable hypotheses.

Only falsity of universal laws or theories can be logically, conclusively deduced, but not their truth. For instance, spotting out even a single Indian at a given space and time, who is not rich, one can deduce conclusively, 'Not all Indians are rich'. If a theory claims more it has more chances for falsification. More falsifiability is always better for a theory. For instance, 'Students of the Hyderabad Central University are hard-working' is more falsifiable than 'Students of Philosophy of the Hyderabad Central University are hard-working'. For, any instance that refutes the latter refutes the former too and the former gives us more information than the latter. Therefore, once a refuting instance occurs one must mercilessly dethrone that theory, however successful and attractive in the past the theory might have been. He states in no unclear terms, "Once a mistake, or a contradiction, is pin-pointed, there can be no verbal evasion: it can be proved, and that is that".[57]

A sophisticated version of falsificationism demands not just a falsifiability of a hypothesis but demands that "A hypothesis should be more falsifiable than the one for which it is offered as a replacement".[58] Sometimes modifications are made in a theory to improve upon the problematic theory. These modifications are not *ad hoc,* as they are independently testable. For the growth of science *falsifying* a cautious conjecture is more important than *confirming* a cautious hypothesis. For "If a *bold* conjecture is falsified, then all that is learnt is that yet another crazy idea has been proved wrong.... Similarly, the confirmation of cautious hypotheses is uninformative. Such confirmations merely indicate that some theory that was well established and regarded as unproblematic has been successfully applied once again".[59]

Conventionality of Methodological Rules

The methodological rules are *a priori,* and not empirical, for they are not accepted on the grounds of any experience. They are not analytic either, for they are not warranted by virtue of the meaning of their constitutive terms. Further, their status is not as that of logic, for they don't have truth-value; there is no place for variables and these rules are accepted on decisions or agreement. There are certainly equally workable rules of methodology, like the game of Chess. It is agreed upon a decision to play in certain ways. However it is possible to have created some other set of rules, entirely different from the present ones. Therefore Popper declares, "Methodological rules are here regarded as *conventions*. They might be described as the rules of the game of empirical science. They differ from the rules of pure logic rather as do the rules of chess, which few would regard as part of *pure logic*: seeing that the pure rules of logic govern transformations of linguistic formulae, the result of an inquiry into the rules of chess could perhaps be entitled 'The Logic of Chess', but hardly 'Logic' pure and simple. (Similarly, the result of an inquiry into the rules of the game of science – that is, of scientific discovery – may be entitled 'The Logic of Scientific Discovery')".[60]

Fallibilism is at the core of any methodological rule. These rules are indeed tentative and at any point of time they can be partially correct: "Whatever we accept we should trust only tentatively, always remembering that we are in possession, at best, of partial truth (or rightness), and that we are bound to make at least some mistake or misjudgment somewhere - not only with respect to facts but also with respect to the adopted standards".[61] Therefore, the notion of falsification is ever present in the enterprise of science and that makes science, in principle, a never-ending game. In the words of Popper, "The game of science is, in principle, without end. He, who decides one day that scientific statements do not call for any further test, and that they can be regarded as finally verified, retires from the game".[62] Thus a theory in science ever remains open-ended. Even a hard-testing provides us with only the corroboration of the theory and never a conclusive proof of it. All these fallibilistic tendencies would make his realistic claims of science to be taken with caution. Perhaps with Hume, he has to accept certainty in practical knowledge, while skepticism and fallibilism in theoretical knowledge.

Popper's Notion of Objectivity

Testability provides not only rationality to science but also it ensures objectivity in science: "For the requirement of scientific objectivity can also be interpreted as a methodological rule: the rule that only such statements may be introduced in science as are inter-subjectively testable".[63] The objectivity is rooted in the critical tradition of the community, which often makes it possible to critically evaluate a dogma, even if it is well-accepted for a very long time. Thus the objectivity in science does not depend on an individual scientist, rather it is a social phenomenon, resulting from mutual criticism

Popper's framework of objectivity always has room for non-rational aspects. This is because scientists, unlike mathematicians, do act upon many non-rational elements in the process of proposing new hypotheses, which would be tested in due course. He clarifies that science is rational and objective not because of the individual scientists; in fact they are also like poets, often inspired by non-rational insights; however this fact need not diminish their scientific vigour, because, "...Kepler's beautiful Pythaogrean dreams of the harmony of the world system did not invalidate the objectivity, the testability, or the rationality of his three laws – nor the rationality of the problem which these laws posed for an explanatory theory".[64] So, one can understand that "Scientific discovery is akin to explanatory storytelling, to myth making and to poetic imagination".[65] The notions of rationality and objectivity can never be allowed to be based on individual scientists. If it were to be so, it would even destroy the very progressive and revolutionary nature of science. Objectivity is rather safeguarded by the convention of the scientific community, a sort of inter-subjective objectivity.

Difference between Verification and Falsification

The process of Verification assumes *the truth of the theory* and therefore it looks for confirming instances. It is like the innocence of the accused is assumed and when the evidence is in her favour, she is declared innocent. Observations are taken to be direct and uncontaminated by any kind of theoretical background. From these observation-statements theories are inductively arrived at, and thereby, the theories are also shown to be true. The final result of the verification process is that *the theory is established to be true*. This constant drive for verification was taken to be the source of their being scientific but, Popper realizes, "It began to dawn on me that this apparent strength was in fact their weakness".[66]

On the contrary, the technique of falsificationism differs at every point. *The falsity of the theory* is assumed. It is like the accused is taken to be a criminal and when there is no evidence in his favour it is regarded as the 'crime has not yet been proven'. Similarly when the evidence fails to falsify it is held that the theory is not yet falsified, not that the theory is true or verified. Again, observations are not the starting point. Theories are not abstractions from observations; rather they are pure conjectures, just creations of a free mind. The initial probability of the theory being true is '0'. When the theory passes a test of falsification the final result is that *the theory is just corroborated* and if a theory fails the test of falsification it is *completely falsified.*

Popper's contribution in the area of rationality in science is very interesting and crucial. I'll come back to Popper's ideas in the next chapter for evaluating them. Now let me proceed to the ideas of Thomas Kuhn, another milestone in philosophy of science.

SECTION 2
PROPONENTS OF NON-RATIONAL MODELS OF RATIONALITY IN SCIENCE

The salient features of Thomas Kuhn and Paul Feyerabend are introduce here in this Section. With Kuhn's epoch making work, *The Structures of Scientific Revolutions* (1962), the scientific community was led to a new awakening to realize that science is fundamentally a social enterprise, and one cannot have an appropriate picture of science without its socio-historical considerations. He looks at science as a paradigm-oriented activity, as a paradigm at a given time governs everything of science, though in his later stage he reduced his stress on paradigm. He stressed on the conceptual aspect of incommensurability in his early stage but he softened it to the level of its linguistic consideration only, at his later stage, to claim that "incommensurability equals untranslatability". He explains the non-rational factors in choosing a theory. However there are five important features of a good scientific theory, as we shall see shortly, which prevent the scientific enterprise from becoming irrational or relative enterprise and they make science still rational. Feyerabend, a vigorous critic of the rational models of science, denies any method in science and shows that 'anything goes' in science. He recommends a sort of 'anarchical situation' in science, which makes growth and development possible. Arguing for a stronger version of incommensurability he shows the need for plurality of theories, which goes against Kuhn's demand for a single paradigm at a given time. He points out the shortcomings with the claims of classical empiricism and cautions that science can really grow only when the basic claims of classical empiricism (like Consistency condition and the Meaning Invariance Condition) are overlooked or modified. Even the theory of falsification does not make science very rational as he shows many instances in the history of science where falsification was deliberately overlooked. He demands that the rational picture given to science needs to be questioned and a 'human face' needs to be given, rather restored, to it. These ideas of non-rational approach are critically comment on later in the next Chapter.

Thomas Samuel Kuhn: A *Paradigm* for Paradigms

After World War II there was a big shift in the understanding of science. The highly glorified picture of science was shown to be untenable with the better awareness of history of science. Science began to be seen in its proper perspective, that is, as a social enterprise and therefore essentially fallible like any other human endeavour. Philosophers of science too needed to change their outlook of science and scientists. As Robert Klee puts it, "Philosophers of science have often had a tendency to misrepresent workaday scientists as 'little philosophers,' incorruptible investigators who are maximally rational and worried about epistemological and ontological issues",[67] while in reality, the scientists are not so.

Historicist theories of rationality in science are proposed as an alternative to the traditional theories of rationality. The historicist theories of rationality demand that a good theory of rationality must in some way fit the history of science and show that most of the episodes in the history of science are rational. Historicists make a useful distinction between methodology and meta-methodology. As Carl Matheson explains, "a methodology for scientific rationality is a theory of rationality: it tells us what is rational and what is not in specific cases… On the other hand, a meta-methodology provides us with the standards by which we evaluate the theories of rationality which constitute our methodologies."[68]

Kuhn (1922–1996) is given the credit of bringing the trends of Historicism into the Philosophy of Science. He received his Ph.D. in theoretical physics from Harvard University in 1949. Soon he became interested in History of Science. In 1956 he joined the University of California at Berkeley to teach history of science. His world-shattering work, *The Structure of Scientific Revolutions* (*SSR*), 1962, evoked controversial discussions in extremely diverse fields. It has been translated into twenty languages. He moved on to Princeton University in 1964 and later to MIT. He suffered from blood cancer and he died in 1996. Some of his papers both on the history and the philosophy of science are given in *The Essential Tension*, 1977.

The notion of scientific rationality saw a radical change with Kuhn's *SSR*. Till then the most predominant theories were confirmationism (verificationism) of Logical Positivism and falsificationism of Popper. While the Positivists held that to be rational a theory needs to be confirmed or verified, Popper insisted upon falsification. However Popper himself was aware of the limitations of his own theory. The critics of falsificationism clearly brought out the major lacuna of the theory of falsificationism, that is, Popper too does not fare better than verificationists, as after all what one needs is to *verify* the counter instance.

The most important contribution of Kuhn to philosophy of science is that he clearly brought out the need and relevance of the knowledge of history in understanding of science. Up until the arrival of *SSR* the scientific community had been thinking the fundamental unit of science is theory. But he insisted that structure (or paradigm as calls it) is the foundational unit, and it sustains minor theoretical changes. And also, "Kuhn's work highlighted the real problems that historically aware accounts of rationality face: when all is said and done, there may be no trans-historical rule for rational scientific procedure".[69]

In this section I see Kuhn in two broad categories – *Early Kuhn* and *Later Kuhn*, where important Kuhnian concepts like Paradigm, Incommensurability and Theory Choice.

Early Kuhn (Till late 1970s)

Kuhn's first book, which was originally published in the International Encyclopedia of Unified Science, *The Structures of Scientific Revolutions*, appeared in 1962. Nearly a million copies in 16 languages were sold. It became the most eagerly sought after book in diverse disciplines, like Psychology, Education, History, Sociology, History and Philosophy of Science.

Science as a Cycle: Science, for Kuhn, is basically a long period of normal science, occasionally interrupted by crises which lead to short periods of revolutions. History of Science presents regular patterns like this: Normal science → anomalies get intensified → crisis → revolution → emergence of a new paradigm → normal science → anomalies get intensified → .... The main preoccupation of normal science is the puzzle-solving activity. A scientist is primarily a puzzle-solver, who is always busy trying to apply the known solutions to the problems at hand, and not so much of an innovator. "Under normal conditions", opines Kuhn, "the research scientist is not an innovator but a solver of puzzles, and the puzzles upon which he concentrates are just those which he believes can be both stated and solved within the existing scientific traditions".[70] When the existing theories get into conflict with the fresh observational data anomalies arise and mostly anomalies are overlooked. Not that every anomaly is expected to lead to a crisis. A normal scientist must learn to ignore them, or else she can't carry on with her research. So Kuhn says, "... if an anomaly is to evoke a crisis, it must usually be more than just an anomaly.... The scientist who pauses to examine every anomaly he notes will seldom get significant work done".[71] She needs to have a basic faith in the puzzle solving ability of the present theory; for "without that faith their work would be wasteful of time and talent".[72] But when anomalies grow too many in number and too strong in their intensity to tolerate a crisis sets in leading to a revolution. After this short period of revolution, once again a long period of normal science ensues. Sometimes anomalies demand some revisions in theories to understand a new phenomenon, or it may require a revision of techniques / instruments to circumvent the anomalies, without necessarily explaining it. Kuhn, therefore, looks at an anomaly, not necessarily as an obstacle in the research, but a source of growth. Therefore as Rouse put is, "They (anomalies) need to be resolved to the extent necessary to get on with research. If an anomaly is sufficiently obstructive, or interesting in its own right, it can replace the original research topic".[73]

With the emergence of a new paradigm lots of changes in the belief system take place. It is not a piecemeal or a gradual change. Change is always a whole-sale change, not just in bits and pieces. For instance, when the phlogiston theory of combustion was replaced by the oxidation theory of combustion, a great change in the world of chemistry took place. As James Ladyman points out, "Up until this time (of oxidation theory) it had been thought almost universally that there was really only one kind of 'air', although it could be of different degrees of purity; after this revolution it was accepted that oxygen is but one among the constituents of ordinary air".[74]

Paradigm - A Paradigmatic Concept!: Paradigm is almost synonymous to a scientific community, in the sense of both the logical and physical closeness. In a very broad sense all natural scientists can be brought under one paradigm. Here are a few examples for a paradigm: Aristotelian physics, Copernican astronomy, Newtonian physics, Ptolemaic astronomy, the phlogiston and the oxygen theories of combustion, (Daltonian Chemistry)[75], the fluid theory of electricity, the caloric theory of heat, particle physics, wave optics, relativistic physics, quantum physics and so on. In a narrower sense the communities of physicists, chemists, botanists can be said to belong to different paradigms. Within the paradigm of chemistry, major sub-groups, like Organic chemists, protein chemists or physical chemists are possible.

The personal value systems of the scientists do affect their scientific taste too. For instance, as Einstein had a strong inclination of realistic view of the world, he held that the scientific theories gave the real picture of the world; whereas some of the quantum scientists had only an instrumentalist view of the world and hence believed that the theories were only the instruments to describe or predict the world. Kuhn also insists that *the sociological and psychological factors* too play a role in the formation of the disposition of the scientists. Some societies may be ready to take risks and others may prefer to be safe by following the prevailing system. The latter may not be willing to sponsor a research and make the infrastructures available for the research. Therefore, as the nature of knowledge does depend also upon the dispositions of the society, science cannot properly be studied without taking note of social elements.

Paradigms are constitutive of nature. Each paradigm imposes its own ontology on scientific thinking and has its own empirical data, so much so that one scientist working in a paradigm cannot understand the data of another paradigm. Each paradigm has its own facts: "Led by a new paradigm, scientists adopt new instruments and look in new places. Even more important, during revolutions scientists see new and different things when looking with familiar instruments in places they have looked before. It is rather as if the professional community had been suddenly transported to another planet where familiar objects are seen in a different light and are joined by unfamiliar ones as well".[76] *Paradigms are constitutive of science.* Each paradigm has its own problems, its solutions and methodology. It binds the scientists so deeply that they "disagree about what is a problem and what is a solution; they will inevitably talk through each other when debating the relative merits of their respective paradigms".[77] *Paradigms are constitutive of meaning.* Paradigms determine the meanings of the terms used in scientific theories. And finally, *paradigms are constitutive of subject matter of discourse.* Within a paradigm there is a strong consensus-thinking. It is present in mathematics and astronomy from the very beginning. But in modern science such consensus is only a recent phenomenon.[78]

Paradigms, thus, play an essential role in the process of scientific inquiry. Kuhn demands a very strict adherence to the paradigm on the part of the scientists. They would do anything to be faithful to it. They seem even to be 'immoral' in maintaining the value of a paradigm. As James Ladyman puts it, "According to Kuhn, scientists are often very much attached to a paradigm, and sometimes particular individuals will do almost anything to retain it in the face of contradictory evidence, including perhaps, distorting experimental data, using institutional power to stifle dissent, using poor reasoning and bad arguments to defend the status quo, and so on".[79]

Incommensurability – A Move to Relativism and Irrationality?: Since science is a paradigm-bound activity fundamental theories become incommensurable. Kuhn hit upon this great idea as early as a graduate student in theoretical physics, when he had to give lectures on Galileo. When he compared Galileo's concepts with those of Aristotle he was stunned to see how Aristotle, a great mind, saw them in such a primitive way; but soon he realized that the problem lay with himself, not so much with the genius mind of Aristotle, because he confused it with his own understanding of the concepts: "All at once the text made sense to me, but only if I changed my way of reading, altering some of the concepts – the meanings of some of the words – that I, coming from a later age, had brought with me to the text".[80] From then on, at various occasions, not only with ancient thinkers, but also with modern ones like Planck and Bohr, Kuhn realized that there is a significant breakdown between older and newer currents of thought.

The proponents of two fundamental different theories, even though they use the same vocabulary, often speak from different viewpoints. There is no common measure among two rival theories; there are no common meanings and so they can't be translated into a single language. For instance, Kuhn, compares Newton and Einstein's theories: "the physical referents of these Einsteinian concepts [space, time, and mass] are by no means identical with those of the Newtonian concepts that bear the same name".[81]

Each paradigm has its own standards /criteria to assess the merits/demerits of a theory; when two theories emerge from two different paradigms, they can't be meaningfully compared or evaluated. Adherents from two different paradigms "will inevitably talk through each other when debating the relative merits of their respective paradigms... (since) each paradigm will be shown to satisfy more or less the criteria that it dictates for itself and to fall short of a few of those dictated by its opponent".[82] (From this 'evaluative incommensurability', later on Kuhn moved on to lay more stress on 'linguistic incommensurability'; by 1983 he completely distanced himself from evaluative incommensurability, as we will see later).

Many critics, like Dudley Shapere (1966), Stephen Toulmin (1972), and Kenneth F. Shaffner (1970), see Kuhn's incommensurability leading to relativism and irrationality. Edwin Hung[83] points out *four theses of incommensurability*, emerging from the nature paradigms: a) Paradigms do not share facts in common; b) Scientists belonging to different paradigms live in 'different worlds'; c) Paradigms do not share problems or standards of solution; and d) paradigms don't share terms (with the same meaning), or statements or subject matter. Similarly, according to Gerald Doppelt, Kuhn's portrayal of paradigms lead to *four types of relativism*: "Kuhn variously argues in different contexts that rival paradigms are incommensurable i) because they do not speak the same scientific language, ii) because they do not address, acknowledge, or perceive the same observational data, iii) because they are not concerned to answer the same questions, or resolve the same problems, and iv) because they don't construe what counts as an adequate, or even legitimate explanation in the same way"[84] and Malone adds the fifth source of relativism: v) because they pertain to worlds consisting of incomparable entities, processes, etc.".[85]

But Kuhn clarifies that it is not the case that theories can never be compared; they can be compared through concrete measurements of phenomena, which are described in terms shared by two theories. For instance, in spite of the problematic term 'planet' in Ptolemaic theory and Copernican theory "The quantitative superiority of Kepler's Rudolphine tables to all those computed from the Ptolemaic theory was a major factor in the conversion of astronomers to Copernicanism".[86] There are also other criteria for comparison; e.g. "there are arguments... that appeal to the individual's sense of the appropriate or aesthetic – the new theory is said to be 'neater', 'more suitable', or 'simpler' than the old".[87] Hence there are many grounds for comparison, despite incommensurability, including "accuracy, scope, simplicity, fruitfulness, and the like".[88]

The Essential Tension: Kuhn agrees with the normal expectation that a scientist essentially needs flexibility and open-mindedness. But he points out another need, which is equally important, if not more, which he terms as 'convergent thinking'. He observes, "'Convergent thinking' is just as essential to scientific advance as is divergent".[89] He elaborates further that only the traditional concepts of flexibility and open-mindedness are not enough for the revolutionary growth of science. It badly needs a firm rootedness in the present tradition: "Only investigations firmly rooted in the contemporary scientific tradition are likely to break that tradition and give rise to a new one".[90] From the very beginning a sort of training in convergent thinking has been taking place in the world of science and without that the present developments of science are well nigh impossible. The convergent thinking is instilled in the fresh minds through the students' text books, which contain the accepted problem solutions. The student is then asked "to solve for himself problems very closely related in both method and substance to those thought which the textbook or the accompanying lecture has led him".[91]

The convergent thinking is ensured by paradigms; therefore a paradigm can be said to be a blessing in disguise, in the sense, for though it does not easily allow new creative ideas, it certainly ensures the steady movement of science. Edwin Hung gives an analogy: "Perhaps taking paradigms as railway tracks would be a better analogy. The rails in one sense constrain the movements of the train, but in doing so guide it smoothly in a definite direction".[92] Having said all these, therefore, one realizes that a scientist, worthy of her name, must be an enthusiastic innovator and a faithful traditionalist at the same time! She needs to possess the qualities of openness and conservatism at the same time. So Kuhn observes very clearly, "the productive scientist must be a traditionalist who enjoys playing intricate games by pre-established rules in order to be a successful innovator who discovers new rules and new pieces with which to play them".[93] And precisely this is the *essential tension* Kuhn speaks of.

Paradigm – Re-visited: Kuhn received varied criticisms against his notion of paradigm. He himself points out that Shapere's criticism (1964) on the concept of paradigm is the most thought provoking and thoroughly negative account[94]. Masterman (1970) finds about 22 usages of paradigm in Kuhn's *The Structure of Scientific Revolutions*, ranging from "a concrete scientific achievement" to "a characteristic set of beliefs and achievements including instrumental, theoretical and metaphysical commitments together".[95] With his 2nd edition of *SSR* (1970) he preferred a *disciplinary matrix* to 'paradigm'. In this edition also he tries to clarify the doubts and answer the queries raised by the critics. With 'disciplinary matrix' Kuhn finds lots of advantages: he says, "... less confusion will result if I instead replace it with the phrase 'disciplinary matrix'. 'Disciplinary' because it is the common possession of the practitioners of a professional discipline; 'matrix' because it is composed of ordered elements of various sorts, each requiring further specification".[96] The disciplinary matrix includes these important elements: a) symbolic generalizations, b) models, both ontological and heuristic, c) values, and d) exemplars – which are essential for the scientific community to obtain or validate genuine knowledge. All sciences have their own set of symbolic generalizations, which are used without any qualms for justification. He gives an example from physics: f = ma. A student of physics uses this formula without any question about its validity or so, because everyone in the community at a given time accepts it without any difficulty. Also the same formulation may assume different forms according to the applications in various domains within physics.

Disciplinary matrix exhibits a fundamental agreement over a lot of things – of fundamental building blocks of the universe, of our possible interactions with them, of what it means to have a solution for a problem, of the meaning of evidence for such a solution etc. When a new problem is shown to be analogous to a problem that has been already solved, the members try to apply the same solution to the present problem. These exemplars make the newcomers learn about the fundamental and already accepted concepts of the particular community, therefore, "Without exemplars he would never learn much of what the group knows about such fundamental concepts as force and field, element and compound, or nucleus and cell".[97]

The scientific community agrees upon certain assumptions which can't be tested (like giving preference to field theory over particle theory). These assumptions show the direction of further research. They all agree with the characteristics of a good scientific theory[98]: a) *Accuracy* – that is, a theory's consequences should be in demonstrated agreement with the results of existing experiments and observations; b) *Consistency* – a theory must be consistent, not only internally with other theories of the paradigm, but also with other currently accepted theories applicable to related aspects of nature; c) *Broad Scope* – a theory's consequences should extend far beyond the particular observations, laws, or sub-theories it was initially designed to explain; d) *Simplicity* – the theory must be able to bring order to phenomena that in its absence would be individually isolated and, as a set, confused; and e) *Fruitfulness* – it should disclose new phenomena or previously unnoticed relationships among those already known.

Later Kuhn (Early 1980s Onwards)

Most of the criticisms against Kuhn, brought out till 1977, may now be softened, in my opinion, as Kuhn comes out with his *The Road since Structure*, in 1990. The same phrase is the title of a well-thought out collection of Kuhn's essays produced between 1970 and 1993 (also an interview with him held in 1995) and it was posthumously published in 2000. Most of the essays here portray Kuhn in a new outfit, which we call 'Later Kuhn'. One can come across a whole lot of series of parallels between Kuhn and Wittgenstein especially in the area of linguistic analyses. Further, as Rupert Read points out in his review of *The Road since Structure*, "It (the book) also evidences a wish to be comprehensible (on their own terms) to the 'Analytic' philosophers by whom Kuhn was most harshly criticized (e.g. Davidson)".[99]

Thus the Later Kuhn seems to be a completely modified version of the earlier version, as he himself confesses in 'the Robert and Maurine Rothschild Distinguished Lecture in 1991': "In this more constructive portion of my talk, I'll be drawing on bits and pieces of a far larger project, the book on which I'm currently at work. Drastic condensation and simplification of even those pieces is going to be essential, and the central part of my project – a theory of what I once called incommensurability – will have to be omitted entirely".[100] Further, "If I were now rewriting", he admits in his paper "Commensurability, Comparability, Communicability" (first published in 1982), "*The Structure of Scientific Revolutions*, I would emphasize language change more and the normal/revolutionary [science] distinction less. But I would still discuss the special difficulties the sciences experience with holistic language change, and I would attempt to explain that difficulty as resulting from the sciences' need for special precision in reference determination".[101]

From the essays in *The Road since Structure* we learn the four fundamental themes emerging out: [102] a) Science is undoubtedly a cognitive empirical investigation of nature, endowed with a special sort of *progress*. But this progress is not to be understood as though science moves towards the fullness of truth, but progress in ever-improving technical puzzle-solving ability, functioning under the strict criteria of success and failure; b) Science is to be understood as basically *a social enterprise*, where the scientists work within the community and only occasionally they move out of it to find solutions for certain anomalies; c) Modifying the earlier notion of scientific development as a series of long period of normal science, occasionally punctuated by short periods of revolution, now Kuhn maintains that *science is period of development within a coherent tradition divided occasionally by periods of 'speciation'* (like the biological evolution of species of Darwin's theory), into two distinct traditions with somewhat different areas of research; and finally d) distinguishing between *commensurable languages* (where translation is possible, between two languages, and therefore, what is said in one language can be said in another) and *incommensurable languages* (where only paraphrasing is possible between the languages!) he gives *a linguistic twist to the understanding of incommensurability.*

Incommensurability – A New Perspective: Kuhn himself agrees that there is a sort of uncertainty and a need for refinement in his understanding of incommensurability. He acknowledges that earlier his views on incommensurability not unambiguous and sometimes even contradictory: "Initially, recognizing the close resemblance between my experience with Aristotle and that of Quine's radical translator with the community he studied, I tended to see translations as a way to resolve incommensurability". But at a later stage he realized "that translation cannot bridge the gap between the language a reader brings to a text from an older time and the passages in that text which involve incommensurable terms. If there is recourse it must be of another sort".[103]

The intensity of the notion of incommensurability seems to soften over Kuhn's long intellectual career. In the beginning he, along with Feyerabend, was for strict incommensurability though there were some differences[104] between them. Howard Sankey (1993) figures out three clear stages in Kuhn's development of the idea of incommensurability – *the earlier position*[105], *the transitional phase*[106] and *the later position.*[107] In the beginning (1962) he holds that the proponents of different paradigms find it impossible to meaningfully communicate with one another as their theories are incommensurable. But a few years later (1970), in the *Postscript* he softens the stand to confine incommensurability to the translation across paradigms. Further, down the line (late 1970s and beginning of 1980s), he restricts it to semantic differences, subjecting it to Quinean indeterminacy of translation. Direct comparison of theories necessitates their formulation in a common language. Unfortunately such a language is not available and also the exact translation between languages of theories is impossible. Finally (from the early 1980s till the end), he has a rather refined conception and claims that there is a translation failure between localized cluster of inter-defined terms of the languages articulating the theories.

In other words one can see Kuhn's understanding of incommensurability, at four levels. From a global incommensurability he moved on to a local one: "Incommensurability is always local, restricted to small sets of interrelated terms, ordinarily terms which must be learned together".[108] He is not happy with the gestalt-switch metaphor, for incommensurability does not depend on the ways of 'different seeing' but on "the conceptual constraints imposed by language and on the way these vary from one language to another".[109] Then he relies on the distinction between the process of learning a language and that of translating it (though unfortunately both these two are conflated in the philosophy of language). Our natural human capacity enables one to learn many languages; but this ability does not imply a complete translation from one language to another; it is our experience that there are propositions which can be said in one language cannot be clearly expressed at all in another language irrespective of their truth value: "It is rather that some proposition which may be true (or false) in one language cannot even be formulated in another. It is not truth value but affability that varies with language".[110] Finally he goes to the very roots of the problem of incommensurability. He realizes that the problem is due to the very nature of language itself. Incommensurability being a relation between parts of languages is not a problem only for scientific language, because, "Its roots lie in the nature of language itself. In science, however, it shows with special clarity with special consequences… For the science, indeed, I take it to be constitutive".[111]

Kuhn still maintains a big degree of incommensurability across revolutions, where we don't have common measure, where meaning and reference of terms get modified. But now he is not very much disillusioned with the fact of incommensurability as this does not rule out the possibility of comparison between two theories. As David Boersema clarifies, "Kuhn emphasizes in various essays that no 'common measure' does not entail no measure at all; no one-to-one correspondence does not entail no relation. Incommensurability does not entail incomparability".[112] Further Kuhn prefers to look at incommensurability *metaphorically*. With the change of theories not all the terms change in their meaning. While most of the meanings remain the same only some of the terms change in their meaning. "The terms that preserve their meanings across a theory change", says Kuhn, "provide a sufficient basis for the discussion of differences and for comparisons relevant to theory choice".[113]

Incommensurability is now confined only to the linguistic domain. Therefore, "Incommensurability thus equals untranslatability".[114] He now distances himself from the non-linguistic aspects of incommensurability. Though originally his discussion involved both linguistic and non-linguistic forms incommensurability he feels "that I now take to have been overextension resulting from my failure to recognize how large a part of the apparently nonlinguistic component was acquired with language during the learning process".[115] From a global (or paradigmatic) understanding of incommensurability he narrows down to local incommensurability: "incommensurability becomes a sort of untranslatability, localized to one or another area in which two lexical structures differ".[116] Therefore, the incommensurability of two theories implies that there is no neutral or common language into which they can be translated without any remainder.

Kuhn makes a distinction between translation and interpretation, though every translation involves some elements of interpretation. The process of translation requires the knowledge of both the languages, i.e. the original one and the one into which something is translated. In translation the same matter is almost repeated in another terminology, whereas, interpretation is an exercise executed by historians, anthropologists, literary authors and so on. It is sufficient for them to know only one language. Difficult concepts and unintelligible notions are taken up to explore their meaning and relevance in the present context. In the process of interpretation the interpreter may end up learning that new language or be aware of the earlier version of his / her own language. So when a historian interprets a particular term, say phlogiston, of the past to a modern audience, in fact, he "is teaching the language which eighteenth-century chemists used in describing, explaining, and exploring that world"[117] to the present generations. While translating a term into another language one may not find the exact equivalent for a given term. For example, the French word, *espirit*, can be replaced by 'spirit', 'aptitude', 'mind', 'intelligence', 'wit', 'attitude', 'judgment' etc. according to the context. Kuhn clarifies that "these are not cases of ambiguity, but of conceptual disparity between French and English",[118] and therefore every translation has to involve an element of author's intention and that thus there is no chance for perfect translation.

But Kuhn highlights two major difficulties in translation and interpretation: *Translating interdefined terms* – for example, phlogiston – its full meaning is embedded with other terms used in the theory; therefore not possible to translate into terms of later chemical theory. Though new word (neologism) is introduced to replace such old terms yet when there are clusters of such terms it is more cumbersome; and *Conceptual disparity among terms* – terms are conceptually so pertinent and relevant to one particular linguistic system, culture etc.; therefore, as noted above, it is not possible to translate the concepts exactly into another language; it is not merely due to the ambiguities in the meaning, but in the concept itself.[119] Therefore among scientists there is no side-by-side or point to point comparison; a scientist has to learn from scratch as a learner of a new language does.

Theory Choice – Without any Actual *Choice*!: Kuhn makes theory choice something which is obtained by the collective decision of the community. All our observations are theory-dependent. Therefore, no hypothesis can be purely objective. So there is no objective way of theory-evaluation. Due to this he faces lots of criticism from different quarters. Lakatos is so strong in his criticism that he labels Kuhn's science as "a matter of mob psychology".[120] Kuhn declares once again that all these are due to a total wrong reading of his ideas: "Reports of this sort manifest total misunderstanding, and I have occasionally said as much in papers directed primarily to other ends".[121] Kuhn feels that the theory choice cannot be purely on rational grounds. It very much depends on the decisions of the scientists in the field, and choosing a theory is like a 'religious conversion' or a 'leap of faith'. Scientists do have different levels of liking for various desiderata. Therefore, Kuhn claims that theory choice cannot be settled on the basis of "logic and experiment alone"[122] and he insists that the final outcome of the theory choice is not bound uniquely by rational procedure but by many other factors.

Kuhn, however, identifies five, among many other, important characteristics of a good scientific theory: *accuracy, consistency, scope, simplicity and fruitfulness of the theory.* He chooses them "because they are individually important and collectively sufficiently varied to indicate what is at stake".[123] Nevertheless he is not unaware of the problems with these criteria. For, *they are vague if considered in isolation and they might contradict one another, if taken together.* The existing theory may have highly positive value in any one of the criteria while the new competitor may have a high score in another criterion. Now the dilemma arises as to which of the criteria to choose. Sometimes in spite of the absence of a criterion the community seems to still accept it. Kuhn gives the example of the theory of Copernicus. When it was proposed it was not more accurate than Ptolemy's. It gained more accuracy only with the Kepler's sixty years' of slogging in radically modifying it. If, for the lack of accuracy, Copernicus' theory had been completely ignored, we would not have remembered him today. Each of the five criteria can be understood or interpreted differently by different scientists. Therefore though two scientists may use the same criteria and yet can reach different conclusions. Kuhn calls them *shared criteria.*

Kuhn prefers to call the shared criteria *as values, rather than rules*. Each creative discipline is characterized, among other things, by different sets of shared values. At times two men committed to the same values may choose different theories. Unfortunately these values may prove to be confusing in actual applications. He explains with an example from the traditional proverbs that are contradicting each other: "He who hesitates is lost" and "Look before you leap" or "Many hands make light work" and "Too many cooks spoil the broth". Though contradictory yet these are taught to the children in the school. Treating the shared criteria as values, not as norms, has some advantages. It explains rationally even those aspects of scientific enterprise, which the tradition normally labels as irrational or non-scientific. These values are very active at the early stages of theory choice, while the tradition denies any meaningful role for them in those stages. After Copernicus proposed heliocentrism, Kepler and Galileo accepted it, though they were treated as idiosyncratic. They accepted it not because they were convinced of the validity of the theory but because they esteemed the shared value, and proceeded to improve upon the theory. Therefore, as Paul Hoyningen-Huene puts it, "each of these values can be interpreted somewhat differently by different members of the same scientific community. For example, what simplicity means exactly and which aspects of a theory are primarily meant is not uniquely fixed by the commitment of a community to this value" and "two cognitive values, can, in their application, contradict each other which makes relative weighing necessary. But their relative weight is, again, not determined by the commitment of the community of the list of values".[124] Thus an adoption of the new theory cannot be regarded as conscious choice on the part of that person. So declares Kuhn, "*No process quite like choice has occurred,* but they are practicing the new theory nonetheless"[125] (emphasis mine). Along with these there are many personal or subjective criteria shaping their theory choice: one's previous experience as a scientist, the duration of his/her research, extra-scientific persuasions (like philosophical or religious convictions), one's personal preferences, for instance, one may prefer originality to coherence (and therefore ready for risks!), while another may prefer vice versa and end up not taking risks. So Kuhn concludes that the theory choice is a complex procedure, because "every individual choice between competing theories depends on a mixture of objective and subjective factors, or of shared and individual criteria".[126]

A scientist cannot just base her theory choice on a matter of her personal taste, and in which case no one would take her seriously. She needs to intelligibly explain her choice to the peers. Even with great Einstein it so happened that towards the end of his life he had to remain isolated due to his view towards quantum physics, which he based only on his taste, whereas Bohr was ready to discuss the bases of his claims. The critics accuse Kuhn of subjectivism on the ground that he bases the theory choice on personal likes and dislikes. Due to this subjective element playing a big role, Kuhn, unfortunately, is accused of making theory choice (scientific enterprise) an irrational enterprise.... Therefore Kuhn's position is generally projected to be "scientific revolutions are events whose outcome is exclusively determined by propaganda, conversion and the dying out of opponents, and not by sober comparative evaluation of the achievements of theories, or other arguments that articulate good reasons for theory choice".[127]

But Kuhn denies the charge as a misunderstanding. He claims that the theory choice is still objective (though with some limitations!), as it can be analyzed in terms of criteria like accuracy and consistency. Though theory choice is affected by "idiosyncratic factors dependent on individual biography and personality",[128] there are also shared (or objective) criteria. Thus for Kuhn, "Analysis of theory choice of an individual scientist thus results in two kinds of active values; cognitive values to which the whole community is committed, and individual values varying within the community".[129] The communal cognitive values may grant rational basis; but individually differing values may not give rational base. But not all of the individual values are irrational; professional experience will surely add to better functioning; this consideration may lead to different choice by different scientists; it does not indicate irrationality, "rather, it may be a sign of different information bases used in the decision".[130] He sees the descriptive account of science as normative also, because these desiderata assure rationality of the scientific enterprise. Therefore, in spite of the ambiguity and vagueness in the desiderata Kuhn affirms that science is very rational. He is convinced of rationality in science so much that he argues: "if history or any other empirical discipline leads us to believe that the development of science depends essentially on behaviour that we have previously thought to be irrational, then we should conclude not that science is irrational, but that our notion of rationality needs adjustment here and there'.[131] He emphasizes the same when he comments on the ideas of Lakatos: "Scientific behaviour, taken as a whole, is the best example we have of rationality".[132]

Further, Kuhn argues that this situation of 'theory-choice' is not a weakness for science; any theory- choice will always be a risk; when the full implications, both positive and negative, are not fully known, it is vital for scientists to give a try; different scientists must try different theories, so this disagreement among them is very significant for the growth of science. The individuals may disagree but the consensus is based on the collective system of cognitive values: Kuhn says: "To understand why science develops as it does, one need not unravel the details of biography and personality that lead each individual to a particular choice, though that topic has vast fascination. What one must understand, however, is the manner in which a particular set of shared values interacts with the particular experiences shared by a community of specialists to ensure that most members of the group will ultimately find one set of arguments rather than another decisive".[133] Therefore, the *question of rationality of theory choice* must be raised in the context of the community, not as individual scientist.

The rationality of decision of community is decided by seeing whether they base on cognitive values; but we have to see whether those values themselves are 'good, rational reasons' to make this choice a rational one. In the extensive discussion with Hempel[134], Kuhn seems to agree with him on the possibility and the means of justification of cognitive values; that is the 'ultimate goal' of science guarantees the justification of the theory choice. For Hempel, "Science is widely conceived as seeking to formulate an increasingly comprehensive, systematically organized, world view that is explanatory and predictive... And if the goals of pure scientific research are indicated by the desiderata, then it is obviously rational, in choosing between two competing theories, to opt for the one which satisfies the desiderata better than its competitor".[135] If the goal of science changes, cognitive values also will change, because these values tell us how to realize that goal of science.

Finally, the idea of theory-choice implies *progress* in science. The notion of progress was widely talked about in the modern times, starting with Copernicus of 16th century. The idea that science always progresses forward got increasingly reaffirmed up until 20th century's logical empiricism. Science was seen to be cumulatively growing, uncovering the layers of truth with each advancement. Science was taken to be realistic and objective. Only with Kuhn's *The Structure of Scientific Revolutions* the cumulative growth of science towards the realization of full truth out there was forcefully questioned. He proposed a non-continuous model of scientific progress, instead of the traditionally-cherished scientific realism's model. Further, "In abandoning the notion that scientists search for truth, Kuhn also abandoned scientific realism, thus challenging a defining characteristic of modern science since the scientific revolution of the sixteenth and seventeenth centuries".[136] For, Kuhn makes it clear: "We may have to relinquish the notion, explicit or implicit, that changes of paradigm carry scientists and those who learn from them closer and closer to the truth".[137] As the editors of *The Road since Structure* point out, for Kuhn, "Progress takes the form of ever-improving technical puzzle-solving ability, operating under strict ... standards of success or failure. This pattern of progress, in its fullest exclusive to science, is prerequisite to the extraordinarily esoteric (and often expensive) investigations that are characteristic of scientific research...".[138] Scientific progress is like Darwinian evolution; it is not a process that pulled towards some pre-determined goal, but it is a process driven from behind. Thus Kuhn's new idea of science undercut the traditional notion of science's goal as finding 'truth' (or getting nearer to it) in nature; rather the progress of science consists in its better puzzle-solving ability.

Popper vs. Kuhn

Kuhn, in his paper, "Logic of Discovery or Psychology of Research", compares and contrasts his views with those of Popper. It seems that both of them have developed similar ideas about scientific enterprise almost simultaneously. For as Kuhn himself claims he has not read Popper's work until the publication of *The Logic of Discovery* (1959), though he has heard that Popper's ideas are widely discussed. *Kuhn enlists a number of claims or attitudes that are common to both of them.* For instance:[139] a) both focus on the dynamic nature of science, rather than the strict logical structure of science; b) both give equal importance to facts and the spirit of actual scientific life, for which both turn to the actual history of science and result in similar conclusions; c) both emphasize upon the revolutionary process of scientific growth, where the new theory replaces the old one and they focus on the failure of previous theory in facing challenges posed by logic, experiment and observation; d) both reject the cumulative growth of science; e) both are found in one company in rejecting most of the tenets of logical positivism, and insisting upon an intimate relation between scientific observation and theory; f) both of them seriously doubt the possibility of any neutral observation language; g) that the aim of scientists is to arrive at explanatory theories is agreed upon by both of them; and finally, another significant similarity between Kuhn and Popper is that h) both of them give due importance to the concept of tradition in the enterprise of scientific knowledge. As early as 1948, Popper boldly confessed, "I do not think that we could ever free ourselves entirely from the bonds of tradition. The so-called freeing is really only a change from one tradition to another".[140]

Nevertheless there are indeed *lots of areas where Kuhn and Popper differ*. For example, Kuhn demands a deeper commitment to tradition on the part of a scientist, while Popper would appreciate if the scientist is always geared to refute the tradition! Further, Kuhn is highly unhappy with Popper's demarcation criterion of science and non-science. But Kuhn prefers to concentrate on the central region of agreement, instead of focusing on the disagreements. Though the disagreement over the demarcation is one of high intensity, Kuhn still feels that the disagreements between them must not be blown out of proportion, as the disagreement is not really a disagreement, rather only a difference of viewpoint. Both seem to see the same lines and diagrams on a sheet of paper but interpret them differently. So Kuhn explains, "That is why I call what separates us a gestalt switch rather than a disagreement and also why I am at once perplexed and intrigued about how best to explore the separation".[141]

A significant disagreement between them is with regard to the idea of demarcation of science and non-science. The notion of 'normal science', which is central to Kuhn, seems to be missing in Popper's framework. The period of normal science basically consists of puzzle-solving activity. Kuhn maintains that in this period when a test is done it is on an individual scientist, not on the current theory as such, whereas for Popper the current theory as a whole is tested. If it fails the test it is dropped. But Kuhn feels that a Popperian scientist seems to be always anxious to disprove his own theory, as if a dagger dangerously dangles upon his head. All that Popper claims regarding testing etc. happens only in revolutionary periods of science, which is quite rare; they are only interpolations in the long course of normal science. If at all there is a distinction between science and non-science, it can be traced only in the normal science. In the words of Kuhn, "it is normal science, in which Sir. Karl's sort of testing does occur, rather than extraordinary science which most nearly distinguishes sciences from other enterprises. If a demarcation criterion exists (we must not, I think, seek a sharp or decisive one), it may lie just in that part of science which Sir Karl ignores".[142] When one focuses too much on testing of theories one is sure to lose the major part of science, for most part of the actual science lies in the period of normal science. Kuhn laments, therefore, "To rely on testing as the mark of a science is to miss what scientists mostly do and, with it, the most characteristic feature of their enterprise".[143] Popper, therefore, seems to make the exception a rule! Kuhn suggests that "Sir Karl has characterized the entire scientific enterprise in terms that apply only to its occasional revolutionary parts".[144] A beginner in science is primarily trained to work in normal science not to throw away the existing theories outright!

Hacking thinks that Popper and Carnap, by 1930s, mummified science. Though both had deeper difference, like the demarcation of science, they agreed upon many important aspects of science, like the rational foundations of science. They disagreed on many things, precisely because they agreed upon the fundamental things about the image of science.[145] As Hacking puts it, "If two people genuinely disagreed about great issues, they would not find enough common ground to dispute specifics one by one".[146] Kuhn has problem with both Carnap and Popper. He denies their images of science; Popper's image of science finds favour with Kuhn, though. He undercuts mercilessly all the claims, eight of them as listed by Hacking,[147] fondly cherished by both Popper and Carnap. Kuhn claims that scientific growth is not cumulative, with no sharp distinction between observation and theory, nor between context of discovery and context of justification. There is no particular method is used in science and there is not strict deductive structure in science. Science is very much social and context-bound and therefore essentially historical.

Therefore, The fundamental difference between Popper and Kuhn is that while former views science to be always in constant revolutions as the basic principles are being always put to test, the latter demands a serious commitment to the existing theories and revolutions or shifts occur only occasionally. A Popperian scientist is always anxious to falsify his/her own theory, whereas a Kuhnian scientist needs to exhibit a deeper commitment to uphold his/her theory.

Now I would like to move on to the exposition of Feyerabend's contribution, the last point of consideration in this Chapter.

Paul Feyerabend: Methodology – Not a *Corner Stone* for Science

Feyerabend (1924 – 1994) was born in Austria. His studies (Physics) were interrupted by the World War II, in which he served an infantry officer in the German army. A bullet injury in his spine caused severe damages which troubled him till his death. After the war he switched over to philosophy of science. He spent a year in working with Karl Popper and for the rest of his life he lived in the USA.

No other critic of science has meted out such a vigorous attack against the received view of science as Feyerabend has done. Strongly criticizing the positivists' view of science, he argued that science was neither unique nor rational. He knocks down the whole concept of rationality from the realm of science. He argues that science has been given an undue importance on two grounds: the methods of science and the results that it comes out with. For him *what makes science rational is its methodology* and *what makes a methodology special is its aim and results*. And because there is no such special method which is unique to science, to be followed in all the domains of science, so he argues, *there is no rationality in science;* therefore he claims that science too need not be unique. He propagates this conviction, mainly in his celebrated works: *Against Method* (1975) and *Science in a Free Society* (1978). In this short presentation I elaborate his significant ideas regarding the rational (or non-rational!) aspects of science and his sincere invitation to free human society from the clutches of science.

Two Phases in Feyerabend's Intellectual Contribution

One can roughly divide Feyerabend's works, as John Preston does in his *Feyerabend – Philosophy, Science and Society,* into two major sections: *from early 1950s till 1970* and *from 1970 onwards, till his death in 1994*. In the first part he was engrossed in developing an all-embracing model for acquisition of knowledge, while in the second part he lost interest in this project and became highly critical of rationalism. One of the significant factors that led him to the shift is the intellectual freedom at the campus of the University of California, Berkeley. It made a considerable impact on him. *Radical Movements* of the students and the *Free Speech Movement* influenced him so deeply that he questioned the hegemony of the Western rationalism. He wondered how grave the mistake it would be to impose Western rationalism upon the students from different parts of the world, who had come to Berkeley for education. Though his rationalist friends thought it was a wonderful opportunity to spread rationalism to all the corners of the world and to enlighten them, Feyerabend felt very differently. He somehow realized that so far he had been promulgating and imposing the ideas of a very small group, who had the hegemony over the others to enslave them. So he asks himself: "Who was I to tell these people what and how to think?".[148]

Being won over by the ideas of Lakatos, Feyerabend decided to hold on to relativist views, denying single paradigm of rationality or any unique way of acquiring knowledge. He was deeply convinced of the big role of social, personal and aesthetic elements played in the history of science. So he came down heavily upon the traditional rationalist themes in his *Against Method* and *Science in a Free Society,* which earned him more popularity and a bigger audience, even from the quarters of non-philosophy-of-science. By opposing Western imperialism he stressed on relativism not only in epistemology but also in politics.

In the beginning feyerabend had a great admiration for Popper and earnestly tried to develop a model for science. He accepted the crucial views of Popper: normative epistemology, falsificationism and inductive scepticism. A rational model of science must provide the aim for science and a *modus operandi* to attain that aim. For the earlier-Feyerabend the aim of science is knowledge-gathering and the methodology to achieve that aim is 'falsificationism'. Any methodology must fulfil two requirements – the testability of the theories and the explanation of the known phenomena. Going along with Popper, he thought that epistemology of science was certainly a normative one, and not a descriptive one. He held that scientific method was to furnish normative rules to be followed strictly, and not just to describe the actual activities of the scientists, and to that method, "actual scientific practice will correspond only more or less closely".[149] In most of his papers (1981) prior to 1968 he supported Popper's ideas of scientific methodology.

I now concentrate more on the second phase which is more relevant to the topic of rationality.

No Method, and therefore No Rationality!

Feyerabend's strong attacks on scientific rationality and scientific method are based on his claim that no rule has survived in history which is applicable in all contexts; great breakthroughs in science took place only when the giants of science (Galileo, Newton, Einstein) went against the so-called long-cherished norms of scientific rationality. They deliberately violated rational guidelines and often acted counter-inductively. The scientists also ignored the demands of rationality to achieve greater heights in science. He gives case studies to prove all these claims and to show that great scientists put forward theories even without any support of evidence, which are often even inconsistent with the well-established prevalent theories, and they develop *ad hoc* theories to explain away anomalies.

Feyerabend identifies *the demand for one single method* (i.e. *methodological monism*) as the core of rationalism. But history shows that there is no *the* method, suitable for all the domains of science. He brings in the case-study of Galileo to show how the latter systematically went against each of the rules of the methodological monists. This is not to show that Galileo did not follow any method in his science at all, rather, Galileo emerges out as a strong sign for the need for *methodological* and *theoretical pluralism*. Similarly, He comes out with many other instances from History of Science to astutely argue that strictly insisting upon any single prescriptive scientific method is not only impossible and not factual, but also it positively hinders the scientific enterprise. For instance, he elaborates the Copernican Revolution, which is usually taken to be the clear and undeniable instance to show the rational and systematic growth of science, which follows a fixed methodology. But he showed that the prescriptive rules were deliberately violated in similar circumstances and claimed that efforts to blindly follow those rules would have even prevented scientific revolutions. Theoretical pluralism is preferable as it does not impose rigid rules on the scientific community and it is necessary for the growth of science, because, "Variety of opinion is necessary for objective knowledge".[150] Therefore he advocates *methodological* and *theoretical pluralism*, which is not only more humanitarian but also more favourable to the growth of science.

Scientist vs. Non-scientist: Since no method has been proved to be successful in all domains and such a concept of methodology seems to have restricted the growth of science, it must be the case that 'any thing goes'. But again it must be noted that it has limitations: that is, because 'any thing goes' we must not think that 'anybody' becomes a scientist, or that anybody can propose any theory. The distinction between a reasonable scientist and a crank does not lie in the fact that the former puts forward a successful and useful theory while the latter comes out with an unsuccessful theory; for, the success or failure of a theory can be ascertained only in course of time, not at the time of proposal. Therefore Feyerabend distinguishes thus: "The crank is usually content with defending the point of view in its original, undeveloped, metaphysical form, and he is not at all prepared to test its usefulness in all those cases which seem to favour the opponent, or even to admit that there exists a problem. It is this further investigation,... the recognition of objections, which distinguishes the 'respectable thinker' from the crank".[151]

Anarchy – Need not Lead to chaos: Anarchy, Feyerabend proposes, is a requirement for any intellectual enterprise. It may be objectionable in politics but certainly not in epistemology. Feyerabend is for epistemological anarchism, to liberate science and society from wrong understanding that scientific rationality has a special and superior character. In fact this is the conviction, as he himself acknowledges, that led him to write *Against Method*: "The following essay is written in the conviction that *anarchism*, while perhaps not the most attractive *political* philosophy, is certainly excellent medicine for *epistemology*, and for the *philosophy of science*".[152]

Feyerabend agues that the rules, which the methodological monists propose are naïve and simplistic, fixed in advance, without any consideration for historical elements. The types of rules that he has in his mind are a) that one must not propose *ad hoc* hypotheses to explain away the noncompliant data; b) one must not yield to hypotheses which go against the hitherto well-accepted experimental outcome; c) one must not accept hypotheses with lesser empirical content compared to the prevalent theories; and d) one must stay away from introducing hypotheses that are self-contradictory.[153] He claims that these rules can neither be descriptive nor prescriptive and goes on to systematically deny each of those claims.

Since many historical episodes, like Galileo's and Copernican cases, show that not even one single rule, however greatly valued and cherished it might have been, does come to stay permanently in science. It is violated at one point or the other. The scientists sometimes not just violate the traditional rules they even choose just the opposite rules. Such violations are *factual*, *deliberate* and even *absolutely necessary* for further growth in science. The newly adopted rules could even be *backward looking.* Circumstances may even demand "to introduce, elaborate and defend *ad hoc* hypotheses, or hypotheses which contradict well-established and generally accepted experimental results, or hypotheses whose content is smaller than the content of the existing and empirically adequate alternative, or self-inconsistent hypotheses, and so on".[154] It is not the case that a new theory has to clarify all the doubts raised by the previous theory and rationally prove its superiority and a far-advanced empirical content before it is used. Theories often obtain clarity only after a prolonged usage. To become empirically successful they need non-rational and non-sensical acceptance in the first place. Such unmethodical 'push-ups' are needed for the new theories.

Feyerabend assures that because we don't adhere to fixed rules or methods in an anarchistic science we need not worry that it will land us in chaos. The less stress on strict law and order in science would not lead to anarchism in science, for he believes in the well-ordered nervous system that the humans have developed over the millennia.[155] Therefore he denies strict, unchanging and all-appealing methodological rules in science; he does not encourage a chaos-like situation, in both theoretical pluralism and methodological pluralism. *For what he has in mind is not the non-existence of such rules but their emergence and dependence only within the given research agenda.* As "there are standards, but they come from the research process itself, not from abstract views of rationality",[156] a scientific research need not be capricious and whimsical.

A Tendency to Follow False Reasoning: True, it is not easy to go against the traditionally well-appreciated practices. For we are generally determined to faithfully follow the *tradition of reason* but actually we fail to realize that what we consider the 'voice of reason' is only a creation of the social conditioning, a *causal after-effect* of the training. Only the ignorance of the potentialities of humans and the complexities of the social surroundings make us cling to "...intellectual security in the form of clarity, precision, 'objectivity', 'truth'...".[157] Thus, when we dispassionately analyse the intellectual processes in history, there seems to be only one rule that is protected and acceptable under all situations and at all times of human development, and that principle is nothing but *anything* *goes*. In fact, he says, a rationalist who has a closer study of the historical episodes cannot but throw up his arms in the air and sincerely exclaim that 'anything goes'. [158]

Feyerabend made a scathing attack on *the stability thesis* (i.e. taking rationality to be more or less invariant over time). There are no absolute rules in science and, in fact, accepted rules are very often violated and even just the opposite ones are chosen. He does not see any 'rational' basis for choosing Newtonian science over Aristotelian science. *Science prevailed magic, or Newtonian science prevails Aristotelian science, not because of any rational superiority on the part of the latter.* Of course, science has a sort of rationality but it is not as robust as we think it is; it is one among many forms of rationality. A scientist or an artist is at the same footing in terms of rationality. They don't need, as Feyerabend puts it, "papa methodology or mama rationality to give them security and direction, he can take care of himself, for he is the inventor not only of laws, theories, pictures, plays, forms of music, ways of dealing with his fellow men, institutions, but also of entire world views, he is the inventor of forms of life".[159]

Falsificationism – Not a Solution: Feyerabend is also critical of *Popper's Falsificationism*. For, no important theory is ever in harmony with all the relevant facts. But naïve Falsificationism demands that a scientific theory must be rejected if it does not agree with the available, known facts. But theories are kept up and even encouraged for further development with *ad hoc* methods, which are essential to the progress in science. For example, Galileo could not give the necessary optical theories to support his observational claims with telescopes; he and his followers had to use *ad hoc* rules until the field of optics could justify their telescopic claims with an optical theory. Feyerabend agrees that falsifiability provides a way to replace invalid inductive thinking with deductive, falsifiable reasoning. Nevertheless he argues that doing so is neither necessary for, not conducive to scientific progress; in fact it hinders it. Thus he argues that science proceeds not by induction, but by *counter-induction*.

The Need to 'Unmask' Science: As humanity learns more and more from the history of science, it comes to realize that science is not so simple as it seems to be – rather, science consists, not just facts and unbiased interpretations of those facts, but "also contains ideas, interpretations of facts, problems created by conflicting interpretations, mistakes and so on".[160] Moreover, we learn another hard-to-believe lesson that there are brute facts provided by science, as whatever enters our knowledge-realm has been filtered by certain tools and coloured by certain ideas, and therefore the scientific knowledge is fundamentally *ideational.* To study science, which is full of ups and downs, many wrong decisions and choices, and lots of mistakes and misunderstandings, one needs a complex system of analysis, not a simple one, based on some fixed rules and methodologies. But today's tendency is to make science appear as a system, which is very simple, highly objective, extremely successful and strictly methodological. Stringent lines of demarcation are drawn between science and non-science, by erecting walls around science to keep it uncontaminated by one's metaphysical, religious, social and cultural background. But unfortunately such a tendency to keep science aloof from all the other social institutions turns out to be harmful for the total welfare of the humanity. Such a move is *not possible*, as the world which science explores is largely unknown and too complex to be framed into a few rigid rules; and it is *not desirable* either, as it ignores all the other disciplines of humanities, which are essential for a holistic development of human beings.

In short, no special methods for science, therefore no special status for science either. If at all one looks for a universal rule for science, that would be, 'anything goes'. The special status of science, if any, is obtained from the social and physical values of the results of science, rather than its methods. The supremacy of science is artificially thrust upon it. Science has forcefully pushed itself up! Not by logical arguments, but because of its power it came to occupy the ruling position. As in the middle ages the religious leaders of Christendom imposed rule on their subjects, now scientists use science as power to push their ideologies upon people.

Feyerabend's Lenience towards Relativism

In Feyerabend one can trace out two possible routes to relativism, through *methodological pluralism* and *incommensurability*:

Firstly, given no fixed method, science has no way of ratifying the choice of one method over another. There seems to be no way of calculating the merits of the different methods or paradigms. Secondly, since there is incommensurability in terms of truth-content, in a very hard-to-ignore fashion, there emerges relativism.[161]

In his "Rationalism, Relativism and Scientific Method" (1977) he combats against rationalism, He differentiates between naïve and sophisticated versions of 'cosmological', 'institutional' and 'normative' rationalisms. He finally gives his doctrine of relativism that there are many 'rationalities', not just one, and it is up to the society to choose from those many. In the same paper he presents a crucial problem what has is known as the 'dilemma of ultimate commitment' – i.e. both the rationalists and the irrationalists make a sort of 'leap in faith' in committing themselves to their respective ideology. Their choice is beyond rational justification; For, he explains: "Assume we want to judge action A by standard S. We apply S to A and render our judgment. But the application must also be rational, so there must be standards S` which judge the pair [A,S] and so on *ad infinitum*, unless we admit that at some place we simply act without being able to provide the standards which make this action rational.[162]

With the insistence of anarchism, the privileged position given to reason is displaced. It becomes equally problematic and abstract as the other monstrous notions of Obligation, Duty, Morality, Truth and 'their more concrete predecessors, the Gods'. All these notions have been very badly threatening humanity and systematically hampering their freedom and joy in developments of every sort. In short, reason just "withers away".[163] Thus he argues that whether rationalism or irrationalism there is no rational basis and it is only a 'leap of faith' which surely lands them in relativism.

Empiricism – Ineffective in Fulfilling its Promise!

According to Feyerabend, empiricism is not able to fulfill even its fundamental promise. However, empiricism has become so powerful that every intellectual enterprise is keen to take it as its official position. All swear upon empiricism that the only way to escape whimsical speculation and meaningless metaphysics is to rely upon a *thorough observational procedure*. But unfortunately modern empiricism seems to establish to a sort of metaphysical dogmatism, which in fact it aims to oppose. It is ironical to note that empiricism constructs its own position on such an edifice which is not refutable by experimental procedures.[164] *Empiricism, if it is less and less tolerant of theoretical pluralism then it becomes more and more dogmatic.* If any intellectual enterprise likes to be objective, it must make an essential room for pluralism. It is this pluralism of theories that enables a sharp criticism of the existing theories, however well-accepted they have been in the past. One must not make the domain of 'observational facts' as the *only* yardstick for evaluating theories, as we don't have any observation, independent of theoretical considerations. Science cannot afford to deny a theory, because it sounds too general or metaphysical. At every stage of science some sort of metaphysics or the other is unavoidable in order to stop science from becoming dogmatic: "A science that is free from *metaphysics* is on the best way to become a *dogmatic* metaphysical system".[165]

Empiricism insists upon two conditions – *Consistency Condition* and *Meaning Invariance*. Feyerabend explains them as follows: [166] a) Only such theories are then admissible in a given domain which either contain the theories already used in the domain, or which are at least consistent with them inside the domain and b) Meanings will have to be invariant with respect to scientific progress; that is, all future theories will have to be phrased in such a manner that their use in explanations does not affect what is said by the theories, or factual reports to be explained. He finds serious fault with both these criteria. Episodes from the history of science show that both these conditions have been violated, especially at those crucial points where great scientific revolutions took place. The inconsistency between Newton's theory on the one hand, and Galileo's law of the free fall and Kepler's laws on the other, and the inconsistency between statistical thermodynamics and the second law of the phenomenological theory and that between wave optics and geometrical optics are widely known violations of the consistency condition. Similarly, the relativity theory is shown to be violating both the conditions of consistency and meaning invariance. The first criterion, the *consistency condition,* increases the tendency to retain the older theory as one can retain many cherished assumptions of the past. At times it forces the scientific community to overlook even the observational evidence in the choice of theories. A theory is rejected, not so much for its counter-evidence at the face of new observational or experimental results, but just because it goes against a well-cherished theory at hand. Here, the age and familiarity become the basis for the theory elimination. So he even wonders: "Had the younger theory been there first, then the consistency condition would have worked in its favour".[167]

Even to know the strength of our own opinion we need to contrast it with counter opinions and trends, as a dark colour can be much appreciated against the background of a light colour and vice versa. Mill, in *Utilitarianism, On Liberty and Considerations on Representative Government*, succinctly puts this point across: "But on every subject on which difference of opinion is possible the truth depends on a balance to be struck between two sets of conflicting reasons. Even in natural philosophy, there is always some other explanation possible of the same facts… and it has to be shown why that other theory cannot be the true one: and until this is shown, and until we know how it is shown, we do not understand the grounds of our own opinion".[168] Many facts are unearthed only with the availability of alternative theories. To refuse to consider the alternatives leads the scientists to the danger of abolishing the potentially refuting facts. These facts may be capable of showing the absolute inadequacy of the theory in question. By ignoring them the scientists are at a loss. They have to go on with the ignorance of holding on to a theory which is, in fact, deficient and incompetent.

Under this 'blissful ignorance' of the refuting facts, one tends to be mistakenly convinced of the specialty and uniqueness of the current theory and reject any single theory that does not go along with the present theory. From here one automatically moves on to the state of metaphysical dogmatism. Here its success is assumed and artificially created by us. The success is not because it corresponds to the truth and nature, nor because it has been scrutinized under many possible alternative theories. Given such a background, that particular scientific theory becomes a sort of a *myth*, which is kept up for many personal, social, political and religious reasons, like one's dread, unfounded inclinations, preconceived ideas, and religious convictions. Efforts are made to reinforce the *myth*, not necessarily by religious priests but by the custodians of science and Nobel laureates themselves. This *myth* permeates the whole life and it becomes almost impossible to criticize, because the possibility of considering any other "set of equally all-embracing principles... has been excluded from the very beginning".[169] *Therefore the insistence upon consistency and uniformity of theories weakens science by removing its critical power; it seems to lead even to opposite effects.* For, the great power of imaginative and innovative ideas is under a threat. It tells upon the realization of one's deep insights. If such a situation is allowed to prevail for a period of time it will slowly and steadily affect the quality and the vigour of education.

Reductionism - To be Rejected!

Feyerabend criticizes Nagel's theory of reduction, which claims that a secondary theory must be logically deduced from the primary theory (or *explanandum*, i.e. the conclusion to be arrived at, must be logically explained by the *explanans*. i.e. the premises from which the conclusion is to be deduced). Further, the meanings of the observational terms don't vary in the process of reduction or explanation. Nagel declares: "The objective of reduction is to show that the laws, or the general principles of the secondary science, are simply logical consequences of the assumptions of the primary science".[170] He gives an example for reduction: Galileo's science can be reduced to Newtonian science. Along the same lines, modern empiricism makes similar assumptions; one finds Popper, Hempel and Oppenheim claiming that *explanandum* is logically deducible form the premises of the *explanans* and meaning are invariant in that process of explanation. Following this guideline, in science also it is assumed that the new scientific theories or hypotheses must be consistent with earlier ones. In other words, as Feyerabend puts it, "only such theories are admissible (for explanation and prediction) in a given domain which either *contain* the theories already used in this domain, or are at least *consistent* with them".[171]

Feyerabend shows that both of these demands are defied by the actual practices and the reasonable methodological procedures of science. Both the conditions fail because theories in universal character cannot be fully comprehended by any set of observations at any time and also the truth of even an observation statement can be guaranteed only with a margin of doubt. Thus we are led to learn that "*one and the same set of observational data is compatible with very different and mutually inconsistent theories*".[172] He takes up the examples of Nagel one by one to show that both the demands cannot be proved on empirical grounds: for instance, while Nagel argues for the deductive relationship between Newtonian celestial mechanics and the Galilean physics, Feyerabend shows that "it is impossible for quantitative reasons, to establish a deductive relationship between T (Laws of Newton's celestial mechanics) and T' (Galilean physics), or even to make T and T' compatible. This shows that the present example is not in agreement with"[173] both the demands. Similarly, while Nagel claims *meaning invariance* of the concept of motion in both the Aristotelian and Newtonian frames of references, Feyerabend explains that motion is understood in terms of *impetus* in the former system and it is *inertia* in the latter system. For Aristotle the state of rest is the natural state of an object, whereas for Newton the natural state can be either rest or motion.

Another instance Feyerabend gives is the notion of mass in Newtonian science and the relativity theory. Hard empiricists would say that under certain conditions classical physics could be shown as a special case of relativity physics. For example, when the velocity is small, then we can show that the classical physics is a special case of relativity physics. Hence the logical positivists argue that there is no big difference between classical physics and relativity. For them, in fact, classical physics is a special case of relativity and hence consistency condition is guaranteed. Einstein's role was only to broaden the domain of classical physics. This is not true because mass is a property of a material body in classical physics. But in relativity, mass is a relational property. The mass of a material body will depend on which frame of reference it is placed. On the surface of the earth I may weigh 140 pounds but in a fast moving space shuttle, my weight will be anything more than 140 pounds.

Suppose I am on a vehicle that is moving with the speed of light, then I will have infinite mass. Thus we learn, Feyerabend argues, that reductionism does not seem to hold water. Consistency Condition is both *unfair* and *not viable* in science. It is unfair, because it eliminates the competitor not because the competitor is in disagreement with the fact but because it happens to come late. What kind of rationality is there? 'First come, first served' is not right in science.

Incommensurability – Too Strong to be Overlooked!

Incommensurability is one of the salient features of Feyerabend's contributions. It is originally a mathematical concept. For instance, it is used in Mathematics, relating the side and the diagonal of a square; the side is measured by a rational number and the diagonal by an irrational number. But "since an irrational number cannot be represented by a point on the rational number line, the quantities are said to have no common measure; they are literally incommensurable".[174] This concept was made popular by Kuhn (in *The Structure of Scientific Revolutions*), as we saw earlier and Feyerabend as well (in "Explanation, Reduction, and Empiricism") in 1962. Interestingly both came up with the idea in the same year, but for different reasons.

Incommensurability originates from the issue of 'theory-dependence of observation'. No interpretation of observations can be done without some theoretical assumptions. Therefore any evaluation or description of an observation is unattainable without a theory. Any paradigm, within which a given theory or a scientific practice is undertaken, operates under many assumptions. These assumptions, which lie at the basis of the paradigm's structure, make the comparison of two paradigms greatly difficult. Feyerabend goes even further to demonstrate that the very idea of incommensurability itself is beyond a total comprehension by logic: "As incommensurability depends on covert classifications and involves major changes it is hardly ever possible to give an explicit definition of it".[175]

Feyerabend's favourite example to explain the notion of incommensurability is, as we saw above, comparing the impetus theory of the medieval period and Newton's classical mechanics, where, he claims, the usage of the term impetus differs beyond comparison. He argues that when the transition takes place from one theory to another, the later theory not just incorporates the earlier theory, but the very ontology is changed. In the new ontology, the meanings of the descriptive terms (if they are used in both theories) also change. For instance, classical mechanics, interpreted realistically, holds that "properties (shape, mass and volume) exist in physical objects and can be changed as a result of physical interference", while relativity theory, interpreted realistically, affirms that "properties such as shape, mass and volume no longer exist, but become relations between objects and a reference frame and can be changed, without any physical interaction, by changing from one reference to another".[176] The meaning of a term depends on the theory in which it occurs. It is determined by the theoretical context they are in. Meanings of observational terms are also determined by theories in which they occur. All terms get their meaning in their context. There is no neutral observational language. Therefore, there will be differences in meaning even when the observation-statements speak about such properties and there by leading to incommensurability. Therefore, every theory is governed by universal rules or principles of construction. Another theory would violate these rules so one cannot render the 2nd theory in terms of the first one. Thus both the theories are incommensurable. Taking T and T¢ to be classical mechanics and general relativity theory respectively, he elsewhere explains that the classical notions of absolute mass, distance etc. can never be defined in terms of T and nor the defined terms of T can be incorporated into T¢, because the meanings of those terms change, and thus making "T and T¢ (are) *incommensurable theories*".[177]

Feyerabend maintains 'contextual theory of meaning', by which, meaning of a term in a theory is determined by the context in which it occurs. A genuine alternative competitive theory must substantially differ from the first theory; must differ in the meanings of all the terms in it; if this difference is not substantial, this is not really the alternative theory but only a variant: The general principles of a theory affect the meaning of every term; when theory changes substantially those principles too change, and with that change, meaning also changes. However, large part of the theory of the world remains intact; that is why the transition from Newtonian mechanics to general theory of relativity has not altered culture, arts, ordinary language and perception. Incommensurability leads to a strict subjectivity in the realm of theory choice. Since there is no logical way of comparing various theories, we are left only with the "aesthetic judgments, judgments of taste, metaphysical prejudices, religious desires, in short, what remains are *our subjective wishes*".[178]

Society – In Need of Protection from Science

Feyerabend strongly pleads that any society must be protected from the too much influence of science, as people demand protection of society from the State. Science does not have any right to be kept at the pinnacle of all the processes of human knowledge. Feyerabend is clear that if one attributes any kind of superiority to the scientific enterprise it is not because there is any deserving substance in science, rather because one has arbitrarily decided earlier to give such an importance. Science is just one field of knowledge among many others developed by humans and it is certainly not the best field. "It is conspicuous, noisy, and impudent, but it is inherently superior only for those who have already decided in favour of a certain ideology, or who have accepted it without having ever examined its advantages and its limits".[179] As no special position to science is justified, different disciplines, like astrology, witchcraft, traditional medicines, voodoo etc., must be given equal access in the society. People must be left free to choose any discipline they want. The children must have equal access to various ideologies.

Feyerabend respects and insists upon individual's freedom. One must be free to choose any discipline one wants. The Western society has disentangled religion and state. One is now free to choose any religion one wants. Similarly science has to be freed from the State and a free society must give freedom to its citizens to choose any discipline to study. The state must give equal importance to magic, astrology, voodoo etc. and thereby an individual can make a free decision in pursing anything one wants. So, "In

Feyerabend's ideal society the State is ideologically neutral".[180] However even in this situation one is not totally free as there are many other aspects, like the existing social structures and the accessibility to education and so on, binding an individual. Hume makes a similar point, in "Of the Original Contract", while criticizing Locke's idea of free adoption of social contract by the members of a society:

> "Can we seriously say, that a poor peasant or artisan has a free choice to leave his country, when he knows no foreign language or manners, and lives from day to day, by the small wages which he acquires? We may as well assert that a man, by remaining in a vessel, freely consents to the domination of the master; though he was carried on board while asleep, and must leap into the ocean and perish, the moment he leaves her".[181]

Feyerabend proposes that committees of lay people need to be established to determine, by popular election, the subjects that are investigated by scientists. Those committees must be empowered to supervise the scientific assumptions and conclusions as well. Thus science must be brought under total control of democracy; that is, not only that people should determine the subjects of scientific research, but also committees of lay people must supervise the conjectures and conclusions of the scientific community. It is not proper to expect only the scientists and the philosophers to decide for the community.

Having presented the seminal ideas of the proponents of the rational and non-rational models of rationality in science, I now move on to the critical evaluation of those ideas in *Chapter 2*.

NOTES & REFERENCES

1. For a very brief and precise explanation of these topics as developed by Hempel, please see: "A Brief Guide to the Work of Carl Gustav Hempel," in *Erkenntnis* 42 (1995), 3-7.
2. C.G. Hempel, *The Philosophy of Carl G. Hempel–Studies in Science, Explanation and Rationality*, 2001, xiii.
3. Thomas Kuhn, "Rationality and Theory Choice", in *The Road Since Structure: Philosophical Essays,* 2000, 208.

4. W. Salmon, "Carl G. Hempel on the Rationality of Science", 1983, 555-562, 555.
5. C.G. Hempel, "Scientific Rationality: Normative Versus Descriptive Construals", in Fetzer (ed.), 2001, 357.
6. *Ibid.*, 358.
7. C.G. Hempel, "Valuation and Objectivity in Science", in Fetzer (ed.), 2001, 375.
8. Imre Lakatos, "Falsification and the Methodology of Scientific Research Programs", in Lakatos and Musgrave, (eds.), 1970, 91-195, 178.
9. C.G. Hempel, "Scientific Rationality: Normative Versus Descriptive Construals", 2001, 364.
10. *Ibid.*, 365.
11. For details: Gerald Holton, "Subelectrons, Presuppositions, and the Millikan-Ehrenhaft Dispute", in G. Holton, *The Scientific Imagination: Case Studies,* 1978, 25-83.
12. C.G. Hempel, "Scientific Rationality: Analytic vs. Pragmatic Perspectives," in E.D. Klemke et al., (eds.), 1998, 457.
13. C.G. Hempel, "Valuation and Objectivity in Science", 2001, 388.
14. *Ibid.*, 383.
15. C.G. Hempel, "Kuhn and Salmon on Rationality and Theory Choice," 1983, 572.
16. C.G. Hempel, "Valuation and Objectivity in Science", 2001, 372-395, 389-90.
17. C.G. Hempel, "Scientific Rationality: Normative Versus Descriptive Construals", in Fetzer, (ed), 2001, 366.
18. *Ibid.*, 369.
19. C.G. Hempel, "Scientific Rationality: Analytic vs. Pragmatic Perspectives," 1998, 463.
20. C.G. Hempel, "Scientific Rationality: Normative Versus Descriptive Construals", 2001, 362.
21. *Ibid.*, 361.

22. *Ibid.*, 363.
23. *Ibid.*, 372
24. Quoted by C. G. Hempel, "Valuation and Objectivity in Science", 2001, 380.
25. G. Hempel, "Scientific Rationality: Analytic vs. Pragmatic Perspectives," 1998, 454.
26. C.G. Hempel, "Valuation and Objectivity in Science", 2001, 389.
27. Ibid., 391.
28. C.G. Hempel, "Scientific Rationality: Analytic vs. Pragmatic Perspectives," 1998, 463.
29. C.G. Hempel, "Scientific Rationality: Normative Versus Descriptive Construals ", 2001, 370.
30. Popper brings out this idea in many of his writings. For instance, his *Logic of Scientific Discovery*, 1959, 54.
31. K. Popper, *Logic of Discovery*, 1959, 16.
32. K. Popper, *Conjectures and Refutations: The Growth of Scientific Knowledge*, 1963, 253-92.
33. K. Popper, "On the Status of Science and Metaphysics", in *Conjectures and Refutations: The Growth of Scientific Knowledge*, 1963, 192.
34. Paul Bernays, "Concerning Rationality" in Paul Arthur Schilpp, *The Philosophy of Karl Popper*, 1974, 600.
35. K. Popper, *The Myth of the Framework – In defence of Science and Rationality*, 1964, 14.
36. *Ibid.*, 14
37. K. Popper, "Gewohneit" und "Gesetzerlebnis" in der *Erziehung. Eine padagogiscs- strukhapsychologische Monographie*; as quoted by Stefano Gattei, 2004, 448-466, 449. He quotes other similar sources too.
38. K. Popper, in "A Letter to the Director of the London School of Economics", as quotedin Lakatos, "Popper on Demarcation and Induction", 1974, 261.

39. As quoted by Bipin Indurkhya, "Some Remarks on the Rationality of Induction," 1990, 95-114.
40. K. Popper, *The Myth of the Framework – In defence of Science and Rationality*, 1964, 12.
41. K. Popper, "Falsificationism", in *Scientific Inquiry – Readings in the Philosophy of Science*, 1999, 68.
42. K. Popper, *The Philosophy of Karl Popper*, (ed.) Paul Arthur Schilpp, 1974, 1015.
43. K. Popper, *Objective Knowledge: An Evolutionary Approach*, 1972, 1.
44. K. Popper, *Conjectures and Refutations*, 1963, 42.
45. K. Popper, *The Philosophy of Karl Popper*, (ed.) Paul Arthur Schilpp, 1974, 1015.
46. K. Popper, *Realism and the Aim of Science*, 1983, 28.
47. K. Popper, *The Philosophy of Karl Popper*, 1974, 1516.
48. *Ibid*., 1019.
49. *Ibid*., 1020.
50. *Ibid*., 1021.
51. K. Popper, "Falsificationism", in *Scientific Inquiry*, 1999, 66.
52. Ibid., 67.
53. Paul Bernays "Concerning Rationality" in Paul Arthur Schilpp, 1974, 597.
54. K. Popper, "Falsificationism", in Robert Klee, 1999, 70.
55. Popper elaborates on this idea in Sections 4 & 6 of Chapter I of *Logic of Scientific Discovery*.
56. K Popper, *Realism and the Aim of Science*, 1992, xxvi-xxx.
57. Imre Lakatos, "Popper on Demarcation and Induction," in Paul Arthur Schilpp, ed, 1974, 248.
58. A.F. Chalmers *What is This Thing Called Science*?, 1992, 50.
59. *Ibid*., 55.
60. K. Popper, *Logic of Scientific Discovery*, 1959, 53.

61. K. Popper, *The Open Society and Its Enemies*, 1962, 391.
62. K. Popper, *Logic of Scientific Discovery*, 1959, 53.
63. *Ibid.*, 56.
64. K. Popper, *The Myth of the Framework – In defence of Science and Rationality*, 1994, 13.
65. *Ibid.*, 7.
66. K. Popper, "Falsificationism", in *Scientific Inquiry*, 1999, 67.
67. Robert Klee (ed.), *Scientific Inquiry – Readings in the Philosophy of Science*, 1999, 199.
68. Carl Matheson, "Historicist Theories of Rationality", 1. (See: http://plato.stanford.edu/entreis/rationality-historicist/
69. *Ibid.*, 2.
70. Thomas Kuhn, "Essential Tension", in *Essential Tension*, 1977, 234.
71. Thomas Kuhn, *The Structure of Scientific Revolutions*, (henceforth *SSR*), 1962, 82.
72. Thomas Kuhn, "Essential Tension", 1977, 236.
73. Joseph Rouse, *Knowledge and Power – Toward a Political Philosophy of Science*, 1987, 33.
74. James Ladyman, 2002, 103.
75. *The phlogiston theory* was first proposed by Becher (1635-1682), and developed by Stahl (1660-1734); according to the theory phlogiston is the 'combustible principle', which is present in all flammable objects; it is emitted out while burning. By itself it is unobservable, colourless, tasteless and odorless. But when it is found in a material object it may give that object observable properties. *The Oxygen theory* of Lavoisier was the first step towards the chemical atomic theory of John Dalton (1766-1844); according to which matter consists of indivisible atoms; though there are different kinds of atoms, atoms of the same kind are identical in every respect, like shape, size, weight etc. Most of the things in the world are compounds rather than basic elements.

76. Kuhn, *SSR*, 2nd ed., 1970, 111.
77. *Ibid.*, 109.
78. Kuhn explains this with the example of *Optics*: from antiquity till the end of the 17th century one cannot trace out any single or a set of paradigms for optics; Newton's particle theory of optics ruled the world of optics throughout the 18th century till the beginning of the 19th century; wave theory was accepted during the second half of that century; now in the 20th century there is a consensus in optics, where light is seen both as wave and particles. In these three phases all the problems and solutions in optics were framed only in accordance with the then prevalent theories in mind. [Thomas Kuhn, "Essential Tension", in *Essential Tension*, 1977, 231-2].
79. James Ladyman, 2002, 119.
80. Thomas S. Kuhn, "Remarks on Incommensurability and Translation," in Rema Rossini Favretti *et al.,* (eds.), 1999, 33.
81. Kuhn, *SSR*, 2nd ed., 1962, 102.
82. Ibid, 109-10.
83. Edwin Hung, *The Nature of Science: Problems and Perspectives*, 1997, 379-383.
84. Gerald Doppelt, "Kuhn's Epistemological Relativism: An Interpretation and Defense", in M. Krausz ad J. W. Meiland (eds.), 1982, 113-146, 114.
85. Michael E. Malone, "Kuhn Reconstructed: Incommensurability without Relativism," 1993, 69-93, 74.
86. Kuhn, *SSR*, 1970, p. 154.
87. Ibid., 155.
88. Kuhn, "Reflections on my critics," in Imre Lakatos and A. Musgrave, 1970, 231-78, 261.
89. Thomas Kuhn, "Essential Tension", 226.
90. *Ibid.*, 227.
91. *Ibid.*,229.
92. Edwin Hung, 1997, 358.

93. Thomas Kuhn, "Essential Tension", 1977, 237.
94. Thomas Kuhn, "Second Thoughts on Paradigms", 1977, 459-482.
95. *Ibid.*, 460.
96. *Ibid.*, 463.
97. Thomas Kuhn, "Second Thoughts on Paradigms", 1977, 471.
98. Kuhn, *The Essential Tension*, 1977, 321-22.
99. Rupert Read, in his book review of The Road Since Structure (See: http://www.uea.ac.uk/~j339/ kuhnreviewbjps.htm).
100. Kuhn, *The Trouble with the Historical Philosophy of Science*, 1991, 4.
101. Thomas Kuhn "Commensurability, Comparability, Communicability", in *The Road Since Structure*, 2000, 57.
102. The editors have done a comprehensive job in identifying the themes in the Introduction. See: in *The Road Since Structure*, 2000, 1-9.
103. Thomas S. Kuhn, "Remarks on Incommensurability and Translation," 1999, 34.
104. Some of the differences are: Feyerabend confined incommensurability only to the semantic factors, while Kuhn at first included non-semantic elements; Feyerabend proposed it only as a reaction against the reductionists' notion of theory succession. Unlike Kuhn's notion, his idea of incommensurability largely remained the same except for expanding it to include world views too.
105. Thomas Kuhn, *SSR*, 1962.
106. Thomas Kuhn, "Reflections on my Critics," 1970, 231-78.
107. Thomas Kuhn, "Metaphor in Science," (1979) and "Commensurability, Comparability, Communicability," (1983) in *The Road since Structure, 2000.*
108. Thomas S. Kuhn, "Remarks on Incommensurability and Translation," 1999, 34.
109. *Ibid.*, 34.

110. *Ibid.*, 35.
111. *Ibid.*, 36.
112. David Bersema, in book review of *The Road Since Structure*. (See: http://www.humboldt.edu/~essays/boersema7arev.html)
113. James Conant and John Haugeland (ed.), *The Road Since Structure*, 2000, 36.
114. Kuhn, "Dubbing and Redubbing: the vulnerability of rigid designation", Wade Savage (ed.), 1990, 298-318, 299.
115. Kuhn, "Dubbing and Redubbing…", 1990, 298-318, 315, (footnote no. 4).
116. Kuhn, 'Afterwords,' in P. Horwich (ed.), *World Changes*, 311-41, 329.
117. Kuhn, "Commensurability, Comparability, Communicability", 2000, 45.
118. *Ibid.*, 48.
119. Kuhn, "Commensurability, Comparability, Communicability", in P.D. Asquith and T. Nickles, (ed.), 1983, 669-88, 679-80.
120. Imre Lakatos, "Falsification and the Methodology of Scientific Research Programmes," in I. Lakatos and A. Musgrave, (eds.), 1970, 91-195.
121. Thomas Kuhn, "Objectivity, Value Judgment and Theory Choice", in Yuri Balashov and Alex Rosenberg, (eds.), 2002, 422.
122. Thomas Kuhn, "Notes on Lakatos", in R. S. Cohen, and R. C. Buck (eds.) PSA 1970, 1971, 136-146, 144.
123. Thomas Kuhn, "Objectivity, Value Judgment and Theory Choice", 2002, 422.
124. Paul Hoyningen-Huene, "The Interrelations between the Philosophy, History and Sociology of Science in Thomas Kuhn's Theory of Scientific Development", 1992, 487-501, 493.
125. Thomas Kuhn, "Objectivity, Value Judgment and Theory Choice", 2002, 436.

126. *Ibid.*, 2002, 425.
127. Paul Hoyningen-Huene, 1992, 494.
128. Thomas Kuhn, "Objectivity, Value Judgment and Theory Choice", 2002, 429.
129. Paul Hoyningen-Huene, 1992, 487-501, 493.
130. *Ibid.*, 495.
131. T.S. Kuhn, "Notes on Lakatos", 1971, 144.
132. *Ibid.*
133. Kuhn, Postscript of *SSR*, 2nd ed, 1969, 200.
134. C.G. Hempel, "Turns in the Evolution of the Problem of Induction," *Synthese*, 46 (1981), 389-404; and "Valuation and Objectivity in Science", in R. S. Cohen and L. Laudan (eds.), 1983, 73-100.
135. C.G. Hempel, "Valuation and Objectivity in Science", in R. S. Cohen and L. Laudan (eds.), 1983, 73-100, 91.
136. David J. Voelker, "Thomas Kuhn: Revolution against Scientific Realism". See: http://history.hanover.edu/hhr/94/hhr94_4.html.
137. As quoted in: See: http://www.emory.edu/EDUCATION/mfp/Kuhnsnap.html
138. James Conant and John Haugeland, ed., *The Road Since Structure,* 2000, 2-3.
139. Thomas Kuhn, "Logic of Discovery or Psychology of Research", in *Essential Tension*, 1977, 267-8.
140. Karl Popper, *Conjectures and Refutations*, 122.
141. Thomas Kuhn, "Logic of Discovery or Psychology of Research", 1977, 269.
142. *Ibid.*, 272.
143. *Ibid.*, 277.
144. *Ibid.*, 272.
145. For a precise and comprehensive presentation of the 'Battlefield' and "Common ground' between Carnap and Karl Popper, see: Hacking, "The Rationality of Science After Kuhn", in Robert Klee (ed.), 1999, 217-220.

146. Ian Hacking, "The Rationality of Science After Kuhn", in Robert Klee (ed.), 1999, 219.

147. *Ibid.*, 220.

148. Paul Feyerabend, *Science in a Free Society*, 1978, 118.

149. Paul Feyerabend, "Explanation, Reduction, and Empiricism", in H. Feigl & G. Maxwell, eds., 1962, 60.

150. Paul Feyerabend, *Against Method,* 1975, 46.

151. Paul Feyerabend, "Realism and Instrumentalism: Comments on the Logic of factual Support", in M. Bunge, ed, 1964, 305.

152. Paul Feyerabend "An Argument Against Method", in Robert Klee, ed., 1999, 228.

153. John Preston, *Feyerabend: Philosophy, Science and Society*, 1997, 172.

154. Paul Feyerabend "An Argument Against Method", in Robert Klee, 1999, 231.

155. *Ibid.*, 230.

156. Paul Feyerabend, *Science in a Free Society*, 1978, 90.

157. Paul Feyerabend "An Argument Against Method", in Robert Klee, 1999, 234.

158. *Ibid.*

159. As quoted by: Ernan McMullin, "The Shaping of Scientific Rationality: Construction and Constraint," in Ernan McMullin, ed., 1988), 10.

160. Paul Feyerabend "An Argument against Method", in Robert Klee, 1999, 229.

161. John Preston, *Feyerabend: Philosophy, Science and Society,* 1997, 193.

162. As quoted by Ibid 195.

163. Feyerabend, *Against Method, 1975,* 180.

164. Paul Feyerabend "How to Be a Good Empiricist – A Plea for Tolerance in Matters Epistemological", *Philosophy of Science – The Central Issues*, Marci Curd and J.A. Cover ed., 1998, 923.

165. *Ibid.*, 925.

166. Ibid., 926-7.

167. Paul Feyerabend, *Against Method*, 1975, 36.

168. As quoted by Newton-Smith, *The Rationality of Science*, 1996, 132.

169. Paul Feyerabend "How to Be a Good Empiricist...", 1998, 938.

170. E. Nagel, *The Structure of science: Problems in the Logic of Scientific Explanation,* 1961, 354.

171. Paul Feyerabend, "Explanation, Reduction and Empiricism", in *Philosophy of Science – Contemporary Readings*, Yuri Balashov and Alex Rosenberg, eds., 2002, 143.

172. *Ibid.*, 145; *mine.*

173. *Ibid.*

174. Muhammad Ali Khalidi, "Incommensurability", in W. H. Newton-Smith, ed., 2001, 172-180, 172.

175. Paul Feyerabend, *Against Method, 1984, 5th Impression,* 225.

176. A.F. Chalmers, *What is This Thing Called Science?,* 1992, 2nd ed., 137.

177. Feyerabend, "Reply to Criticism: Comments on Smart Sellars, and Putnam," in 1981, (1st pub. 1965), 104-31, 115. (Italics original).

178. Feyerabend, *Against Method*, 1984, 5th Impression, 285.

179. *Ibid.*, 15.

180. A.F. Chalmers *What is This Thing Called Science,* 1992, 2nd edition, 243.

181. *Ibid.*, quoted in 143.

Chapter 2

Assessment of the 20th Century's Theories of Rationality in Science

INTRODUCTION

Chapter two undertakes the evaluation of the 20th century's theories of rationality in science that have been presented in the previous chapter. The first Section of this Chapter focuses on the proponents of the rational models of science, Hempel and Popper, while the second section critiques the proponents of the non-rational models of science, Kuhn and Feyerabend. Though Hempel's understanding of 'goal-means' notion of rationality is very satisfying to the commonsensical approach there are some serious problems with the very concept of goal(s). Some of the serious and popular critiques of Hempel are presented. Similarly, while acknowledging the great contributions of Popper in the field of philosophy of science, I point out that all is not well his theory of falsification. In fact, since he relics heavily on the notion of falsification to make science rational and to relieve science from the problem of inductive reasoning, once that notion is challenged his whole efforts of making science rational and critical come under severe attack. Further, the limitations with Popper's ideas of progress in science and verisimilitude are also laid bare to show that it is not easy to buy his arguments without some modifications.

In Section two, Kuhn's ideas are assessed. While he is given the credit of making the scientific community realize that science is basically a social enterprise, he is to be taken with a caution as he might easily land science in a trap of relativism. Science cannot just be equated to mob psychology. Even after the revision of his notion of paradigm certain aspects of it are not clear. Shapere's notion of 'chain-of-reasoning connections' is presented as a way-out for Kuhn's problem of incommensurability. Turning to Feyerabend, the Section begins with presenting the views of both his admirers and the opponents of Feyerabend. While recognizing his constructive critique against the highly rational face of science, one needs to be cautious in not agreeing with him that 'any thing goes' in science. It is shown that there may not be unique method, applicable for all sciences at all times, but it is not proper to deny the role and relevance of any methodology; in fact, even to deny methodology, one needs a method.

This Chapter, in short, highlights some of the serious difficulties with both the powerful models of rationality in science, namely, the rational and non-rational. This paves the way to look for a new approach to the whole issue of rationality in science, which would, in a way, tries to absorb the strengths of both models, while correcting their weaknesses.

Section 1
Assessment of the Rational Models of Rationality in Science

This Section directs its attention to the critical evaluation of Hempel and Popper. Both have undoubtedly made significant contribution to philosophy of science. However, there are certain areas that need to be dealt with sufficient watchfulness. For instance, the critiques of Hempel's notion of goal-means rationality, by four important authors, are recalled here: Philip Kitcher argues that rationality does not always demand utility-maximization; Kuhn points out that Hempel's efforts to construct a purely logical program for scientific rationality is only a dream; Robert Nozick, after pointing out many problems with the standard model of science, propounded by Hempel, Popper, and many others, prefers the 'functional model of rationality of science'; and finally, Wesley Salmon indicates that Hempel has not sufficiently addressed the 'crucial problem' of the inductive reasoning. Then, a short analysis of the complex nature of 'goal(s)' is undertaken in order to bring out the shortcomings in confining rationality only to the 'goal-means' relationship. Regarding Popper, one has to appreciate his critical spirit towards knowledge. There are a few serious problems with his bold claim of having solved the problem of induction, because he seems to have watered down the concept of induction. Several troubles can be highlighted with the notion of falsification, to show that it cannot provide the conclusiveness that Popper aims at.

Hempel: Underestimating Goals in Science and Life

Hempel is too an important thinker to be treated lightly. His contributions are significant and thought-provoking that it attracts the attention of many critics. Let me give some of them, which I deem to be directly relevant to our purpose here, before I give my comments on Hempel's ideas.

"Rationality Does not Always Demand Utility-Maximization" - Philip Kitcher

No doubt, Hempel's notion of rationality has two distinct components, namely *normative* (evaluation) and *explanatory* (description) components. Hempel defines, as we have already seen, rationality in terms of the goals of the action and the beliefs available for the agent to accomplish that action. Kitcher calls this as *utility-maximization notion of rationality.* But the problem with this notion of rationality, as Kitcher shows, 'success' can be a matter of luck too. One may attain the maximized utility by a mere matter of luck. So a modification in Hempel's account of goal-means relationship is needed to take care of the aspect that the agent does the actions in the most reliable way, not by luck, to maximize the perceived utility.

Kitcher (2000) tries to show that utility-maximization of action need not always be counted in a materialistic and numerical way. He gives the example that Charles Darwin, before proposing to his cousin, jotted down all the advantages and disadvantages of proposing to her and marrying her. Weighing the relative strengths of both the aspects he finally decided to marry. We cannot totally call this approach a rational one. For, this kind of decision-theoretic approach to such a phenomenon of choosing one's life-partner does not seem to be appropriate: "A person who was self-aware and conscious of the depths of commitment required for a happy marriage would question his own ability to make the numerical judgments and would be profoundly skeptical of attempts to compute the 'utility-maximization action".[1]

More studies in psycho-neuro related fields reveal many striking elements with regard to the process of decision making in our ordinary day-to-day lives. Very often we don't make any elaborate cost-and-effect analysis of our actions before we perform them; we need to act almost impulsively having no time for strict rationalization – for example, jumping into the river to save a drowning child, or jumping quickly away at the sound of leaves' rustling fearing a snake, and later finding out that it is just dry leaves. Decisions can very well result from our emotional responses, like being moved by pity and sympathy helping someone, even by going out of one's way and perhaps even though we don't seem to like him/her.

Kitcher prefers the phrase 'being reasonable' to 'being rational'. The one who is reasonable gives considerable importance to reason and emotion. Such a person goes far beyond mere decision-theoretic calculation. Such a person goes beyond strict rationality. She is aware of the possibility of being swept away by momentary impulses, who knows how to handle this danger, who can distance oneself and suspend from action. From this realization, one can see History of Science as not a product of pure rational actions which slowly but steadily take us nearer to *truth*; nor is it a net result of socio-politico and economical upheavals. Rather history of science is moulded by, "the real actors (who) are those psychologically complex beings, reasonable people".[2] [*Note*: I do find a great sense of relevance in his distinction between *reasonableness* and *rationality*. I take this up at length in evolving my account of rationality in science in chapters three and four].

"Hempel's Efforts - An Impossible Dream" - Kuhn

Kuhn tries to show that Hempel tries to construct a purely logical program for scientific rationality. Both Hempel and Kuhn agree that theory choice has to presuppose a prior specification of the goals – say, to make the theory more efficient or better 'puzzle solving'. A theory is more efficient, if, say, there is a better fitness between theoretical predictions and the results of experiment and observation. Suppose, given this goal, one chooses a theory which does not provide a better fitness then we can straightaway see that one is irrational. Yet, Kuhn differs from Hempel, when the latter, following Carnap, favours constructing a formal inductive logic. Hempel wants to build a program of theory choice based upon logic and experiment, whereas Kuhn affirms that logic and experiment alone can't give sufficient grounds for theory choice. Kuhn, as well as other historians like him, is of the opinion that this very program is basically untenable. However it is good to note, as Salmon clarifies, that "It is not that Kuhn wants to regard science as irrational; instead he would offer a more 'naturalistic' conception of rationality".[3] Further, Salmon argues that a common understanding of the notion of 'logical structure of confirmation' has to be arrived at in order to resolve the differences between Kuhn and Hempel. Therefore he queries, "Unless we have some fairly clear conception of the logical structure of confirmation, how can we hope to come to any intelligent resolution of the kind of issue that divides Hempel and Kuhn on the nature of scientific rationality?".[4]

Kuhn explains that verification and falsification are not compatible with the actual practice of science. They can't work as the basis of normative construal of scientific inquiry. It is not only due to the perennial problem with justifying induction (because Kuhn recognizes, as Hempel also does, that the problem of induction is very deep, a thorn in the flesh so to say, yet he wants a solution: "Together with most of you, I share Hume's itch. Preparing this paper has made me realize that the itch may be intrinsic to the game, but I am not ready for that conclusion"[5]), but also because any comprehensive account of scientific rationality cannot afford to ignore the actual practice of science.

Logical positivism derived their methodological rules for the notion of rationality from certain a priori models. But there are no a priori determinations or norms to which scientific activity is expected to conform. Theory of rationality must be formulated by looking at the actual practice of science. Kuhn also rejected the clear-cut distinction between descriptive and normative accounts; if at all there is a distinction, *only the descriptive account can work like a normative account*. Normative account has to take note of the actual practice of science. As essential functions of science-practices lead science to improvement and progress. So they can be considered as normative. Scientific progress is, as we mentioned earlier, like Darwinian evolution: "Scientific development is like Darwinian evolution, a process driven behind rather than pulled towards some fixed goal to which it grows ever closer".[6] All these lines of thinking make one thing very clear, that is, Hempel's efforts for a logical program for rationality of science always eludes philosophy of science.

"Functional Model of Rationality of Science is preferable" - Roert Nozick

Nozick points out many problems with the *standard model of science*, which has been cherished by Popper, Hans Reichenbach, Hempel, Ernest Nagel and so on, for the last seventy years of the twentieth century. Their main thrust was the idea that the origin of scientific theories might be anything but they should be empirically testable for justification. But problems are many with this standard model of science. For instance, scientific hypothesis is never a single isolated hypothesis but a combination of a few auxiliary hypotheses (so called Duhem problem); theories are not thrown outright unless some sort of alternatives are available; no clear-cut demarcation to differentiate between genuine, wise modifications and mere *ad hoc* reformulations; theory-ladennes of observation; underdetermination of theory by the data and so on. As a response to these problems there have been two trends: some take the *radical step* of denying the objectivity of science totally, and others take *defensive step*, to maintain that "although there are not precisely agreed-upon rules of theory choice, rationality does not require such explicit rule-following" and therefore, "science can be rational and objective, *despite* the factors which complicate the standard model".[7]

As Hempel is aware that due to the problem with the nature of desiderata, a totally objective account of scientific rationality is not possible; therefore he accepts, as we see later, 'a relaxed but an objectivist construal of methodology'. But this is due to a sort of later thinking. For, till late 1970s Hempel maintained a strict list of requirements for scientific rationality. For instance, in "Scientific Rationality: Normative vs. Descriptive Construal,"[8] he demands that scientific procedures to be precise to make a theory choice which is unambiguous at all times; and these procedures follow strict and explicit norms and these rules are followed for advancing scientific knowledge. He also believed in the continuous growth of science: "(the new) theory does not simply refute the earlier empirical generalizations in its field; rather, it shows that within a certain limited range defined by qualifying conditions, the generalizations hold true in fairly close approximation".[9] He views the growth of science a sort of cumulative growth, as the latter theory is more comprehensive than the previous one. However by the middle of 1980s we see a sort of liberalization in his demands, as he maintains that the methodological principles which don't refer to psychological or sociological aspects are not desirable[10] and he goes in for a relaxed objectivity construal.

After analyzing the two common models, Nozick proposes a third trend with the model, namely, the *Functional Model.* It transcends both the radical and defensive models. He boldly claims that science is rational, not in spite of, but because of the problems we have in the desiderata. For they add to objectivity and rationality. For science always builds upon what has gone ahead; we don't, cannot and even should not begin always form the beginning: "We do not start from scratch unless we have to. It would be irrational to always begin at the very beginning".[11] May be realizing this, Newton seems to have said, 'Only when we stand on the shoulders of our ancestors we are able to see a little farther'. Given our neural wiring, which is the product of a long evolution, we cannot have a pure observation with zero theoretical content. First of all, is it possible to have an observation with zero theoretical content? Even if so what will it be like? Is it desirable?

Nozick is convinced that however we try we can never make the theoretical content nil while making observations, because the way we have evolved has placed us in a position where we don't have zero content. However when we say that we always have to work upon what is already there, this does not mean that science is always path-dependent. A total shift from one paradigm to another is possible, and even desirable. "The lack of such fixed and fully determinate rules of acceptance, then, is a virtue, not a defect".[12]

Underdetermination of theories by data (i.e. from a set of data more than one theory can be arrived at, which are not incompatible) need not affect the rationality of science. Compared to the data we have about the world and the world out there, there is always a big gap. This gap is that which makes science interesting and even possible. Therefore, the "Epistemological or ontological underdetermination.... propel(s) the existence of science; we would not need an organized activity of science if our observations reached all the way to basic processes... The fact that our theories reach farther than our data shows how far extended the reach of our theories – a cause for celebration rather than for lament".[13] However Nozick is confident that the underdetermination of theory can be addressed by creating new theories, incorporating the successful theories of the past, and by getting the criteria provided by the paradigm in which one is working. Similarly, the theory-ladenness of observation can be tackled to an extent by using several distinct theories to derive predictions, so that we don't allow any one theory to dominate or to determine our observation.

Nozick concludes that this kind of model, *functional model*, is "neither radical nor defensive but one that delineates the role of these factors (underdetermination of theories, theory-ladennes etc) in contributing to the rationality and the objectivity of the scientific endeavour, thereby giving us a better understanding of precisely how science *is* rational".[14]

"Hempel Is Missing Something Crucial" – W. Salmon

Hempel recognizes the age old nature of the problem of induction which is *the* problem for him: "That is the problem of justifying the way in which, in scientific inquiry and in our everyday pursuits, we base beliefs and assertions about empirical matters on logically inconclusive evidence".[15] He identifies the crux of the problem to be the inability to provide decisive proof for the claims made in conclusion as they go beyond the known instances.

Though Hempel claims that he has not solved all the problems related to inductive reasoning, Salmon argues Hempel has missed something very central to the whole issue; he has not done enough to show the justification of empirical predictions through inductive reasoning. Salmon points out that Hempel even ignored an important aspect undiscussed; namely, for Salmon it is "odd that nowhere in the entire discussion does he (Hempel) address the question of whether inductive logic contains the resources for us to infer the existence of unobservable entities from evidence about observables. That is, I Think, *the* crucial question".[16]

Nevertheless, Salmon himself is not unaware of the crux of the inductive reasoning: i.e. on the one hand, inductive reasoning seems to go beyond a conclusion and on the other hand, many of the philosophical issues centre around it. Therefore he acknowledges, "the logic of confirmation – inductive logic – is an extremely difficult and subtle subject. We still have much to learn about it. Issues that are absolutely fundamental to philosophy of science – such as the rationality of science, the status of scientific theories, and scientific significance in general – still hinge upon the answers (to the questions of induction)"[17] and as long as this problem is not solved we are always "left with an uneasy feeling about the cognitive status of theories".[18]

Salmon proposes a probabilistic account of theory of appraisal along Bayesian lines. Theory choice is actually a consideration of one theory from a whole lot of other theories. From the fact of theory underdetermination we learn that there can be an indefinite number of theories for consideration. From these very large number of theories we segregate some of them on the basis of prior plausibility, say ($p = 0.01$), for further consideration. This prior plausibility depends on personal motives, political affiliations etc. However, scientific rationality need not be threatened due to this;

for prior plausibility is also a part of rational process. Therefore he argues that "(if) Bayes's theorem is correct, then just such plausibility judgments are a *necessary component* in *rational theory choice*".[19]

Difficulties in Confining Rationality to Sheer Goal-Means Relationship

Complex Nature of Goal-Means Relation: Often we find the goals of human actions are not clearly or fully defined or expressed. One may verbally express a particular goal for one's action but implicitly may have another goal in one's mind. For example, a student asks for an appointment to meet his Supervisor in the School to discuss her thesis with him. But the student tells herself, 'after the discussion I'll wish my supervisor for his forth-coming birthday'. Specifying strong, one-to-one relation between goal and means, can't account for such unexpressed goals of human actions which are also part of the action.

Stephen Nathanson (1994) also has some relevant comments to make regarding the means-end conception of Rationality. This type of understanding rationality is very close to common sense understanding. Critics of rationalists can no way deny this means/ end conception of rationality; even those who deny the elements of logicality, objectivity, deliberation, detachment etc. can't deny this concept of rationality; to do so will make that person highly irrational or insane, for even to deny all these is an *end* of their rejection.

Nevertheless there are serious issues to be settled with goals-means notion of rationality. It is understandable that an action can be shown to be rational if it has the highest probability of bringing about the desired end; but this can be decided upon only when we know about the agent's desires; we can't easily judge the aims of the agent in an unambiguous manner. Emotions play a big role in life, it can't be denied at all; without desires, passions there can't be any aim at all in life. It is likely that one is under the power of strong emotions one thinks that a particular thing to be achieved but on a second thought the agent may change his or her goal. If reason requires acting for the best, many other questions arise – what is the best? What things are of value? Whose best? Of the agent's? Of the society? Of all humans? Of all sentient beings? – all these cannot be settled in a definite manner. Further, there can be cases where an action can succeed in realizing an agent's goals and *still be irrational.* The end must be coherent with many other goals in life; a rational action is the one which enhances the prospects of satisfying our central goals in life, rather than interfering with it. Since all our goals in life are often inter-dependent, it is not proper to take a particular action in isolation to study it in terms of its means/end relationship. More discussion on his ideas is given in Chapter III.

Goal Vis-à-vis Science: The goal of science is to be understood in a more complex manner, not as an attempt to attain fixed goal. The goal in science is not like the goals of the activities of day-to-day life. Even if there is one grand pre-fixed goal for science, at no point of time one can claim that the goal of science has been achieved; suppose that goal is achieved one day, the further activities of other scientists seem to be redundant. Down the centuries science shows that it is an on-going activity; it is never able to reach the already one fixed final goal. For, a) one fixed goal is not given to the scientific community; dimensions of goal change in relation to the realms of actual practices. One can't realize some fixed goals in a political realm also. For instance, justice or equality - which is not possible (not only practically speaking) to attain in its fullness in a political realm; it is impossible also in principle, because the yardsticks or requirements of justice may change from time to time within political systems and across various political systems; and b) the goals of scientific inquiry are such that they can't be attained by one or two groups in a specified time-frame. It requires a very long process and it needs the collaboration of diverse groups, from diverse regions for a long period of time.

Science - Not Headed to One General Goal: a) Science is not purely a goal-governed activity; usually when the repeated attempts fail to achieve a goal, that action is renounced, but in science on the contrary, they try more hard or try in different methods to attain that goal; b) Scientific activity is somewhat different from fixed goal; the activity itself generates new dimensions and aims for its development. [Galileo, Newton and others replaced the hitherto held practice of explaining reality in terms of occult qualities and adopted a mechanical view of reality; but Newton himself in order to explain gravity (i.e. action at a distance) had to rely on a sort of occult elements though he had not renounced the general goal of providing a mechanical view of nature, yet he has to shift / change his goal for a specific purpose. The context itself imposes certain requirements or conditions (like the need to abide by the religious principles in vogue at that time)].

No Pre-fixed Goal in Science: In actual practice goal-action relation is not one-to-one and not straightforward; it is a very complex activity. Scientific activity can't be seen as something motivated to achieving one clear goal; it is a complex activity which has to satisfy many requirements as per the circumstances of the scientific activity. It may not always be directed to a specific goal; For, specific scientific activities are functioning within a certain structure (community), and that structure often directs the scientific activity towards a goal. Institutional requirements include influences of circumstances; it is not therefore directed purely to achieve a certain goal. Moreover goals can be of two types: *achievable* and *unachievable*. Even among the achievable goals there is no one-to-one relation; so rationality can't be decided upon analyzing the activities which are not directly related to the goal; and if the goals are unachievable fully, it can't guide the action on a rational path; a) then the action is not directed towards one goal, rather divided towards many aims and b) the contexts redefine the goal(s) continuously.

Goal of the Institution, Not of an Individual Activity: Science is not just one single homogenous inquiry, rather it is an institution. Therefore it has its own requirements and restrictions. Not an individual activity but the institution as a whole is directed by the proposed goal. Thus, not the conditions of goal but the condition of the institution is to be focused on as desiderata. Conditions of institution can work independent from each other, while condition of goal work together to characterize the goal, in the latter case, say, if a certain theory is simple, it does not mean that it satisfies the goal of scientific activity. When we look at science as an institution (structure) the whole institution as such represents the goal, not an individual activity. The rationality of science depends on not only the conditions of goal, but also the institution.

However science is not to be seen as an activity to be carried away by the requirements and influences of the circumstances alone. Science activity itself has a determinative power, which is unaffected by the requirements and influences of the circumstances. This determinative power, Hempel claims, is the necessary condition. This determinative power does not govern all the process. These necessary conditions are also the product of an evolutionary process, as new conditions and aspects of goal continuously develop over the period. "... the idea that the desiderata, which were taken here as reflecting the goal of scientific research or the idea of scientific progress, should be viewed, not as fixed once and for all but as themselves subject to change in the course of the evolution of science".[20]

Therefore rationality has to be decided not just by looking at the conditions of goal, but the conditions/requirements of the institution within which science functions. In a complex activity the failure (or difficulties to achieve) to achieve a goal does not imply the activity is irrational. For instance, a good government, a harmonious society, removal of poverty and various discriminations in the society may not be satisfactorily achieved in a society, but this does not mean that actions with that aim are irrational. Often each activity is assured not with the realization of its goal but whether it sets the step for further practice towards the realization of that goal. Not that any condition is accepted as a necessary condition of rationality. Through the interactions with the structure the necessary condition is accepted. These interactions develop the structure of science also. *Therefore instead of 'goal-action rationality', it is more appropriate to speak about' goal-institution rationality' in science.*

Authority of Science to Determine the Goals: An empirical science can't advise anyone what one should do, though it can help one to clarify for oneself what one can do or want to do. Goals are related to the likes and dislikes, ambitions and anxieties of an individual or a society. As Max Weber points out the issue of determining goals for an individual or a community does not lie with natural or empirical science. Perhaps this is the area where one can highlight an obvious difference between social and natural sciences: that is, the difference between them is not in any of their special methods of investigation inherent to them, but they differ only in the aims and the interests of the scientists. There are, in fact, lots of similarities between these sciences, like, total explanation is not possible in both sciences, and both involve abstraction from the multidimensional aspects of reality. The choice of problems is always 'value relevant'; "There is no absolutely 'objective' scientific analysis of culture or... of 'social phenomena' in dependent of special and 'one-sided' viewpoints according to which – expressly or tacitly, consciously or unconsciously – they are selected, analyzed and organized for expository purposes".[21] Weber insisted that science can help her to critically judge or analyze and understand value judgments; it may check its internal consistency of her actions and desired ends, but "As to whether the person expressing these value judgments should adhere to these ultimate standards is his personal affair; it involves will and conscience, not empirical knowledge".[22]

Hempel – A Bridge between Logical Positivism and Kuhn

Hempel reformulates logical positivism's account of rationality in order to include its socio-cultural elements and tries to correct Kuhn by showing that considerations of goal (and means to achieve it) of the scientific activity lays down restrictions upon the desiderata so as to make them necessary and sufficient norms for scientific rationality. He rightly shows that the accounts of logical positivism and Popper are incompatible with actual practices of science. Therefore normative account has to take note of the actual practice of science. He corrects Kuhn to show that no purely descriptive account can be taken as normative. Hempel charges Kuhn for not making a distinction between the explanation of behaviour of the scientists and the issue of justification. However Hempel is not ready to accuse Kuhn of irrationality, though he knows that other critics do so. It is true that desiderata are insufficient in theory choice; they don't provide any algorithmic procedures for science practices. However they are conditions of goal; the goal lays down restrictions upon the scientific procedures and acts as motivating reasoning or direction. Therefore science is not an inquiry where 'anything goes'. We need context to reconstruct the rational account of science. He comes out with a new account of rationality, as we saw earlier, which he names as *objectivist but relaxed rational reconstruction,* which is seen as a bridge between logical positivism and Kuhn. Though Hempel finds some commonalities between the normative and pragmatic methodologies yet he seems to be inclining more towards a normative methodology. As Fetzer makes it clear, "The most important lessons of Hempel's enduring legacy must be that scientific standards cannot be derived from descriptions of its practice alone".[23]

'Relaxed but an Objectivist Construal of Rationality' – More Relaxed than Objective?

Due to the inconclusive nature of the desiderata (as enlightened by Kuhn), but the determinative propel of the scientific activity by the goal, Hempel opted for a 'relaxed but an objectivist construal of methodology'. It is true, the desiderata provide only a relaxed notion of rationality. But it is debatable how the conditions of goal are able to provide an objectivist construal of rationality. With all the problems with the assumed one pre-fixed goal, the goal-means relations and so on as we noted earlier, one can see that the goal-thesis of Hempel does not seem to do the job what he attributes to it. Therefore I tend to look at his relaxed objectivist construal as something 'more relaxed than objective'.

Hempel argues further that the goal of science cannot be truth, for not all desiderata are connected with truth, except logical consistency and testability. Therefore true theories are not the goal of scientific activities. In the words of Hempel, "If scientific theory is to be viewed as a systematically goal-oriented enterprise, its goal surely is not the attainment of true theories... Scientific theorizing is oriented, we might say, not at the ontological goal of truth, but at the epistemological one of optimal epistemic integration, or at epistemic optimally, of the belief system we hold at any time".[24] But

Nozick suggests that science is the best available tool for humanity; we don't have anything else to obtain truth: "But if there is no better way than science, then perhaps true theories is the goal of science... Perhaps rationality requires that we pursue the most effective means to our goal... but it does not require that we always pursue a goal for which our means are quite effective".[25] So the scientific enterprise does give some truths about nature out there. They may not be the mirror-pictures of reality. One cannot deny that within all the limits, practical and in-principle ones, science enables to have a rather 'true' picture of the world. Therefore Hempel seems to stand in need of a correction when he says that science does not aim at truths.

Karl Popper: A Fallible Falsificationist!

It is interesting to note that there is a wide spectrum of reactions to Popper's ideas, from high appreciation to strong objection. Some praise him very highly, like Stephen Hawking who uses Popper's idea of demarcation to define science, and others, like G.J. Warnock (1960), find very serious faults with him (e.g. the issue of induction). Popper is known for his simple style in presenting his deep ideas and profound arguments, unlike his contemporary scholars. Bazhenov has high regards for the literary style of Popper: "Sir Karl converses with the listener-dilettante on the most complex and subtle problems of epistemology in a way that makes these problems comprehensible to the dilettante while he says things of enormous interest to the professional".[26] Simple and clear presentations, Popper is convinced, are the signs of great and serious scholarship. He charges the then scholars to have created "the cult of un-understanability, the cult of impressive and high-sounding language"[27] and overcomes that very cult of making presentation of ideas much more complicated and cumbersome.

Many consider Popper's contribution a philosophical breakthrough. For, while traditionally it was considered that only established theories deserve the status of scientific, Popper held that any hypothesis which was capable of being tested experimentally would be scientific; similarly, tradition thought that repeated experiences would lead to generalised theories, but Popper held that we would learn from experience by repeatedly speculating explanatory hypotheses and falsifying them by experiments and getting to the nearness of truth by stages.

On a bit deeper analysis we find Popper's claims for scientific rationality run into many difficulties, so much so Newton-Smith calls him 'the irrational rationalist'.[28] It is because a) Popper's method of *Conjecture and Refutation* is not able to produce theories with greater *verisimilitude*; in fact, falsification is not only helpless it even "involves a grave distortion of actual scientific practice"[29] and b) the relationship between the goal of science and the method of science, which is essential to claim rationality in science, can't be established. Moreover, he claims that the methodological rules in science are a priori, while they are conventional as well. They are also tentative. This seems to be inviting trouble given his realistic and rationalistic approaches. Let us elaborate on some critical remarks on Popper's views.

Appraisal of Popper's Views on Critical Rationalism

In general, Popper's critical attitude is highly appreciable. Today the tendency to be dogmatic, and thereby intolerant, seems to be alarmingly increasing, not only in the matters of faith and religion, but also in the matters of intellectual pursuits. The failure to criticize oneself leads to intellectual stagnation and spiritual arrogance. Overconfidence in one's theory is not at all a sign for intellectual growth. As Claude Bernard makes it clear, "Those who have an excessive faith in their ideas are not well fitted to make discoveries".[30] Popper's critical attitude towards rationalism enables him to realize the limits of reason. As W.W. Bartley has it: "Popper thus argues that human reason is unlimited with regard to criticism, yet limited with regard to its powers of prediction; and shows that both the lack of limitation and the limitation are, in their respective places, necessary for human rationality to exist at all".[31] Popper, being realist assesntially a, criticizes subjectivism and idealism and upholds objective realist approach to knowledge. In his three-volume *Postscript to the Logic of Scientific Discovery* (1982/3) – *Realism and the Aim of Science, The Open Universe: An Argument for Indeterminism and Quantum Theory* and *The Schism in Physics* – one finds each volume attacking some approaches or the other of subjectivists or idealists, highlighting some aspects of objective realists' approach.

In Popper's schemata of knowledge enterprise *criticism* plays a very crucial role. In fact the ability to critically evaluate their own cognitive activities characterizes humans as rational. As Boris Salazar explicates, according to Popper, the rationality of humans depends not only on their ability to judge a situation and devise proper methodologies to solve them, but also on their ability to critically assess their solutions and advance their corpus of knowledge.[32] Of course, the understanding of rationality itself needs to be scrutinized critically.

In inductive arguments the conclusions don't follow from their premises. Further, inductive reasoning cannot provide sufficient justifications. Therefore such reasoning is of no significant value in critical thinking. For Popper, as Stefano Gattei puts it, rationality is not a distinguishing property of knowledge, it is the mission of the researchers to achieve. Rationality does not refer to the content of a theory but to the manner in which the theory is held. Thus, "Being rational means nothing but being willing to appeal only to reason and arguments to resolve our disputes. Rationality requires no foundation, no grounding upon expert or consensus. It needs only critical dialogue".[33] For instance, causal determinism, which is very predominant in the classical physics of Newton and Laplace, loses its grips at the subatomic levels. Laws of causality need to be replaced by laws of probability. This is the characteristic of progress in science. Any drastic change can take place but it needs to be accepted after critical analysis. If this process is objected or denied, science will lose its rational and empirical character.

However, more than one critique charge Popper that he is not giving any clear idea about methodology. The *Critical Rationalism* (*CR*), which he claims to be the method of philosophy and natural sciences as well, does not seem to be well-defined. *CR* as an overarching methodology suitable for all disciplines, seem to remain the almost contentless ruling. For instance Robert Nola argues, "Critical rationalism may be no more than the general directive to criticize our theories, or always to attempt to show that our theories are false...CR then becomes an unobjectionable but rather empty set of rules which at best excludes dogmatism but says little more that is positive".[34] In order to substantiate this claim Nola quotes the passage from Popper, where the latter proposes one method which is common for both philosophy and natural sciences as well.[35] Hence his methodology of critical rationalism does not seem to do the job of guaranteeing the foundation of rationality to science in a very definitive matter; since it is applicable to other disciplines as well the force of critical rationality as the methodology of science seems to be watered down.

The Legitimacy of Fallibilism in Science

Popper sounds very practical as he is a man of 'scientific fallibilism'. Indubitability, absolute exactitude, absolute precision always evade the domain of scientific knowledge. To make science infallible is to block the progress of science. Any theory in science can be refuted at any time, and that is the strength (and may be weakness too!) of science. Empirical testability is very essential for scientific knowledge and at the same time testability does not give us absolutely conclusive and precise theories. For such scientific theories can be questioned any time in future.

Popper seems to have great regards for Darwin's theory of Evolution, so much so that Popper tries to see the process of knowledge from the evolutionary perspective as well. Popper's well-known picture of the development of knowledge is, as L. B. Bazhenov puts it:[36] Problem (P_1) à tentative theory (TT) à error elimination (EE) à new problem (P_2). The evolution of both living organisms and our knowledge proceeds according to this schema. In biological evolution organisms that fail tests perish [Failing the tests in biological evolution imply the inability to survive unfavourable and inclement environmental conditions]; in cognitive evolution hypotheses that fail tests are eliminated.

Popper distinguishes between *logical aspect* and *practical aspect* of falsifiability. But he is sad that many of his critics seem to be unaware of this distinction. For instance, Kuhn, though he is aware that Popper is not a naïve falsificationist, can't help label him as a naïve falsificationist. In *The Philosophy of Karl Popper* (ed. by P.A. Schilpp) Kuhn writes: "Though he is not a naïve falsificationist, Sir Karl may, I suggest, legitimately be treated as one". For this Popper replies, "This passage is really astonishing. It is exactly like saying: Although Popper is not a murderer, he may, I suggest, legitimately be treated as one".[37]

The Limitations of Falsificationism

The theory of Falsificationism might seem to be acceptable at the first sight as it claims to guarantee conclusiveness (in falsifying), unlike verificationism which can give us only probability. However on a bit deeper analysis, one becomes aware that there is an ocean of concerns from all possible sections. As one commentator puts it, "there are many problems with his (Popper's) account of progress through Falsificationism, (for) ... some kinds of scientific evidence, including probabilistic evidence, are unfalsifiable and... in the lab, scientists often ignore falsification".[38] There are lots of problems and conflicts that infest the theory of Falsificationism. Let me consider some of them here:

a. One of the things that Falsificationism upholds is theory-ladenness of observation. But Falsificationism itself seems to assume that perfectly assured observation-statements are available. For though it rejects *confirming the truth* of the given theory, all the time it makes efforts *to establish the truth of observation statements that are inconsistent* with them. Therefore conclusive falsification is also *not* possible as the observation statements upon which it is based themselves may be proved false in the course of further developments.

b. A scientific theory or a statement is a very complex one in reality. So it is not that easy to falsify it. As Duhem-Quine show that a single falsifying instance can't falsify the whole theory, as any theory is a collection of many hypotheses. Not all the theories are as simple as 'A pen is on the table';

for, theories are interconnected and no theory would be established or disproved in isolation and hence the conclusive falsification is made very difficult. So given a falsifying observation statement it may be the case that one of the auxiliaries or initial conditions is false. The falsity cannot be automatically attributed to the hypothesis in question. So the hypothesis can be retained in spite of a falsifying test by modifying the auxiliaries or the initial conditions. This, in fact, can keep on going and at no point one can strictly falsify a hypothesis. Therefore, "A theory cannot be conclusively falsified, because the possibility that some part of the complex test situation, other than the theory under test, is responsible for an erroneous prediction cannot be ruled out".[39] The same idea is echoed by Maxwell as well. No theory is in fact falsifiable. Any theory can be 'saved' by somehow cooking up some unknown or unobservable factors of the initial conditions that produced the observed results.[40]

c. We have many examples in the history of science to show that even after falsification a particular theory is not just discarded. After a while it may be revived and gets accepted, tentatively though, by the scientific community. There are instances to show that scientists normally don't reject a theory as soon as they come across an anomaly. There are no quantum jumps in the history of science; one finds only a gradual and evolutionary growth of knowledge. Further, even after the so called 'falsification' they keep the theory. They just temporarily keep it aside and hope for a better explanation in future. Meanwhile they make *ad hoc* modifications to keep the theory going. Therefore it is not proper to demand an immediate rejection of a theory when it is falsified. As Lakatos quotes the example, Mercury's anomalous perihelion was not taken as a falsifying instance of Newton's theory of our planetary system. As a result Newton's theory was not rejected, rather they kept aside the anomalous instance to be considered later and ad hoc solutions were given to save the theory.[41] Therefore Lakatos feels that the falsification is not crucial, rather any theory with a *better* empirical content would ensure progress in science: "And for this replacement of one theory by a better one, the first theory does not even have to be 'falsified' in Popper's sense of the term. Thus progress is marked by instances verifying excess content rather than by falsifying instances".[42]

d. In history, there are many episodes, which can be adequately accounted for, neither by Inductivists nor by Falsificationists. For instance, the Copernican Revolution took about 150 long years to be established. In 1543 Copernicus published his helio-centrism theory and it was confirmed by Galileo's telescope only in 1609. Hence the acceptance of the Copernican theory did not take place at one instant of falsification as Popper would have us to believe.

e. The experimental testability, which is shown to be very crucial for science, seems to have a limit; that is, at a certain level the acceptance or the rejection of a theory depends on the general consensus of the investigators; therefore the test of a theory must stop at a kind of statement, "...about whose acceptance or rejection the various investigators are likely to reach agreement".[43]

f. The scope of falsificationism is limited in the sense it cannot be employed in the case of probability statements and analytic statements. For the technical falsifiability seems to be unhelpful in such cases. Popper[44] makes the distinction between *logical Falsificationism* (referring to the logical possibility of falsification in principle) and *technical Falsificationism* (referring to a conclusive and practical proof of falsity). Probability statements are neither falsifiable nor verifiable in the logical sense. They are not verifiable, for the usual reasons as any other universal generalizations and not falsifiable either, because in a throw of a coin the probability for 'head' is ½; that is, both head and tail have equal frequency. We have a series of throw and in all of them we get head; yet the probability of getting tail is not ruled out. So with a finite number of throws of a coin we can't falsify the probabilistic statement. Probabilistic hypotheses don't rule out anything observable. Probabilistic estimates can't contradict, or be contradicted by basic statement, nor by a conjunction of any finite number of basic statements, and therefore, not by any finite number of observations either. We learn that they have no falsifiable consequences and they can't themselves be the consequences of verifiable statements either. Only infinite sequence of events can contradict a probable estimate. Probability hypotheses are unfalsifiable, as their dimensions are infinite. So they are empirically uninformative, without any empirical content. But science (Physics) has grown successful in the predictions derived from the hypothetical estimates of probabilities by a method called 'practical falsifiability'. That is, though the hypothetical probabilities can't be fully falsified, practical falsification is achievable by a methodological decision to rule out highly improbable events.

g. Problem with having falsifiability as the criterion of demarcation of science and non-science: any statement which is falsifiable need not automatically be scientific. As Putnam gives the example, "if anyone puts a flour sack on their head and raps the table 99 times then a demon will appear"[45] – this statement clearly has epistemological character, logical property and falsifiability too. But it is hardly scientific!

h. Falsification is not very different from verification. Popper shuns verification as it is impossible to conclusively verify a universal theory. The same difficulty seems to arise in the case of falsificationism too. Rom Harre and Michael Krausz clearly bring out the problem of asymmetery between the process of confirmation and falsification, when they analyse Popper's ideas: [46] Furthermore it (Popper's ideas) suffers from the general problem of methodological fallibilism that the supposed asymmetry between confirming a hypothesis as true and disconfirming it as false is difficult to defend. While the former seems to depend on the inductive assumptions that the future will be like the past in relevant respects, itself in need of inductive support, the latter seems to depend on a related assumption that the future will not be like the past in relevant respects, an assumption equally in need of inductive support.

i. Above all, as per the demands of Popper, a theory becomes scientific only when one is able to specify experimental conditions which would falsify that theory. But when this demand is applied to the *theory of Falsificationism* itself, he seems to fail to give those conditions. As Lakatos points out, "He does not answer the question: '*Under what conditions would you give up your demarcation criterion*".[47] (Italics in original).

One further wonders how far Popper is proper in making falsification the bed rock of his philosophy of science. "Testability is falsifiability"[48] demands Popper. These negative instances are very important because positive instance give only inductive support or probabilistic conformability. Further, because falsifiability takes science nearer to truth (greater verisimilitude), he persuasively claims, the method of bold conjectures and refutations must be *the* scientific method. However one does find that many important theories in science and otherwise are glaringly non-falsifiable.

Therefore, on a closer analysis, we find Falsificationism to be infested with some insoluble and intrinsic problem. The powerful words of Anderson seem to sum up the difficulties well: "It (falsification) has failed both on a methodological level, e.g. albeit rejecting induction Popper nevertheless reasons inductively, as well as on a meta-methodological level, e.g. does not correspond to contemporary methodology nor past ones. Further, Popper fails to postulate a rational criterion of refutation".[49]

Popper's Idea of Progress – Is it Progressive?

Popper's idea of scientific progress is also not without problems. As the growth is cumulative he claims that the latest theory is nearer to the truth, having more correspondence to reality than the previous theory. But this claim cannot be justified as we don't have any other direct access to reality. Popper would bring in the concept of verisimilitude, which is truth content minus falsity content of the theory (V = Tc - Fc). Thus the empirical content is more with the latest theory than with the previous one. But unfortunately Popper uses verisimilitude in two senses: one is measurable but not epistemic (Tc - Fc) and the other is epistemic but not measurable. The relation between the latest theory and reality is in the epistemic sense which is not measurable. Measurable verisimilitude and epistemic verisimilitude are different, cannot be bridged. For instance, two students can be compared, say in terms of their attendance and their character; the attendance can be easily measured from the records, while there is no such easy and straightforward measurement for their character.

The Role of Positive Instances

In the beginning Popper did not see any importance for positive instantiations. He even encouraged the scientists to always look only for negative instances to falsify, claiming that the positive evidence has no value for science. Popper treats only the negative instances with utmost importance; for, as Joseph Agassi beautifully explains it, "refutation opens the road to innovation. Nothing is more potent heuristic than refutation. Nothing is more conducive to progress than criticism of the current situation, nothing more likely to herald the new than discontent with the old. Criticism is liberation. The positive power of negative thinking".[50] But by 1960s, Popper could see the significance of the positive instances as well; they are not just a few more corroborations. Corroboration began to be seen as even essential; a theory needs to be refuted but at the same time not to be refuted so soon, as it gives encouragement to the scientific community. Thus his earlier view of science as 'conjectures and refutations' is now changed into 'conjectures, corroborations and refutations', where the corroborations are the third requirement. As Stefano Gattei puts it, "His [Popper's earlier] first two requirements were: a) 'The new theory should proceed from some *simple, new, and powerful, unifying idea* about some connection or relation...

between hitherto unconnected things.. or facts.. or new 'theoretical entities'..." b) 'the new theory should be independently testable. And the third was: 'the theory should pass some new, and severe, tests".[51] Thus the confirming instances do have a value and with corroborating instances the probability of the hypothesis goes higher. They are added to the bulk of background knowledge. But the problem is that this view obviously involves inductive argument.

Popper's Understanding of Induction: Over-simplified?

Popper seems to almost misportray the understanding of induction, when he says, "The fundamental doctrine which underlies all theories of induction is *the doctrine of the primacy of repetitions* (italics original)... According to this doctrine, repeated instances furnish a kind of justification for the acceptance of a universal law".[52] However Adolf Grunbaum points out that some post-Baconian inductivists recognize that not all positive instances have the same kind of automatic 'instantive support' and they are not exponents of *unbridled instantionist* inductivists. So he argues to show that "*The ability of a theory T to 'explain' and/or predict certain phenomena deductively* (with the aid of suitable initial conditions) *is generally only a necessary and not a sufficient condition for qualifying these phenomena to count as T-supporting instances over and above being merely positive instances of T!*".[53]

For a few decades in the recent past there have been many attempts to evaluate Popper's solution to the problem of induction. Both the proponents as well as the opponents seem to argue that the other camp miss the point. Andrew Swann (1988) raises the question and tries to answer whether Popper really misses the point. A rationalist has to go along with a realist in her metaphysics. A realistic position is expected to have *ontological components* (which tell us that there is a world independent of us), *causal components* (which show us that "at least some of the theoretical terms of a theory denote real theoretical entities which are causally responsible for the observable phenomenon that prompts us to posit their existence"[54]) and *epistemological components* (which assure us of some reasonable grounds to grant likelihood of more truth than another theory). But Popper's falsification method lacks in the epistemological aspects, as he repeatedly denies any criterion of truth and any privileged set of observation statements, which can be surely shown to be true or false. As we don't have any criterion of truth sometimes it is possible that we actually possess true theories, but we are not able to establish that they are true. It might also be the case that, because we lack any criteria, we may establish some statements as true by a matter of mere luck rather than rationally applying any criterion.[55]

Is Popper Proper in Denying Induction?

Popper indeed held on to Inductivism in his early stages. Stefano Gattei (2004) has dealt with this aspect in great details. However very soon Popper denied or ignored inductivism. Now let us see how far he is right in denying induction altogether. Popper ignores inductive reasoning as unnecessary because, "There is neither a psychological nor a logical induction"[56] and science needs only a deductive inference. Inductive reasoning will not be of any help as it would not lead us to any conclusive truth. Nevertheless, though he claims to have solved the problem of Induction, in the theory of corroboration it makes a slow reappearance. Is there any real difference between the view that a theory depends for justification on the occurrence of confirming instances *and* the view that it depends on the failure of the falsifying ones to occur?

It is true, inductive reasoning is not without problems either. But going to the extent of effacing inductive reasoning (if at all possible!) from the very enterprise of science will land us in scepticism. Of course scepticism has its own strings of difficulties. Nevertheless without induction it is well nigh impossible to show that there is a growth in science and that science is a rational enterprise. It is highly likely that science would be empty, if inductive reasoning is completely ruled out. It would even be suicidal for science to reject inductive process. As Salmon puts it, "... that science is inevitably inductive in matters of intellectual curiosity as well as practical prediction. It *may* be possible to excise all inductive ingredients from science, but if the operation were successful, the patient (science), deprived of all predictive import, would die".[57]

The very notion of verisimilitude, which is very dear to Popper to affirm that science, is rational, progressive and worth the while, itself is in danger without some sort of *inductive metaphysics* (i.e. a framework of metaphysics where one is allowed to have the inductive reasoning as a valid form of reasoning). Lakatos takes pains to elaborate on the disturbing implications of Popper's verisimilitude for the problem of induction. The actual dividing line between classical scepticism and Popper's fallibilism is rather thin. So Lakatos thinks, "The difference between total scepticism and humble fallibilism is so small that one frequently feels that one is engaged in a mere verbal quibble…".[58] So it would do well for Popper to realize that some form of inductive process is crucial in the scientific process.

In Chapter III we have more discussion on the notion of Induction, as a positive legacy from the traditional views of rationality.

Indelible Mark of Popper

No one can brush aside Popper's contribution to philosophy of science and especially to the ideas of rationality in science. He has left an indelible mark on philosophy of science. His views are so powerful and important that one may either accept or reject his ideas but can never ignore them altogether. L. B. Bazhenov puts it very succinctly: "Even if you do not agree with Popper, you must accept three points unconditionally. First, his conception meets what Einstein called the criterion of 'inner perfection' as distinguished from that of 'external justification'. Second, the schema presented satisfies fully Popper's requirement of accessibility to critical discussion… And finally, third, the conception of evolutionary epistemology places the problematic of theoretical knowledge in a completely new light".[59]

For Popper rationality is critical debate. The core of Popper's philosophical breakthrough is, as Stefano Gattei has it, "rationality requires not foundation, only critical dialogue: it is the end of foundation philosophy".[60] In the 20th century, the debate over foundationalism plays an important role; people either deny it (and hence scientific knowledge is unjustified and therefore irrational) or hold that scientific knowledge is a justified belief, while weakening the idea of truth as correspondence with reality, or the idea that justification shows that a statement is true. Gattei falls on Socrates, who said that we are living in the twilight zone between knowledge and ignorance, to claim that "the failure of foundationalism does not drag epistemology with it. The twilight zone between knowledge and ignorance is an ideal well worth holding on to".[61]

Having seen the critical remarks of the proponents of rational model of science, now in the next section I'll elaborate the critical evaluation of the proponents of non-rational model, namely, Kuhn and Feyerabend.

Section 2
Assessment of the Non-rational Models of Rationality in Science

This Section is concerned with the evaluation the ideas of Kuhn and Feyerabend. Kuhn, in spite of all his efforts to retain and reaffirm the rational character of science, because of the problems with the desiderata for theory-choice, he is not able to achieve what he aims at. The danger of slipping into the snare of relativism cannot be denied. However, one can expect to overcome strong relativism, basing on some of the common elements of human nature, like the ability to reason out and to have language. Later Kuhn predominantly stresses on the linguistic aspects of incommensurability, though these aspects are also found in the early Kuhn. Paul Hoyningen-Huene points three misunderstanding of Kuhn and if they are corrected Kuhn will not appear irrational. Some elements of incommensurability are bound to be there, in terms of conceptualization or ontology, but they are not insurmountable. All these can exonerate Kuhn from the accusation of making science relative and irrational. One needs to understand that science is essentially social but that in no way makes it irrational or 'disturbingly' relative. Then, mixed and extreme reactions against Feyerabend are presented. His critical views on the highly glorified and rational picture of science and scientists are appreciated. He insists upon the social and moral responsibilities of science. However one cannot deny rationality totally; rules may have been broken in great discoveries but this cannot be made into a rule to say that rules need to be broken. Authors, like Laudan, point out Feyerabend makes hasty conclusions from some of the historical case-studies. Though his passion for methodological pluralism is a welcome move and his invitation to be aware of some incommensurability-problems, one cannot allow his principle that 'anything goes' in science. This section does not fail to laud his concern for humanity, as he desires the world to remember him for his 'love' for humanity.

Thomas Kuhn – A Rational Relativist!

Kuhn was undoubtedly one of the creative and original thinkers. He attributes his capacity for creative and open mindedness, as he acknowledges in an interview, to the association with the Progressive education, a movement which "really sprung from some proposals of John Dewey's. It emphasized subject matter less than it emphasized independence of mind, confidence in ability to use one's mind".[62] Kuhn has to be given the credit of giving a new dimension to understand science and its rationality. He popularized such a view though we do find some pre-cursors of such notions in the past. He has made it clear that the communitarian, psychological and the human dimensions of scientific methods play an important role in the development of science. Kuhn's contribution to philosophy of science is significant. His *Structure* was indeed an eye-opener for humanity. It made its strong ripples in both the quarters of supporters and opponents as well. "No other philosopher of science is," claims Klee, "as widely misconstrued by so many different kinds of thinkers in so many different disciplines as is Kuhn's work".[63]

Alexander Bird (2000) presents a much precise and balanced evaluation of Kuhn's ideas. The very cyclic understanding of science, as normal science – crisis - revolution – normal science, is disproved by history. If normal science and revolutionary science are taken as the two extremes, most of science is in between, neither normal nor revolutionary. Most of the revolutionary sciences did not involve any bloody- fight, for instance perhaps we can label Crick's unraveling of the DNA structure as a bloodless coup. Many of

Kuhn's critics, like Hoyningen-Huene (1992), find fault with Kuhn that he does not give any explicit theory of meaning in his *Structure*.

Suppe (1977) tends to think that Kuhn loses his influence in contemporary philosophy of science due to many reasons: such as, Kuhn's modifications and clarifications of his positions seem to exhibit his inclinations towards a sort of neo-positivistic view; the limited role for rationality in his understanding of science; his distinction of normal science and revolutionary science seems to be increasingly becoming more and more untenable; and above all his epistemological and ontological outlook sadly reduces scientific knowledge to nothing more than just a collective beliefs of a given scientific community.

It seems we can't expect from Kuhn a 'yes' or 'no' answer to the question, 'Is science rational?' For him science is not rational and not irrational either. The rationality of science is said to be under threat under Kuhn's scheme mainly because of his following claims: a) the paradigm shift is like a religious conversion or a gestalt switch; b) the notion of incommensurability, whereby we don't have a neutral language to express the old and new theories of science and therefore scientific growth is a radical break from the existing framework and c) the theory-choice is due to shared values of scientific community. We shall focus on some aspects of Kuhn for our critical remarks.

Science – Basically a Social Enterprise

It was Kuhn who brought out the social aspects of scientific knowledge to the fore. Issues of rationality and those of reality, though not unrelated, are the main pre-occupation of the contemporary philosophy of science. Kuhn gave the wake-up call to humanity from their so called 'rational' slumber. The foundations of science and there by the foundations of rationality are put into a big test. As Hacking reasons out, "The book (*Structure of Scientific Revolutions*) itself produced a decisive transformation and unintentionally inspired a crisis of rationality".[64] A large part of the credit goes to Kuhn in making us realize the real picture of science. It was a shock. For, "Kuhn famously depicts the ugly underbelly of science when he describes the sometimes unfair ways scientists fight with each other professionally during a period of revolutionary crisis".[65] Kuhn shows science as value-based enterprise, yet it produces powerful new techniques for prediction and control. He himself wonders how it is possible. In other words he agrees that the problem of induction is still an unsolved problem. This lacuna does not disturb him much, because even if science progresses by some fixed, strictly scientific and rational algorithm, still its success could not be explained.

Kuhn is the first one to bring out the need of Philosophy of science for history of science and vice versa. Mutual enrichment is needed but must be two different disciplines: "Though a new sort of dialogue between these fields is badly needed, it must be inter- not intra-disciplinary"[66] and in fact, history of science can produce more fruits by bringing philosophy of science and science together. History for Kuhn was not just a collection of data and anecdotes. It is not merely descriptive of the facts but of the connections among them. Historians of science need philosophy, as much as they need knowledge of science, for the simple reason up until the 17th century there was no separation between philosophy and science. Scientists, consciously or unconsciously, were involved with researches and ideas that were actually in the realm of philosophy. Therefore, historians need the knowledge of philosophy for any fruitful criticism of science. He insists upon the need for philosophers to learn from history too. They need to take the narrative constructions of the historians of science seriously to learn more about science. Before Kuhn the prevalent idea of science was that each new theory in the successive periods was better refinements of the older theories, adding more to the knowledge content. As the latter theories were more comprehensive the scientific growth was a continuous one. The goal of science was to arrive at the whole truth, whatever it might have meant. But for Kuhn it is not so. Arriving at Truth, whether full or partial, is not the goal of science. Truth does not pull science forward but only the puzzles push it ahead. Puzzle solving is the goal of science and to achieve this goal all need to work together.

Kuhn shows the need of interrelation between philosophy and sociology of science and strongly criticizes the distinction between the context of discovery and the context of justification, which has been well-cherished by logical positivism and critical rationalism (of Popper). It is not that each and every question in sociology of science can be answered by philosophy of science, but the mutual dependence of sociology of science and philosophy of science needed as some deeper levels: "the treatment of *some* allegedly pure philosophical questions, such as those about the dynamics of theories or about scientific progress, necessarily involves sociological aspects" and "*some* allegedly pure sociological questions, such as those about the values governing the behavior of scientific communities, necessarily involves epistemological aspects".[67]

For Kuhn the subject of science is the community of the scientists, committed to values, whereas for logical positivism and critical rationalism, the subject of science is the individuals following rules (algorithms). For them theory choice is done with well-defined rules; so anyone does it rationally will eventually come to the same decision; it is a rational decision.

Paradigm – Still Problematic

Paradigm can be said to be blessing in a disguise, in the sense, for though it does not easily allow new creative ideas, it certainly ensures the steady movement of science. However, as pointed out earlier, after 15 years of publication of *SSR*, Kuhn seems to realize that the concept of paradigm grew out of his control. He agrees, paradigms were supposed to be held just as exemplary problem solutions, "but they expanded their empire to include, first, the classic books in which these accepted examples initially appeared and, finally, the entire global set of commitments shared by the members of a particular scientific community".[68] All such expansions in the limits of the notion of paradigm led to lots of confusion. The notion of scientific community for the early Kuhn was very restricted and limited. Perfect consensus and unity prevails there. Now he removes the idea of consensus among the members. In fact, this consensus may stifle the dynamism of scientific growth.

In spite of all his efforts to clarify the doubts of the critics still many think that Kuhn is not able to solve all the puzzles or answer all the queries. On analyzing the new concepts of disciplinary matrix and the related elements, one still has a doubt whether Kuhn is able to really solve all the problems of the earlier concept of paradigm.[69] It is true that this new proposal clarified much of the misgivings that were in association with the notion of paradigm. Yet many of the lacunae associated with old notion of paradigm were still glaring at one's face. For instance, Carl Matheson opines that many interpretations of Kuhn's ideas, very divergent at that, are possible, perhaps "because the internal consistency of Kuhn's positions still stands in some doubt".[70] Kuhn's distinction of normal science and revolutionary science is not so problematic as the description of the transition from one paradigm to another. The way he deciphers the transition projects him to be non-rational, if not irrational! The transition is not something which is rationally deliberately chosen, rather a sort of conversion experience.

Kuhn claims that a paradigm is not rejected unless another paradigm is simultaneously accepted by the scientific community. The decision to reject one and the decision to accept another must go hand-in-hand. However Mark A. Stone (1991) points out that this is not the case; Stone differentiates three types of discoveries: spontaneous, implicit and directed and argues to show that Kuhn's claim may be true for the first two types of discoveries, but not for the directed ones. He shows Kuhn's own example of Copernican Revolution does not subscribe to this claim, because, by Kuhn's own admission, Copernicus has rejected the existing paradigm even without any successor at hand. Paradigm is not a fixed-mould, in which as though you insert a problem form one side and you get the ready-made solution from the other! Paradigm is only a tool with which one can handle different types of problems. As Rouse puts it, "Accepting a paradigm is more like acquiring a skill than like understanding and believing a statement. Actually, it involves many multiple skills simultaneously: applying concepts, employing mathematical techniques... using instrumentation and other apparatus, and recognizing opportunities for varying or intervening in particular theoretical or experimental situations".[71]

Theory Choice – Not a Cake-Walk!

Kuhn's insistence upon theory-ladenness of observation and meaning variance thesis and further his denial of any role for correspondence rule in scientific knowledge had naturally led him to view knowledge as a matter of shared judgment of a given community. Because of this many (for instance, I. Scheffler, 1967, Shapere 1964 and 1972) accused him of making science a totally irrational enterprise. In the enlarged edition of the *SSR*, he tries to soften his doctrine of theory-ladeness of meanings, whereby he maintains that disciplinary matrix is that which determines the meanings of terms in such a way some shared interpretation of some of the terms found in both theories is made possible. This naturally paves the way for a common vocabulary and some sort of rational assessment of the theories. In spite of Kuhn's sincere efforts to make science more rational in his outlook he does not seem to have succeeded much. For rational assessments a scientific community needs a body of basic elements outside the normal science. But, as Suppe points out, "Kuhn's analysis implies the absence of such scientific domains outside normal science and denies the need for such assessment within normal science, hence cannot accommodate such forms of rational hypothesis and theory evaluation".[72]

Due to the subjective element, introduced by the desiderata in theory choice, Kuhn is accused of making theory choice (scientific enterprise) an irrational enterprise.... Therefore Kuhn's position is generally projected to be "scientific revolutions are events whose outcome is exclusively determined by propaganda, conversion and the dying out of opponents, and not by sober comparative evaluation of the achievements of theories, or other arguments that articulate good reasons for theory choice".[73] But the communal cognitive values may grant rational basis; but individually differing values may not give rational base. He is, however, convinced of rationality in science. The *question of rationality of theory choice* must be raised in the context of the community, not as individual scientist. The individual may disagree; work with various theories; only in course of time a consensus may emerge; only when there is a decision reached the question of rationality can be raised. The consensus is based on the collective system of cognitive values.

This notion of consensus in the scientific community makes science social and at the same time it prevents science from becoming irrational. We will turn to this aspect later in Chapter III.

Relativism – Kuhn's Inevitable Consequence?

Though Kuhn claims himself to be a rationalist, other rationalists don't seem to be happy about it, because, unlike other rationalists, Kuhn does not see rationality as a rule-governed process and he denies the cumulative growth of science. These two principles land him on doubtful grounds of rationality and make the accusation of relativism possible. Kuhn[74] lists and discusses in some detail five desirable values for theories: accuracy, consistency, scope, simplicity, and fruitfulness. These criteria enable one to differentiate two paradigms to see which is better. Therefore the *threat of relativism* is removed.

I think, some reflections on the important aspect of relativism are in order. Relativism seems to be an accepted way of life in the 21st century. The diversities in cultures, religions, morals, languages, social and psychological behaviours seem to demand the acceptance of relativism. It is quite attractive and enticing to most of the moderns, for, among many other, it reinforces the long cherished freedom from moral and social bindings. There are varieties of relativism, depending upon its focus. Rom Harre and Michael Krauz spell out four types of relativism: *Semantic, Epistemic, Ontological* and *Moral Relativism*. They also point out two fundamental attitudes infesting relativism – *a sense of skepticism*, which holds that "no point of view is privileged, no description is true, and no assessment of value is valid" (malign relativism); *and a sense of permissiveness*, which declares that "all points of view are equally privileged, all descriptions are true and all assessments of value are equally valid" (benign relativism).[75]

Various elements like, the difficulty, even impossibility, in translating languages completely, the incompatibility and the incommensurability between two theoretical structures, the irreducibility of one theory to another and so on, are stressed in different ways by different relativists. Various combination of these different elements lead to different types of relativism. However, the broad-spectrum of philosophical claims for relativisms may be summed up into *four negative theses*: "a) There could be no universal language adequate to a common human world; b) There could be no unified theory universally applicable in all contexts at all times; c) There could be no definitive system of descriptive categories picking out a common range of natural and social phenomena and d) There could be no universal moral/aesthetic principles viable in all cultures at all times".[76] Of all the four (b) seems to be directly involved with the ideas of Kuhn, as he is blamed of relativism, and even irrationalism, due to his ideas on incommensurability.

As we know strong relativism has many problems. The ancient form of relativism, Protagoraean relativism, itself can be shown to be a paradox. It claims that a belief, which is true for one person, may be false for another. But this belief is demanded to be true for all, at all times. If so, at least there is one belief which is true for all, to disprove the original claim. So if the staunch Protagorean relativism would be false if it is true. It is like someone claiming, 'nothing is certain in this world, except this statement!'. The strongest kind of relativism would deny universalism, objectivism and foundationalism.[77] This type of relativism can be easily shown to be either *false* or *unprovable*, precisely because of the need for a non-relative or absolute stand point of view to realize the relativism. Harre and Krausz explicate this point with two theses: "*The thesis of ontological independence*: Entities, states, experiences and so on exist independently of culture for the fact of cultural diversity vis-à-vis these entities to show up; and *the thesis of transcultural intelligibility*: Descriptions of some entity, state, experience etc. must be universally intelligible, if it is to be possible to realize that the entity, state or experience being described is being treated differently in different cultures".[78]

One can trace out some *common elements of human nature* to argue against strong relativism. The ability to feel or reason out seems to be a part of common human nature, the feelings and the reasoning may differ though. The ability to learn an elaborate language-system, which the non-human beings seem to lack greatly, is certainly found with all normal humans. They have to make meaning out of their own lives, intellectually or emotionally, morally or metaphysically within the human conditions in which they find themselves. No doubt, languages are artificially created, based on various semantic and syntactic rules, for the purpose of communication. But to make this language possible, demands Wittgenstein,[79] (1953) *a transcendental pragmatic condition* must be fulfilled. That is, there must be language-like precursors, "natural regularities in form of life in order for the 'artificial' regularities of rule-following to take root. Without the capacity to grasp the import of a rule, language use would be impossible".[80] Language implies mutual trust and the need for honesty to speak the truth.

Further, though what is right and wrong, good and bad, may differ from person to person, culture to culture, region to region and religion to religion; even for the same person it may differ from time to time, yet that there is a distinction between right and wrong, good and bad, seems to be a universal human experience. Any normal thinking person will have this distinction and she experiences a natural inclination to follow good and right. Even if a person does something what is generally considered to be wrong, bad or immoral, that person very often does it not because she does not know it is wrong or immoral, nor that she deliberately wants to cause evil, but because she sees some good (which overrides the bad or immoral aspect!) in doing that. It is a sheer wonder to realize that how and whence this moral sensitivity has come to humans!

Kuhn is said to be a relativist because of the notions of paradigm shifts as gestalt switch, and the incommensurability. Gestalt interpretations don't seem to suit the examples he gave; all those cases are non-perceptual ones, and can't be limited to just ways of seeing; therefore Malone points out that, "if we do take him (Kuhn) literally, his own examples, which are non-perceptual cases of 'seeing…as', refute him".[81] Kuhn treats gestalt switch as more of a metaphor rather than a psychological impulse it can't be taken in a literal and purely perceptual sense. So if at all he is accused of relativism it would be a sort of anti-objecitvism or a weak relativism. Though we can argue against relativism, we cannot decisively prove absolutism either. The arguments for anti-absolutism are also quite strong and we cannot take them lightly: "It seems that while we cannot prove that all knowledge is relative to culture, we cannot prove either that we could obtain knowledge that was absolute, that is universal, foundational and objective!".[82] Similarly, incommensurability need not imply deep relativism. As Malone points out, "Kuhn's incommensurability thesis requires only these two features: (a) *concepts and correlated ontologies sufficiently different that the world depicted in one theory cannot be fully represented by the rival theory*, and (b) *an inherently vague theory-fact distinction.* This incommensurability is not insignificant, but it is also not radically relativistic as implied by the standard reading, and is not obviously a matter of differences in perception".[83]

Should this relativism make us irrational? I think not. Various paradigms or view points of various people at various times add something to the body of human knowledge. All our human efforts are like arranging the various pieces of zigzag puzzle to arrive at the final and whole picture. Unfortunately we don't have any other access to arrive at the final product except the one that humanity is engaging in now. But that need not depress us rather that adds life and beauty to the whole process of seeking knowledge. The rational aspects need to be complemented by emotional, intuitive and psychological aspects, without which humans would be far from complete. Inconsistence and irrationality have even provided unexpected and creative solutions; as history shows many great scientific and mathematical discoveries are the results of sudden inspiration, or thinking differently, violating existing rules, even acting 'irrationally'.

Therefore it is good to note that relativists need not be irrationalists. Relativism and rationalism need not be mutually exclusive. Relativists have their own internal checkpoints to circumvent irrationalism. When it is denied that there is no *single universal* method for science it does not imply *no* method at all! Thus we can show that it is popper to categorize Kuhn as 'rational relativist'.

Incommensurability – Causing Trouble for Kuhn?

Many critics, like Dudley Shapere (1966), Stephen Toulmin (1972), Kenneth F. Shaffner (1974), and Wolfgang Stegmuller (1976), see Kuhn's incommensurability leading to relativism and irrationality. His idea on incommensurability is equivocal and even sometimes contradictory. However, as pointed out earlier, in his latter phase he moved to lay more stress on linguistic incommensurability; by 1983 completely distanced from evaluative incommensurability and claimed, "Incommensurability thus equals untranslatability".[84]

It is not that only the later Kuhn turned to linguistic aspects. On close reading one can find, as Kulkarni explains, the linguistic aspects in the early Kuhn also. Incommensurability, after all, had to deal with meaning variance of central theoretical concepts of two fundamental successive theories: "(a) The changes brought about by a scientific revolution were construed by him earlier in terms of a visual gestalt and later Kuhn finds this construal highly inadequate because visual gestalt does not alter the elements whereas this is not so in the case of a scientific revolution, and secondly, visual gestalt concerns perception which is essentially an experience of individuals whereas a revolutionary change in science is experienced by a community. Hence Kuhn construes scientific change as a change in the linguistic framework; and (b) A scientific community could be defined, he thought earlier, in terms of certain sociological features related to education and other matters of professional life. Later he construes scientific community as essentially a language community".[85] However in the later Kuhn concern with language became more radical.

Though Kuhn moved on to the linguistic aspects of incommensurability, he is misunderstood regarding incommensurability. Paul Hoyningen-Huene[86] points out that there are three misunderstandings of Kuhn: a) Kuhn is misunderstood to be proposing 'radical meaning change' or 'total' incommensurability, that all concepts used in two theories have different meaning; but he has advocated only 'local' incommensurability; b) he is said to be denying any sort of continuity between succeeding paradigms; but the fact is already in *SSR* (1970) he speaks of experimental and theoretical ones, otherwise how to meaningfully speak of better puzzle-solving ability of the latter theory; and c) he is said to be saying that there is no rational comparison between theories is possible, because two incommensurable theories deal with different worlds, with mutually untranslatable vocabularies; and therefore there can't be any empirical friction between such theories and there can't be any real competition between them. But due to the 'local' nature of incommensurability, *some* of the local consequences of the two theories can be compared *immediately*, where the mutually incommensurable terms are not involved. Further, a proponent of the theory has to identify and learn those parts of new conceptual vocabulary that are different from his own.[87] And above all, the two theories can be compared globally with respect to their simplicity, accuracy, fruitfulness, predictive power etc. This form of comparison is not at all irrational, and provides a reasonable ground for theory choice.[88]

Putnam and others have proposed, what is known as 'Causal Theory of Reference' to overcome the issues of incommensurability. Davidson focuses only on 'meaning incommensurability' though Kuhn and Feyerabend speak of methodological incommensurability also. Davidson argues that if at all there is a translation-failure it will be on only of a local nature; most part of the language is translatable into other languages. Thus one can safely conclude that no language has radically different conceptual schemes from our own language. Therefore meaning incommensurability is not tenable. He further shows the need for 'generally successful translation' even to realize that the failure of translation: "a background of generally successful translation… provides what is needed to make the failure intelligible"[89]. One needs to translate most of the other languages; one may not be able to realize that the other is a linguistic entity. Therefore totally untranslatable language does not exist. Davidson distinguishes *untranslatability in principle* and *untranslatability in practice* and he agrees with the latter, and not the former. This questions Kuhn's claim that 'incommensurability is untranslatability'.

Shapere's 'Chain-of-Reasoning Connections' – A Way-out for Incommensurability?

Shapere's notion of 'Chain-of-Reasoning Connection', I believe, can help us to address the issue of incommensurability. Shapere introduces, what he calls, 'concept schema' to deal with certain conceptual issues in philosophy of science. To understand the recent sophisticated science, he suggests, it is better that one treats the important concepts employed in it as a 'concept schema', instead of individual concepts. For, the meanings of these concepts, including the scientific ones, like observation, explanation and so on, are not given once and for all as one individual concept; rather they develop over the years, with all related aspects. Concepts can be understood well only in terms of the *content* of background information. He prefers to discuss the notions of observation, meaning, reality, truth, knowledge, and so on as 'concept schema'. Philosophers have failed to arrive at the exact meaning of many concepts and that can be evidence that there is no such exact meaning. Their meanings can be made up only by a chain of ideas, linked by reasons.

Thus a concept schema means:[90] a) A concept schema is formulated in terms of a 'framework' that remains reasonably stable over a significant period of time, or over contemporaneous areas of inquiry; b) The usage of the *content terms* depends on accepted scientific beliefs, which are the elements of the background information; [By content terms he means those terms which refer to the content of the accepted scientific beliefs. These terms are contrasted with meaning-terms and reference-terms]; c) The accepted background information can and does often change from one epoch to the next, and therefore the usage of the content terms of the concept schema (e.g. what is to be observational) also changes. Thus the usage or application of the content terms can vary over the period during which the framework is otherwise stable; of course, the framework itself can be changed; d) The usage of the content terms is changed in the light of new scientific beliefs, which count as reasons for making changes. Thus there is a *chain-of-reasoning connection* between the successive usages of the concept schema; e) A concept schema is treated as part of 'an approach to inquiry' and the interpretation of its results. There can be many rival approaches to an inquiry; one is accepted and the others are rejected. The selection of a given approach is based on legitimate reasons. The approach involving concept schemata is also contingent; and f) A particular schema is used to a sequence of specific usages, not just to a single set of specifications. The members of the sequence are linked by a chain-of-reasoning connection.

Though the usage of a term differs, and the background context differs at successive stages, the usages can be compared due to, what Shapere calls, 'the chain-of-reasoning connection' between those usages. At successive stages some properties or the other of the terms are dropped or added for specific reasons. Let U_1 and U_k are two usages that determined by theoretical contexts, T_1 and T_k. There is a chain of reasoning which helps us to understand why certain properties ascribed in usage U_1 and its successors up to and including U_k were deserted, modified, or replaced; the same chain-of-reasoning explains the possibility of comparing the two usages and their theoretical contexts even though the last stage U_k involves very few or even no features in common with the first usage U_1.[91] Shapere explains with an example of usage of electron over centuries: After a rethink upon Faraday's electrolysis experiment Stóney concluded that electron was an indivisible unit of charge (and not matter); later Thomson added the property of mass to electrons and electrons became classical particles, with mass, position and velocity, and negative charge. But the recent quantum mechanics abandoned simultaneous attribution of position and momentum; and added another property of spin to electron. In the further evolution of quantum field theory, electrons are seen as a variety of leptons. Like all particles, electrons are also seen excitations of a quantum field, possessing a range of quantum numbers. All these developments were even unimaginable to Stoney and Thomson in those days. However, we know all these properties are about electrons, not because we ascribed essential properties to electron, or some sort of necessary and sufficient conditions of applications of the term. We understand the term refers to electrons also not because of an irreducible assertion that we 'refer to electron'. The continuity and comparability of the usage is not due to some common descriptions or common reference, rather, "the rationale for saying that we have referred to the same thing all along is given by the linkage of reasons which gives continuity to the history".[92] The historical usages of electrons *are extremely different, if not radically incommensurable, yet they are not incomparable* due to chain-of-reasoning connection. Thus, I tend to believe, that Shapere's notion of 'chain-of-reasoning connections' offers us a way-out for the problem of incommensurability.

Kuhn's earlier position on incommensurability is more psychological and empirical, whereas that of later is more philosophical in style and a priori in method. His earlier position was more naturalistic approach. Led by Quine the whole trend of philosophy was moving towards naturalism; it began to see the questions about the nature and the possibility of knowledge not as something belonging to a priori philosophy, but as 'empirical cognitive psychology'. Kuhn could have proceeded to give a thoroughly naturalistic account of theoretical change, world-change and incommensurability, instead of turning to linguistic aspects. Therefore, to understand the nature of science and its development, we *must look in the process of reasoning employed in science and not in philosophy of language.*

Having all said and done, strong incommensurability, if it is taken literally, cannot hold. There may be some problems in interpreting or understanding a very old theory, but those problems are not insurmountable as Kuhn (and Feyerabend) makes us believe. For, if it is taken literally, as Newton-Smith points out, "the intelligibility of the thesis is questionable...if the theories are genuinely incommensurable why should I be faced with the problem of choosing between them? Why not believe them all?".[93]

Finally, Kuhn does stand tall in his contributions to philosophy of science. In realizing his enormous contributions to the analyses of science, Gürol Irzik and Teo Grünberg gladly acknowledge that "probably no single philosopher has achieved as much as he has. Both this paradigm, which contains the novel idea of exemplar, and his structured lexicon provide a much richer and more detailed description of normal and revolutionary science than anybody else's".[94] Kuhn's ideas on *science as a social enterprise*, where the *consensus* playing an important role and *the notion of theory-choice* are, as we will see in Chapter III, are very helpful in my construal of rationality of science in terms of reasonableness.

Feyerabend: An Intellectual, but Human to the Core!

Mixed Reactions over Feyerabend's Contributions

Gonzalo Mune¢var gladly declares that Feyerabend is the most important thinker of the 20th century, especially as humanity looks back to evaluate all that has gone in that century. In introducing the volume on *Beyond Reason*, he acknowledges, "Some philosophers think that Paul Feyerabend is a clown, a great many others think that he is one of the most exciting philosophers of science of this century. For me the truth does not lie in between, for I am decidedly of the second opinion that is becoming general around the world as this century comes to an end".[95] It is Feyerabend who made the world aware of the actual situation of science and scientists with regard to the claim of their being the most objective and rational lot in society. As an aftermath lots of conferences and colloquia take place every year on the socio-political dimensions of science and the moral and psychological aspects of scientists.

That science is socially constructed and therefore the human and affective elements are brought in a very decisive way into science is not denied these days. Science can no more be treated as a body of rational disembodied knowledge. One cannot overlook these aspects in any serious study about knowledge or science. That is why, even a decade back Elzinga and others strongly realized that "in the future, graduate programmes as well as undergraduate curricula will have to devote more time to ethical topics, and also to enlighten the students on the ways that political interest intervenes into science".[96] Grover Maxwell introduces his essay on Feyerabend as, "it is a purpose of this essay to praise Paul Feyerabend, not to bury him – not, of course, that he is in need of my praise, or of my defense..." and he goes further to gratefully record: "it is Paul

Feyerabend's influence on my own modest philosophical development has been deep and pervasive".[97]

Some, on the other hand, are very tough in attacking Feyerabend to call him a childish thinker who is not serious about his own claims. For instance Herbert Schnädelbach thinks that Feyerabend is a playful man; he quotes Feyerabend to have diluted the claim about 'Anything goes', as it was never meant to be a principle, but only "a somewhat joking portrayal of the situation of the rationalist who wishes to have universal principles but who has to increasingly empty them of all contents when faced by the material I offer. 'Anything goes' is all that remains".[98] Feyerabend is projected to be a staunch proponent of irrationality in science, or at least he is made to declare that science is not guided by any rational element but power, propaganda, prejudice.

Feyerabend in New Light!

Now there are efforts among the scholars to remove the erroneous picture of Feyerabend, by reconstructing his arguments to show that he (at least the later one!) is not against rationality in science. He argues for rationality in science based on certain values. Robert Farrell (2003) is one among such sympathizers of Feyerabend, who try to show that Feyerabend's rationality is to have a fine balance among the competing (even incompatible!) methodological demands. He projects Feyerabend as having a 'tightrope-walking rationality'. For, though Feyerabend denied any universal 'rules' of rationality yet he recognized temporal values that are universal, like comprehensiveness, empirical accuracy, fruitfulness and testability.

Rationality involves something more than mere adherence to rules; rationality is often identified by a rationalist with always acting in compatibility with certain strict rules and regulations. If following the rules is the only criterion for being rational, then, Herbert Schnädelbach (1991) wonders, whether we must treat programmed computers and washing machines as rational! Feyerabend is against only the strict traditional understanding of rationality; in a strict rationalist point of view, science must fulfil at least one of these three theses - the logical thesis, the factual-foundational thesis and the methodological thesis: the *Logical thesis* claims that all our thinking must conform to logic in providing the structure of our knowledge; *the factual-foundational thesis* demands that there are indubitable sources of our knowledge and they exist; and *the methodological thesis* maintains that a scientific research must be performed according to the rules of an unchanging scientific method. These theses would not allow any room for subjective or rhetorical elements in scientific progress, as these would be considered 'irrational'. But Feyerabend's notion of scientific progress does have room for all these elements and still it can be (and is) rational. He was always trying to show that rationalism would have to be rejected by its own standards.

Science – A Custodian of Rationality?

In the normal parlance science is seen as the *epitome of rationality,* because science uses a special and unique methodology: therefore, to be scientific or to be an expert is to be rational. If any one comes out with a theory or a kind of new invention or a discovery, it is expected to look for its scientific basis. That is why, nowadays every discipline likes to claim its statements or principles are scientific in order to sell it to the people, to make it more acceptable and to make it more rational. The scientific character of the traditional natural sciences, like Physics and Chemistry, is appropriated not only by the Study of Politics and Society, but also by the Study of Historical Materialism, Administration and Library and the like; that is why today we also have, as C. Trusedell lists[99] Library Science, Administrative Science, Speech Science, Forest Science, Dairy Science, Meat and Animal Science, and even Mortuary Science.

But such a received view of science, such a highly elevated picture of science and its methodology are being questioned in the recent decades. Unfortunately many of the scientists seem not to be aware of these critical developments in the aspects of philosophical appraisals of science. Still many top level scientists are convinced that they, as scientists, are the custodians of rationality. Daniel P. Moloney points out, "Stephen Jay Gould, Richard Dawkins, E. O.

Wilson, Stephen Pinker, Daniel Dennett - all these have, in recent years, tried to leverage their credentials as scientists into positions as guardians of public rationality. And all of them want to set the record straight: science is close to figuring out all the answers and anyone who says otherwise is an irrational obscurantist".[100] In the early 1960s, the Nobel laureate physicist Richard P. Feynman affirms the same view in the collections of his lectures, *The Meaning of It All: Thoughts of a Citizen Scientist.* He elaborates that the distinguishing character of a scientist is her ability and desire to 'question everything'. This attitude of questioning very often lands a scientist in the realm of atheism, as she does not find rational answers for the questions regarding the relevance of a religion or the existence of God. Feynman makes a useful distinction between *questioning* and *investigating.* In the process of *questioning* one withdraws any already fixed conclusion and is open to change one's conviction at the end of the process, while in *investigating* one initiates the process to lay a better foundation to the already held conclusions. A student of science needs to realize that she *does not question but investigates* in science, meaning to say, that she already has many assumptions to work upon. At least some of these assumptions cannot be *questioned* in Feynman's sense.

Feyerabend's caution against in absolutizing rationality is appreciable and meaningful. It is true and correct that scientists are not that rational and objective as the popular picture of science makes us believe. But we need to realize that the 'questioning-everything' attitude is not always at work at the scientific enterprise. It is not true to say that science arrives at every theory or inference through a process of scepticism. So, "If we define a conclusion (in science) to be rational only if it can withstand the sort of scepticism Descartes employed, i.e. only if there is no logically possible alternative, then there might not be a single meaningful scientific conclusion that we can call rational".[101] Scientists do act upon many assumptions and suppositions, which cannot be put into a scrutiny of Cartesian methodological scepticism. Nevertheless this in no way makes science and scientists irrational, rather it only teaches us that logical certainty is not possible in scientific enterprise:

"Scientists make as many assumptions as the rest of us. And sometimes, just like the rest of us, their assumptions turn out to be wrong, especially when, like Feynman, they assume they know more than they do".[102]

Feyerabend maintains that the scientific activities are so varied that no common rules can be abstracted. The theories are so incommensurable that the preference of one theory over another is only a question of subjective ideologies and personal orientation; it totally depends on one's ideology and not on any rational basis. He seems to accept that inconsistency can be one of the criteria to objectively decide that a theory is deficient. Therefore at least there is one factor which is not bound by one's ideology. However he does not have any problem with inconsistency, as inconsistent theories have brought progress in many cases. As Newton-Smith objects, "the fact that inconsistent theories have brought progress is not reason to revise logic by dropping the law of non-contradiction. Inconsistent theories have brought progress through their development into consistent theories".[103]

No Method and therefore No Rationality?

It is quite common among philosophers to think that scientific methodologies are theories of scientific rationality; for example, we can cite Lakatos, Musgrave, Worrall and so on.[104] Methodology is identified with rationality. Therefore one tends to impose one's understanding of methodology to evaluate the rationality of the past. But Curtis cautions that a historian must not use one's own modern methodology to evaluate the rationality of the research programmes to scientists of the past years. One needs to consider the scientists of the yester years in their own intellectual milieu. The methodology that the present historian arrived at now is the product of the intellectuai contribution of the long generations of the bygone era. He makes an elaborate study on Darwin's understanding of methodology and rationality and concludes that we can consider the overthrow of creationism by Darwin and his "overthrow of the requirement for proven causes itself" as rational, "only if we decline to use our own methodologies by which we appraise the rationality and correctness of the revolution as a whole. We can see this only if we take into account the rationally changing methodologies of the participants themselves".[105]

Feyerabend claims that great scientists have produced very important theories by violating the well-cherished rules of their times. This is alright even for staunch rationalists; no one claims that all scientists are rational at all the times in all the matters; only a few that too the trend-setters of modern science fit Feyerabend's description; as Laudan says, "It is not scientific charlatans he is describing; rather... the figures he is writing about have always been considered as the folk heroes of our scientific culture. It is the likes of Kepler, Copernicus, Einstein and Galileo who turn out to be the chronic 'cheaters' in the scientific game, for they are the ones who proceed counter-inductively, who ignore falsifying evidence...".[106] It is true, they all went against, or even modern progress would not have been possible had not they done so, but to make this exception a rule, that science always progresses, or has to progress this way, is a bit too much.

Further, some authors strongly argue that the historical case studies that Feyerabend undertakes are not historically accurate (for example: P. Machamer, 1973, and McEvoy, 1975). Yet, Laudan is convinced that "even if all of Feyerabend's historical claims were correct, most of the conclusions which he draws from the history simply do not follow. An analysis of the reasons why they do not follow will expose several key Feyerabendian confusions about the nature of methodological rules".[107] Laudan proceeds further to analyse the situation: A legitimate methodological rule is that which optimizes the opportunities of achieving some cognitive aims. Therefore it can be criticized on two grounds; either a) a rule chronically fails to promote a certain goal, or b) it is less efficient than other rules. All that Feyerabend case studies show is that certain scientists at certain times promote the goals of science though they violated the given rules. A methodological rule is proposed that a particular end will be achieved by following that rule; sometimes it may happen that the end is achieved by violating the rule. So what is the problem about the rule-breaking behaviour of the scientists? When science asks us to follow a rule it is not forcing us to follow it; in fact, it suggests that the rule will help us to attain our goal rather than defying it. Just because in one or two cases the rules are violated, he cannot inductively infer that all methodologies are useless. In fact he himself is dead against inductivism. In such violations we can only say that said rule is not the best for the desired end; or that rule may never work out to get the intended end. Only from a few successful rule-breaking events by a few scientists nó universal conclusion can be drawn against methodology itself. Laudan wonders if in some cases cancer may get cured without any medical intervention but this does not imply that all cancer patients must not seek any medical assistance at all.

Feyerabend not only claims that great scientists have broken the well-accepted rules and methods to develop science and behaved in a way what modern philosophers would call irrational, but also that the rules had to be broken to attain scientific progress. Granted, the prevalent methods may be faulty; they may not be efficient; and one method may not be answering the questions of different disciplines. Science is always in need of new methods and not abolition of all methods. Herbert Schnädelbach explains: "We need methods because only by means of them can science be the enterprise which it should be in a free society, namely subject to public and inter-subjective verification. The fact that certain philosophers of science believe they can put an end to methodological imagination once and for all is no reason to denigrate the methods themselves".[108] Laudan concludes in very clear terms: "The fact that prominent scientists have sometimes succeeded by failing to follow what we regard as sound rules of scientific inference does not challenge the legitimacy of those rules as general strategic guidelines. Moreover his one detailed effort to mount a general epistemic argument against methodology… falls flat on its face, because it will not sustain the conclusions which he draws from it".[109] Thus we can conclude that one cannot completely do away with methodologies and their rationality in science.

Need for the 'Consistency Condition'!

The total denial of the *consistency condition* (i.e. *new hypotheses agree with accepted theories)* by Feyerabend does not seem to be acceptable to the scientific community. Therefore, Newton-smith makes three slight modifications to in order to make it acceptable to the scientific community: "New hypotheses must agree with currently accepted *acceptable* theories" (because they are aware that no theory is beyond any correction in due course), "all things being equal, new theories should agree with currently accepted acceptable theories" (because even a staunch rationalist would not claim that there is a set of strict and changeless rules to be adopted in theory-choice) and more significantly "all things being equal, new theories should agree with the observationally successful aspects of currently accepted acceptable theories".[110]

One cannot afford to ignore science or reason; we need traditions as a spring board for diving deeper, or soaring higher into the explorations of reality. As Herbert Schnädelbach explains: "We, by contrast, wish to nurture and care for traditions and not simply because they are there but because it is reasonable to do so; and in so doing we wish in each and every case to ask ourselves and those concerned whether this is rational. We wish to despise neither reason nor science, and least of all reason in science. However bad it may be, however little it will solve the problems of our lives, there is nothing better to be had".[111]

Does / Can Anything *Really Go?*

Feyerabend precisely seems to be making the mistake of throwing the baby with the bath water! Yes, no fool-proof certainty is available with science and no universally accepted a-historical single methodology is to be applied, but that does not certainly imply *anything goes*! One might even see a sort of rashness in jumping to a conclusion that anything goes! The failure of the inductive methodology and that of Popper's deductive methodology seem to have made him largely disheartened. He is in search of a method or a set of rules without any exception, but because he is not able to come across one he turns out to be highly sceptical to declare 'anything goes!' Moreover, he is angry with the society which gives undue importance to science and makes science 'everything' for the society. For instance, John Preston summarizes Feyerabend's arguments to show our society has ended up adoring science: "First, it is impressed upon us that modern science is a fact of life… Science represented the vanguard of enlightenment, pitted against the forces of authoritarianism and superstition. Second, science is defended on the ground that although it is not simply a body of known truths, it is in possession of a special method… and lastly, defenders of science sometimes urge that although non-scientific traditions may have 'answers' to Western problems, they cannot be taken seriously because Western institutions alone have results".[112] So as a reaction to such a society he goes to the extent of denying any rationality to science at all.

No moderate rationalist would deny that some methodological rules have failed at some time or the other. We are not omniscient. We don't have any direct access to truths of reality other than the tools that are available with us. We need to somehow judge the achievements of science by the comparison of rules and methods available at hand. A historical analysis may find fault with one particular methodological rule but that does not justify denouncing all the rules: "The best we can do through an historical investigation is to take up a single plank of the ship of methodology while the rest remain, for the moment at least, firmly in place. An attack on a particular aspect of method presupposes method".[113]

Methodological Pluralism – A Welcome Move!

Critics tend to think that because Feyerabend is for methodological pluralism he is against rules and norms. But methodological pluralism does not imply the absence of rules and norms in science. Rather, it implies that there are different sets of rules for different sciences. The methodological monists seem to have been misled by the examples taken only from physics of the last few centuries. Actually it is not even conceivable that the physicists, neurophysiologists, geologists and astronomers use the same method(s) in their respective investigations. But radical rationalists are reluctant to agree with this fact, for methodological monism is the bedrock for them. For them, perhaps as they give undue importance to the view of methodological monism, if there is no fixed methodology there is no guarantee for a rational process in science: "If no principles of evaluation are fixed, there is no 'objective view-point' from which we can show that progress has occurred and we can say only that progress has occurred *relative to the standards that we happen to accept now*".[114] Whereas *reasonable rationalists* will make room for the fact that the methodological rules need not be absolutely immune to changes.

True, Feyerabend makes a very valid point: in order to impartially judge a scientific theory or a concept we need to distance ourselves from it, and perhaps create new system or a theory as a tool to evaluate the existing theory. (That's why he finally distanced himself from both Objectivism and Relativism – so he declares, "Objectivism and Relativism are both untenable as philosophies [and] guides for a fruitful cultural collaboration"[115]). The more the new system goes against the traditionally accepted norms and the well-cherished notions, it becomes a more powerful system. One can even "import such a system from outside science, from religion, from mythology; from the ideas of incompetents, or the ramblings of madmen. This step is, again, counter-inductive. Counter induction is thus both a fact – science could not exist without it – and a legitimate and much needed move in the game of science".[116]

The Need and Relevance of Metaphysics

Feyerabend's fresh understanding of empiricism, which attacks the traditional inconsistency condition and meaning invariance condition, does make room for metaphysics and ideologies. He makes it clear not only that our observational claims involve non-observational background elements but also no part of observations is free from it. As Newton-smith has it, "... it is not that our observational judgments may have an ideological component, our observational judgments have no components that are not ideological".[117] Therefore, he is very well aware that without some metaphysical elements science would not be what it is today. In his framework of empiricism we have a different understanding of metaphysics. Metaphysical systems can be looked as scientific theories in the making. Metaphysics, as an alternative, is desirable for a better criticism. If a metaphysical claim contradicts a well-cherished principle, then it shows how metaphysics can be useful; for only in the face of alternatives science can really progress. Therefore if whole of metaphysics is mercilessly ruled out it is very likely that the scientific theories themselves may transform into dogmas. It is not enough that an empiricist checks the knowledge claims only on the basis of observational evidence, rather she has to create many alternative theories, including metaphysical ones, to evaluate the original theory. So in short, "a good empiricist must be a critical metaphysician".[118]

Credibility of Incommensurability

In a way, Feyerabend's vies of incommensurability is stronger than that of Kuhn. Feyerabend himself agrees that it is not an easy job to give a clear definition of incommensurability: "it is hardly ever possible to give an explicit definition of [incommensurability]"[119] Feyerabend sounds more consistent than Kuhn in understanding the concept incommensurability. He stresses only the linguistic aspect of it, but the reasons given by him differ from those of Kuhn. Feyerabend's understanding of incommensurability sounds more global, unlike that of Kuhn, for whom it is more localized. The former does not restrict it to one single problematic terms or cluster of terms, therefore substantial theory change affects all the terms in a theory. Feyerabend views the reason for incommensurability to be a kind of extreme holism regarding the nature of meaning itself, while for Kuhn it arises from the translational difficulty.

However, after criticisms both, Feyerabend and Kuhn, are not ready to soften their demand for total incommensurability; they are determined to look for alternative modes of comparison. For instance, two incomparable theories can be critically compared, he grants, "the use of incommensurable theories for the purpose of criticism must be based on methods which do not depend on the comparison of statements with identical constituents. Such methods are readily available".[120] Commensurability may also be achieved by using 'the pragmatic theory of observation' which focuses on the causes of the production of a certain observational sentence rather than the meaning of that sentence. But a serious difficulty with this idea is that Feyerabend claims even the 'descriptive terms' also change. Finally he follows Kuhn in giving the example of scientist's learning a new theory is like a child learning a new language, starting from scratch, without any help from anywhere, and therefore it is not like translating from one language to another[121].

If the issue of incommensurability is taken to the extremes, Ian Hacking wonders whether crucial experiments are even logically possible: "If an experiment is to decide between theories, would there not have to be a sentence asserting what one theory predicts and what the other denies? Can there be such a sentence?".[122] There are two main trends in responding to the problem of incommensurability, as Khalidi (2001) points out: a) Denying some elements of incommensurability; creating a linguistic comparison among theories; and b) Rejecting linguistic incommensurability to develop other ways of comparing scientific theories. Among the former group comes Scheffler (1982), who tries to show that at least one component as a theory is not affected by untranslatability, and that is, reference; however he is not elaborating a theory of reference to show how the reference of terms belonging to different theories can be compared; and then we have Hilary Putnam (1975), who suggests that the causal theory of references can be used to give a version of the meaning of natural kind terms, and argues that the same can be applied to theories in science as well; and finally Davidson (1985) maintains, going against Kuhn, that all presumed conceptual schemes are inter-translatable, most probably including the scientific theories embedded within them. He claims that incommensurability makes no sense because it is supported by the idea of different and incomparable conceptual schemes. Among the latter group, which focuses on *non-linguistic types of response*, we find further two trends: to express theories in model theoretic terms, so that the mathematical mode of comparison is possible (proposed by authors like Joseph Sneed and Wolfgang Stegmüller);

or to focus on the cognitive approach to science, which tends to see scientific theories to be entities in the minds / brains of scientists (e.g. Paul Churhclnad, Roanld Giere and Paul Thaggard). Thaggard (1992) seems to be more convincing in his arguments in answering incommensurability by using procedures derived from the connectionist research programme in artificial intelligence. However it is interesting to note that another cognitivist, Chruchland (1992), claims that connectionist models in fact supports. Shapere (1966) may be included here who, as pointed out earlier, argues that there is a chain of reasoning to identify the sameness of meaning among successive theories to make the comparison among them possible.

But as we noted above in critiquing Kuhn, it is quite reasonable to allow a certain degree of incommensurability between two theories, belonging two different paradigms. True, comparison between two rival theories may sometimes be logically impossible. Further, to a certain extent a scientist is influenced in theory choice by the elements, which Feyerabend has mentioned, and some other 'external factors such as career prospects and availability of funds' do play a significant role. However, in spite of these subjective elements we cannot conclude that "individual judgements are immune to rational arguments. The preferences of individuals can be criticized, for example, by showing that they are seriously inconsistent or by showing that they have consequences that the individual holding them would not welcome".[123] But all these elements need not undermine the edifice of rationality of science as a whole.

After all these discussions, one realizes that incommensurability is still a live issue. It arises due to the linguistic representation of theories; as long as linguistic elements are there (which will always be there), who looks for comparison of scientific theories point-by-point, will have to face the problem of incommensurability; if one thinks such piecemeal comparison is necessary then some other way of comparing theories has to be proposed.

Feyerabend - a Staunch Humanist

It seems to be better to identify Feyerabend as a humanist, rather than a philosopher of science. He is primarily concerned about liberation of the society from various evils. His concept of liberty is not abstract or superficial. He calls for an *unselfish dedication* to work for the liberation of the oppressed. This liberty is to be manifested in thoughts, in education and in the teaching methods. This kind of liberty is essential for practical purposes of *free development* and which in turn would lead to *happiness*. This liberation of the oppressed is not just for any political or abstract reasons but "to liberate people so that *they can smile*".[124] He is very practical and down-to-earth. Unless one has the sense of liberty and the ability to smile one cannot liberate others and make them smile. One cannot be iron-handed and self-righteous in working for such liberty, as it would create another kind of oppression. Therefore we must realize that "The hardest task needs the lightest hand or else its completion will not lead to freedom but to a tyranny much worse than the one it replaces".[125]

Feyerabend is convinced that because of an undue admiration for science the society mistakenly concentrates only upon the scientific developments investigating plenty of time and energy, money and human power. We need to concentrate on the totality of human welfare. The State has first to focus on the making safe drinking water, sanitation, employment and other basic facilities available for all its citizens. This demands at least as equal if not more, the attention science gets. He voiced out throughout his life his deep concerns for humanity and liberty, in all the realms like politics, science and intellectual pursuits. He was mindful of those human values even at the last days of his life. That is why, in spite of all his contributions to the intellectual world, finally he wanted the world to remember him basically as a simple and ordinary human being, who values love the most. Just a couple of weeks before he breathed his last on 11 February, 1994, he wrote in his touching autobiography: "My concern is that after my departure something remains of me, *not* papers, *not* final philosophical declarations, but love... That is what I would like to happen, not intellectual survival but the survival of love".[126]

Having seen some of the salient features of the proponents of rational model (Hempel and Popper) and the non-rational models

(Kuhn and Feyerabend) of rationality in science in Chapter I, and having critiqued some of those claims in Chapter II, now we turn, to the notion of "Reasonableness - as an Integrated View of Rationality of science" in Chapter III.

Notes & References

1. Philip Kitcher, "Reasonable People", in James H. Fetzer, ed., 2000, 252.
2. *Ibid.*, 261.
3. W. Salmon, "Carl G. Hempel on the Rationality of Science", 1983, 555 - 562, 557.
4. Ibid., 561.
5. Kuhn, "Rationality And Theory Choice," in *The Journal of Philosophy*, 1983, 563.
6. Kuhn, "The Trouble with the Historical Philosophy of Science", in *The Road Since Structure: Philosophical Essays,* 2000, 12.
7. Robert Nozick, "The Objectivity and the Rationality of Science", 2000, 292.
8. C.G. Hempel, "Scientific Rationality: Normative vs. Descriptive Construals", in J Fetzer, ed., 2001, 357-371.
9. Carl G. Hempel, *Philosophy of Natural Science*, eds. Elizabeth and Monroe Beardsley, 1966, 79.
10. C.G. Hempel, "Valuation and Objectivity in Science", 2001, 372-395.
11. Robert Nozick, "The Objectivity and the Rationality of Science", 2000, 296.
12. *Ibid.*, 298.
13. *Ibid.*, 301.
14. *Ibid.*, 303.
15. C.G. Hempel, "Recent Problems of Induction," in James Fetzer, ed., 2001, 29.

16. W. Salmon "Carl G. Hempel on the Rationality of Science", 1983, 555 - 562, 562.
17. *Ibid.*
18. *Ibid.*
19. *Ibid.*, 560; italics original.
20. C.G. Hempel, "Valuation and Objectivity in Science", in James Fetzer, ed., 2001, 391.
21. Edward Shils, and Henry Finch, (eds.), *Max Weber on the Methodology of the Social Sciences*, 1940, 72.
22. *Ibid.*, 54.
23. C.G. Hempel, James Fetzer, ed., 2001, xxxii.
24. C.G. Hempel, "The Irrelevance of Concept of Truth for the Critical Appraisal of Scientific Claims," Unpublished manuscript, as quoted by Robert Nozick, 2000, 295.
25. Robert Nozick, "The Objectivity and the Rationality of Science", 2000, 295.
26. L. B. Bazhenov, "Reflections on Reading Popper," 2002, 7 – 22, 8.
27. As quoted by L. B. Bazhenov, "Reflections on Reading Popper", Fall 2002, 7 – 22, 9.
28. W. H. Newton-Smith, *The Rationality of Science*, 1981, 44.
29. Ibid, 45.
30. As quoted by Popper, *The Myth of the Framework*, 1994, 7.
31. Karl Popper, *Realism and the Aim of Science*, 1992, XV.
32. Boris Salazar, "How Rational is Popper's Rationality Principle?: A Critical Note on Oakley". (See: http://hesta.fec.anu.edu.au/review/ejournal/pdf/32-c-1.pdf).
33. Stefano Gattei, "Karl Popper's Philosophical Breakthrough", 2004, 448-466, 459, foot note 37.
34. Robert Nola, "The Status of Popper's Theory of Scientific Method", British Journal of Philosophy of Science, 38 (1987), 441-480, 455.

35. Karl Popper, *Logic of Discovery,*' 1959, 16.
36. L. B. Bazhenov, "Reflections on Reading Popper," Fall 2002, 7 – 22, 10-11.
37. Karl Popper, *Realism and the Aim of Science*, 1992, xxxiv.
38. Elanor Taylor, "Why Science Matters", 2006. See: http://www.sirc.org/articles/why_science_matters.shtml
39. A.F. Chalmers *What is This Thing Called Science?*, 1992, 64.
40. Grover Maxwell, "Corroboration without Demarcation", Paul Arthur Schilpp, ed., 1974, 296.
41. I. Lakatos, Popper on "Demarcation and Induction", in Arthur Schilpp, ed., 1974, 247.
42. *Ibid.*, 250.
43. Karl Popper, *The Logic of Discovery*, 1959, 104.
44. K. Popper, *Realism and the Aim of Science*, 1992, xxii, Foot Note 46.
45. Hilary Putnam, *Reason, Truth and History,* 1981, 197.
46. Rom Harre and Michael Krausz, *Varieties of Relativism*, 1996, 71.
47. I. Lakatos, "Popper on Demarcation and Induction", in Arthur Schilpp, ed., 1974, 245.
48. Popper, *Conjectures and Refutations*, 1963, 36.
49. Soren Bo Anderson, "Relativism". See: www.phys.port.ac.uk/what/example.htm
50. Agassi, "Sir Karl Popper in Retrospect: The Positive Power of Negative Thinking," in Joseph Agassi, *The Philosophical Art of Philosophical Polemics. Selected Reviews and Comments,* 1988, 479-501, 497-498.
51. Karl Popper, "Truth, Rationality, and the Growth of Scientific Knowledge" 1963. As quoted by Stefano Gattei, 2004, 448-466. 462.
52. Karl Popper, *The Logic of Scientific Discovery*, 1959, 420.

53. Adolf Grunbaum, "Karl Popper Versus Inductivism", in R. S. Cohen *et al.*, eds., 1976, 213-252, 221. Italics original.
54. W.H. Newton-Smith, *The Rationality of Science,* 1981, 46.
55. Arthur Schlipp (ed.), *The Philosophy of Karl Popper*, 1974, 1105.
56. Karl Popper, *Conjectures and Refutations, 1963,* 55.
57. W. Salmon, "Rational Prediction", in Martin Curd and J.A. Cover, eds., 1998, 443.
58. I. Lakatos, "Popper on Demarcation and Induction", 1974, 260.
59. L. B. Bazhenov, "Reflections on Reading Popper," 2002, 7 -- 22, 11.
60. Stefano Gattei, "Karl Popper's Philosophical Breakthrough", 2004, 448-466, 464.
61. *Ibid.*
62. Thomas Kuhn, "A Discussion with Thomas Kuhn," in *The Road since Structure,* 2000.
63. Robert Klee, ed., 1999, 199.
64. Ian Hacking, "The Rationality of Science After Kuhn", in Robert Klee, ed., 1999, 217.
65. *Ibid.*, 200.
66. Thomas Kuhn, "The History and the Philosophy of Science", *The Essential Tension*, 1977, 4.
67. Paul Hoyningen-Huene, "The Interrelations between the Philosophy...", 1992, 487-501, 491-2.
68. Thomas Kuhn, "Essential Tension", 1977, xix.
69. For instance, Shapere comes out with a sharp criticism against Kuhn's attempts to solve the problems of 'paradigm'. See: Shapere, "The Paradigm Concept", 1971, 706-709.
70. Carl Matheson, "Historicist Theories of Rationality", 2. (See: http://plato.stanford.edu/entreis/rationality-historicist)
71. Joseph Ruse, *Knowledge and Power – Toward a Political Philosophy of Science*, 1987, 30.

72. Frederick Suppe, ed., *The Structure of Scientific Theories,* 1977, 647.
73. Paul Hoyningen-Huene, 1992, 487-501, 494
74. Kuhn, *SSR,* 1970, 158, 199, 262; and *The Essential Tension,* 1977, 321-2.
75. Rom Harre and Michael Krausz, 1996, 3.
76. *Ibid.,* 8.
77. Rom Harre and Krausz give us the definitions of these in a nutshell, in *Varieties of Relativism*, 1996, 4-5. a) *Universalism* – discursive variant: there are beliefs (classes of statements) which hold good in all contexts, at all times and for all persons and ontological variant: there are entities (classes of existents) which exist for all persons; b) *Objectivism* – discursive variant: there are beliefs (classes of statements) which hold independently of the point of view, corpus of beliefs or conceptual scheme held to and employed by any particular person or society, ontological variant: there are entities (classes of existents) which exist independently of the point of view, corpus of beliefs or conceptual scheme held to or employed by any particular person or society; and c) *Foundationalism* – discursive variant: there is a common set of basic statements, not capable of further analysis, which serve in each context for each kind of enquiry for the assessment of all judgments of a relevant kind; ontological variant: there is a common ontology or set of basic existents, incapable of further analysis, out of which all other existents are constructed.
78. Rom Harre and Michael Krausz, *Varieties of Relativism*, 1996, 26.
79. L. Wittgenstein, *Philosophical Investigations*, 1953.
80. Rom Harre and Michael Krausz, *Varieties of Relativism*, 1996, 32.
81. Michael E. Malone, "Kuhn Reconstructed: Incommensurability Without Relativism", 1993, 69-93, 81.
82. Rom Harre and Michael Krausz, *Varieties of Relativism*, 1996, 110.

83. Michael E. Malone, 1993, 77. Emphasis original.
84. Kuhn, "Dubbing and redubbing: the vulnerability of rigid designation", 1990, 298-318, 299.
85. S.G. Kulkarni, "Subject in Post-Positivist Philosophy of Science: Some Reflections on Thomas S. Kuhn", 2003, 108-116.
86. Paul Hoyningen-Huene, "Kuhn's Conception of Incommensurability, 1990, 481-492, 488ff.
87. Kuhn, 'Response to Commentaries', in T. Asquith and T. Nickles, 1983.
88. Kuhn, "Objectivity, Value Judgment, and Theory Choice," in *The Essential Tension*, 1977.
89. Davidson, "On the very Idea of a Conceptual Scheme," 1985, 183-198, 192.
90. Shapere, *The Rational Dynamics of Scientific Inquiry*, Chapter II, p. 8-9. (Forthcoming).
91. Shapere, 2001, 199.
92. Shapere, "Evolution and Continuity in Scientific Change", 1989, 429.
93. Newton-Smith, *The Rationality of Science*, 1981, 148.
94. Gürol Irzik and Teo Grünberg, "Carnap and Kuhn: Enemies and Allies?", 1995, 285-307, 301.
95. Gonzalo Mune¢var (ed.), *Beyond Reason*, 1991, p. ix. In fact it is a very good collection of 25 essays on Feyerabend's contributions; it ranges from glorifying him as teh rarest of the rare thinkers of the 20th century and accusing him to be the most dangerous man of humanity. For example, Vine Deloria, in "Perceptions and Maturity: Reflections on Feyerabend's Point of View", acclaims highly, whereas on the other hand Jospeh Agassi, in "As You Like It", goes even to make a scathing personal attack on him accusing him of supporting the Nazi ideology.
96. Aant Elzinga, Jan Nolin et al., eds., *In Science We Trust? – Moral and Political Issues of Science and Society*, 1990, 9.

97. Gorver Maxwell, "Feyerabend's Materialism", in Gonzalo Mune¢var, ed., 1991, 453-463.

98. Feyerabend, *Erkenntnis fur freie Menschen* (Suhrkamp: Frankfurt, 1979), p. 87. As quoted by Herbert Schnädelbach, "Against Feyerabend", in Gonzalo Mune¢var (ed.), *Beyond Reason* 1991, 433-448, 434.

99. As quoted by A. F. Chalmers, in *What is This Thing Called Science?* 1992, xvi.

100. Daniel Moloney, "Questioning Everything?", in http://www.leaderu.com/ftissus/ft9811/opinion/moloney.html

101. *Ibid.*

102. Daniel Moloney, "Questioning Everything?" in http://www.leaderu.com/ftissus/ft9811/opinion/moloney.html

103. Newton-Smith, 1981, 128.

104. For instance, Kulka thinks that for Lakatos, methodology provides a rationality theory ("Some Problems Concerning Elkana and Lakatos", 1977, 325-43); Musgrave argues that various methodologies provide cannons of rationality ("Why Did Oxygen Supplant Phlogiston?", in C. Howson (ed*.), Method and Appraisal in the Physical Sciences*, 1976, 181-208); and in Worrall's understanding a methodology has a direct relevance upon rationality or irrationality ("Thomas Young and the 'Refutation' of Newtonian Optics", in C. Howson (ed*.), Method and Appraisal in the Physical Sciences*, 1976, 107-79).

105. Ronald C. Curtis, "Are Methodologies Theories of Scientific Rationality?", 1986, 135-161, 158.

106. Larry Laudan, "For Method: Or, Against Feyerabend", in J.R. Brown and J. Mittlestrass (eds.), *An Intimate Relation*, 1989, 299-317, 302.

107. *Ibid.*, 303.

108. Herbert Schnädelbach, "Against Feyerabend", in Gonzalo Mune¢var, ed., 1991, 433-448, 438.

109. Larry Laudan, "For Method: Or, Against Feyerabend", 1989, 299-317, 313.

110. Newton-Smith, 1981, 130.

111. Herbert Schnädelbach, "Against Feyerabend", in Gonzalo Mune¢var, ed., 1991, 433-448, 446.

112. John Preston *Feyerabend: Philosophy, Science and Society*, 1997, 204-6.

113. Newton-Smith, *The Rationality of Science*, 1981, 134.

114. J. Worral, "The Value of a Fixed Methodology", 1988, 274.

115. Paul Feyerabend, *Killing Time*, 1995, 152.

116. Paul Feyerabend, *Against Method*, 1975, 68.

117. Newton-Smith, *The Rationality of Science*, 1981, 139.

118. Paul Feyerabend "How to Be a Good Empiricist – A Plea for Tolerance in Matters Epistemological", 1998, 944.

119. Feyerabend, *Against Method*, 1975, 225.

120. Feyerabend, "Reply to Criticism...", 1981, 104-31, 115.

121. Feyerabend, "Putnam on Incommensurability", *in Farewell to Reason*, 1987, 265-72, 266.

122. Ian Hacking, *Representing and Intervening ...*, 1987, 73.

123. A. F. Chalmers, *What is This Thing Called Science?*, 1992, 138.

124. Paul Feyerabend, "How to defend Society against Science?", 1998, 65.

125. *Ibid*.

126 Paul Feyerabend, *Killing Time*, 181.

Chapter 3

Reasonableness – An Integrated View of Rationality in Science

Introduction

In this Chapter an attempt is made to arrive at an integrated view of rationality in science. A few traditional views of rationality are analyzed in Section 1, and their inadequacies are brought out. The idea of equating *rationality with deducibility* is discussed in some detail, because this idea has been quite influential in the past few centuries. It is, of course, not the case that every aspect of all the traditional accounts of rationality needs to be rejected or reformulated. For instance, I explain the role of *inductive reasoning*, in developing an adequate account of rationality in science. Due to the shortcomings of the traditional accounts of rationality, there is a need to revise it. Therefore, in Section 2, I elaborate on the need for such a revision. I support my arguments with the views and alternative accounts of rationality, put forward by Hilary Putnam ('Integrated View of Rationality'), Stephen Toulmin ("Over-reliance on Rationality), Christopher McMohan ('Collective Rationality') Stephen Nathanson ('Reasonable form of Rationalism') and Jurgen Habermas (Consensus as a form of Rationality).

In Section 3 I try to present my account of rationality in science. I tend to look at the notion of *Reasonableness as the distinguishing character of a comprehensive account of rationality in science.* I enumerate some of the features of the notion of reasonableness, like *Imagination, the Role of the agent, Common sense, the rejection of Zero-tolerance, Skeptical attitude, an Enriched notion of Objectivity, a bit of Pragmatism, the role of Judgment, and the Consensus of the scientific community.* I categorize these features, for the sake of more simplicity and precision, under three groups, though they need not be taken as water-tight compartments; namely, *Reasonableness in Scientific Methodology*, *Reasonableness in Scientific Practice* and *Reasonableness at the Personal level of Scientists*. In order to highlight the need and relevance of reasonableness a few other domains, like Judicial System, Environmental Protection, Feminist Approach to Rationality, and Critical Thinking are discussed.

The consensus of the scientific community seems to assume a greater importance in the contemporary scientific research frontiers, like Cosmology, Astronomy, Particle Physics and so on. I explore this aspect of *Consensus of Scientific Community* in Section 4. I begin with the need for consensus in (Contemporary) science. I explore an example for consensus from the domain of Astrophysics, namely, the experiments on Gravitational Wave detection, to show how consensus is important in such experiments. This analysis shows how consensus can be seen as one important way to overcome the Experimenters' Regress and also it underlines the role of non-cognitive factors in science.

SECTION 1
TRADITIONAL ACCOUNTS OF RATIONALITY AND THEIR CONSTRAINTS

While highlighting various traditional views on rationality, this section tries to present some of their inadequacies also. Though some aspects of these traditional accounts of rationality have already appeared in the earlier discussion in the previous chapters, here in the section, I cull them out specifically and analyze them a bit more closely. I discuss in some detail the claim of equating *rationality with deducibility*, as it seems to be more forcefully held in the past few centuries, and show how inadequate that understanding is. Though I show the constraints of the traditional views, I do argue for the need of inductive reasoning, perhaps a positive legacy from the traditional understanding, to show that science is rational in spite of inductive reasoning.

Various Accounts of Rationality and Their Constraints

Rationality – As Logical Consistency

One aspect of rationality is usually taken to be the logical consistency. So it is irrational to affirm both p and ~p at the same time. By expanding this notion, one can hold that it is irrational to try to achieve a goal with unsuitable means, and have a conscious knowledge about the unsuitability of the means. If an agent lacks conscious knowledge of the unsuitability of means she is not irrational. Charles Taylor argues that the concept of rationality is richer than mere consistency. It involves more than avoiding inconsistency; it depends on different judgments, as we are going to see shortly, especially when we compare incommensurable cultures and activities.[1]

Rationality – As Justification

Rationality is often equated to justification. A justified belief will necessitate (entail) a rational belief; and when a belief is unjustified it is irrational. However, in actual practice, the relation between justification and rationality is not very clear and there does not seem to be any direct relation between them. For, our practical life shows that people don't always undertake evaluative process by strict rules, methods, deductive / inductive reasoning to arrive at a conclusion or to evaluate the validity of a conclusion; people are ready to believe even an invalid (logically speaking) conclusion and reject even a valid conclusion. There is a deep interplay among justification, rationality and beliefs; it is a complex phenomenon. Therefore, merely to equate rationality with justification is to do injustice to the notion of rationality, and conversely, unjustified beliefs need not always lead to irrational actions. Justification and rationality need not always go hand-in-hand.

Rationality – In Terms of True / False Beliefs

Generally rational / irrational beliefs are equated to true / false beliefs. However, not all our rational beliefs are true and not all our irrational beliefs are false. False beliefs may even arise from true empirical facts depending on how we interpret a phenomenon. We do entertain false beliefs, we do act upon them, but that action / decision need not be irrational. Often, first we act, and then try to give validating reasons. So our ordinary beliefs can't always be put under the traditional scanner of justification.

Rationality – As Scientificity

Several scientists don't seem to be aware of the last thirty years of philosophical investigations showing the shortcoming of the scientific methods. Considering themselves to be expert on everything, including the question what makes human beings human, they tend to consider themselves to be the custodians of rationality. Rationality has been seen to be synonymous with science. To be scientific is to be rational; and some also hold that to be unscientific is irrational. As science is seen as the best exemplification of human rationality, to question the rationality of science would be like asking whether a bachelor is an unmarried person.[2] But as Putnam observes, "'Scientific' is not coextensive with 'rational'. There are many perfectly rational beliefs that cannot be tested 'scientifically'". He discusses three important areas of our human lives, where science has nothing to do, nor even can it affirm that those facts exist: a) the domain of objective values; b) the domain of freedom; and c) the domain of rationality itself.[3] Popper also shows that "poetic inventiveness" and "the invention of criticisms"[4] are also important components of rationality. Further, Ian Hacking (1983) argues that philosophers have made a mummy of science, but when they finally realized that science constitutes history and sociology they resented that rationality is at stake. When history and sociology of science seem to teach that nothing of science is available in absolutistic, non-historical terms they thought the whole science which stands on the edifice of rationality collapses. So rationality obviously goes beyond the realm of Scientificity. Neither science nor scientists can be the guardians of rationality. Scientists are also in need of assumptions as all of us are. There is no guarantee that their assumptions will always be correct.

*Rationality – Provided by Autonomous (*a priori*) Principles*

The traditionalists claim that: a) there are epistemological principles that are *a priori* and autonomous from science and that it is possible to discover them; and b) these principles provide a warrant for normative claims as to how scientists should act. However, there is a Naturalistic account of rationality, especially of rationality in science. In recent times many, like Churchland (1979), Hooker (1987), Hull (1988), Rouse (1986), are attracted by this view of science, in opposition to the traditional view of the first few decades of the 20th century. *These naturalists argue* that: a) There is no such autonomous realm of epistemological principles; though there are better or worse ways of attempting to do science, and they may give some guidelines regarding the better ways of doing / or not doing science; and b) however, these principles are only instrumental or conditional, and they connect research strategies with goals of research. Ronald Giere, going along the same lines, proposes 'instrumental (or conditional) rationality': "To be instrumentally rational is simply to employ means believed to be conducive to achieving desired goals. It is part of the standard, folk psychological, model of human agents that they are normally rational in this instrumental sense"[5] and those actions that don't fulfill these requirements are 'irrational'. Thus we can have an alternative account of naturalistic view of rationality.

Rationality – As Goal-Orientedness

Rationality is invariably understood as a goal-oriented activity. But Goal-means understanding of rationality does not seem to be sufficient to capture the notion of rationality. In human activities there are several types of actions. There are different types of social action: *rational actions* that are purposeful; *goal-oriented actions* – here goal and means are rationally chosen; *value-oriented actions* – it strives for a substantive goal, which in itself may not be rational (like attaining salvation) but pursued by rational means; *emotional or affective actions* – these actions are rooted in the emotional state of the actors rather than in the rational weighing of means and ends (like religious or fundamental sects); and *traditional actions* – these are rooted in customary of thought, 'the eternal yesterday' (actions of Orthodox Jewish congregation). The modern society shows more importance for 'goal oriented' action, while earlier the society focused on the other actions. In modern society, even in the realms of politics, economics, the law, surprisingly in interpersonal relations, the method of means-to-end has become predominant.[6] Thus, there are several types of human actions, which are rational, but cannot be brought under the 'goal-means' types of actions.

Philip Kitcher points out how 'decision-theoretic approach' to rationality cannot explain the multidimensional aspects of human lives. For instance, as it is already explained in Chapter I, a complete rule-oriented or goal-oriented rational approach to taking decisions to marry someone will be inadequate, as married life involves companionship and commitment in marriage, which cannot be fully explained by this rational analysis. In life there are many more aspects which lie beyond the purview of mere goal-oriented rationality. What numerical value can be given to love, pain of losing one's beloved, joy of reunion with one's beloved, satisfaction, contentment, quality of life etc.?[7]

Rationality in science also cannot be reduced to mere rule-following. Scientists are people with various cognitive traits and abilities. *Choice of problems*, *framing goals, drawing conclusions* are done with lots of *imagination, confidence*, and *exercising caution*, giving much importance *to moral*, social and even *religious concerns*, *valuing loyalty* and *friendship* and so on. Therefore the real actors in science are not just agents who follow rule-procedures, or socially determined performers but, Kitcher puts it, as we noted already, "are those psychologically complex beings, reasonable people".[8] These reasonable people show concern and respect for others and a willingness to participate in dialogues with others. Humans not only reason out, but also love, feel, suffer and desire. All these are not signs of weakness but essential and integrated parts of being human; not that everything that originates apart from reason is bad.[9] A notion of reasonableness, as we will see later in this Chapter, will have room for all these dimensions of human lives.

Rationality – As the Fundamental Assumption of Being Humans

Some claim that human rationality is the fundamental assumption for human existence and for their relationship with others, and it is the 'gift' of the long evolutionary process. Philosophers like Quine (1960) put forth, what is known as, 'the argument from charity', by which they show that unless we assume that the other person is rational, we cannot communicate, nor we can attempt to understand or translate the other's language. The other person, if there is no specific deformity, has to be rational. Hume distinguishes between *performance reasoning* and *competence reasoning*, where one may lack rational characteristics in the former but not in the latter. *The weak version of the principle of charity* claims that unless there is a strong epistemic evidence against human rationality people must be treated as rational; but *the strong version of the principle* demands, as proposed by Daniel Dennett (1987), that people should be always treated as rational, and if there is any deviation, it is only a performance error, and not a competence error. The strong version is intertwined with *the argument from evolution*; it is the design of evolutionary process that humans are rational: "evolution has designed human beings to be rational, to believe what they ought to believe and want what they ought to want".[10] David Papineau (2003) also holds a similar view. Evolution has gifted us with two great cognitive abilities: understanding the other minds (people) and means-end thinking; this enables them to have *theoretical rationality*, i.e. rationality of the beliefs that one adopts; from this emerges *practical rationality*, i.e. rationality of the choices one subsequently makes in actual situations.

It seems plausible to view rationality as the fundamental human experience. But taking rationality as the fundamental aspect of human existence sounds over-simplification of the whole issue. This also sounds highly anthropomorphic and does not seem to encourage critical evaluation of the features that make humans essentially rational. Our Universe is such a multifaceted one that it appears that we need different types of rationality to comprehend it. Michael Polanyi, according to William Scott, categorizes various kinds of rationality which include: the purely deductive systems in mathematics, the deductive systems concerned with elementary entities in physics, geometric systems such as crystal physics and organic chemistry, structural-functional systems in biology..., ordering and operational principles, ... many complex systems of thought, relation and action in the social sciences, and quite generally, the filling of gaps in the ever-growing edge of the web of science. Therefore it is not proper to confine rationality only to any one particular frame work.[11]

Rationality – As Deducibility: An Analysis

I explore this account of rationality as deducibility a bit more as this account has been quite influential view of rationality in the past few centuries.

A Legacy from Plato: Though deducibility has been equated with rationality when one analyses the equation more closely, one understands that the notion of rationality transcends mere deducibility. As Gangavane[12] explains, right from the time of Plato, deductive reasoning has enjoyed a very privileged position. Since knowledge, worthy of its name, is taken to be certain and necessarily true, for Plato the only source of valid knowledge is reason. Sense experiences do vary and cannot give knowledge that is certain and infallible. It can give only opinion. Geometry is the ideal type of knowledge. The unchanging eternal world, the world of forms, alone can be the object of knowledge. This kind of understanding of rationality had a lasting impact in the history of epistemology. However, even within the Greek tradition, not everyone agreed with Plato's ideas. Sophists criticized the concept of certain *a-temporal* knowledge and argued for historicity of knowledge. Descartes, following Plato, demands that a piece of knowledge has to be certain, infallible and true. As Plato, he too considers geometry as the ideal type of knowledge. Therefore, all knowledge system must start from self-evident axioms and derive valid conclusions from premises, with the help of some logical principles. This is made clear in the third rule for the direction of mind. In our inquiries we must not focus on the thoughts of others, and not on our own conjectures; we must rather focus on what we can indubitably deduce with certainty; there is no other way of having knowledge worthy of its name. Whatever is achieved other than this way would lie outside the fence of knowledge and thus *rationality is equated with deducibility*.

Impact on Logical Positivism: Though the empiricists later strongly criticized the Cartesian view of reason as the only source of knowledge, they did not question the pre-eminence of deductive reasoning. They challenged *only* the equation of rationality to deducibility. They asserted that sense experience was also an equally important source of knowledge. It was generally agreed that sense experience gave the raw material for knowledge which was later systematized by reason. In pure sciences, like mathematics and logic, we have reason as the only source of knowledge. The empiricists held that sense experience could also give us knowledge, if not as certain as reason. In the long run, empirical movement ended up in Logical Positivism. The logical positivists held empirical verifiability to be the only criterion for meaningfulness; experience is the only source of knowledge, though an explicit role of reason is acknowledged in the systematization of knowledge. It is to be noted that due to the problem with 'strong' verificationism Carnap proposed a softer version of verificationism, which is now known as the 'Received View'. If one insists on a total and definitive verification of a theory, all the universal statements are at a great risk. So he proposed only a 'gradual confirmation' instead of verification. Through this repeated empirical confirming tests science has been progressing.[13] Positivists insisted upon the *criterion of deducibility* in the methodology of science and they held that an argument would be sound, if it were materially true and formally valid. Thus *empirical provability* (which is to be given by empirical verification) and *logical demonstrability* were two essential features of knowledge. The Logical Positivists identified rationality with scientific rationality, and there too tried for a unity of method. They questioned the rationality and validity of status and importance of social sciences, as these could not fit into the straight jackets of scientific rationality.

However, the critiques of logical empiricism and the lessons learnt from the sociology and history of science by the second half of the 20th century demanded for a reconsideration of the equation of rationality with deducibility. One can grant the persuasive nature of the deductive reasoning; but one still needs to question "the demand for rejecting all other types of reasoning or for considering them to be always inferior to deductive reasoning".[14]

Complexity of a Deductive Syllogism: A deductive syllogism has two premises and a conclusion; it must have three terms, each one must occur in two of them; the validity of the argument just depends on its form, and thus providing only a formal validity. This appears to be simple and uncomplicated. But as Toulmin points out its complication is camouflaged by its deceptive simplicity. In any deductive syllogism, there is a conclusion and to support it the data is given. There is a need for a third element, what Toulmin calls, a *warrant*. This warrant provides the link between data and conclusion. For example:

All planets in the solar system go around the sun (warrant, which needs backing),

Unless it is bigger than the sun (condition of rebuttal)

Earth is a solar planet (Data)

Therefore the earth goes around the sun (Conclusion).

Toulmin points out that the conclusion does not always *necessarily* follow from data and warrant; at times it follows only *probably* or *under certain conditions*. The conditions to which it is subject are given in the conditions of exception or rebuttal. It is to be noted that the context of an argument decides what can be taken as warrant. For instance, a pole of a certain length makes a shadow of particular length at a given time. The height of the pole explains the shadow and not the other way. One can't say that because the shadow is of this sort at a given time, the pole has a certain height. Further, the authority of warrant needs to be established and it is called the *backing* of the warrant. The backing, which is required to establish the authority of a warrant, is different from one domain of science to another; it may refer to a law, or some statistical report or a system of taxonomical classification. The process of ascertaining the backing is also a part and parcel of the deductive argument itself. One can't close one's eyes to this aspect and take up only deducibility as rationality. In the traditional notion of the major premise (universal premise) the distinction between a warrant and backing is not paid proper attention to.

Thus it is clear that the structure of a syllogism is *internally complex* and never simple. Any argument can be presented in the form of 'data – warrant – so conclusion', and thus made it to appear formally valid. Therefore, Toulmin cautions that the notion of formal validity refers only to the manner of phrasing the arguments and "nothing about the reasons for its validity. These reasons are to be understood only when we turn to consider the backing of the warrant invoked".[15]

Toulmin also makes a distinction between *analytic arguments* (where the backing of the warrant openly or implicitly contains the information provided in the conclusion) and *substantial arguments* (where the backing does not contain the information provided in the conclusion. In analytic arguments, unlike the substantial arguments, by verifying the backing the truth/falsity of the conclusion is also automatically established. Logicians tend to accept the substantial arguments as only the second best because they try to force the analytical ideal of universality in all fields of knowledge, irrespective of their peculiarities. The inconclusive or probable nature of substantial arguments need not be a deficiency; it is only the peculiarity of the subject matter in question, and not the weakness of the argument. In fact, in natural sciences, social sciences, history, ethics, religion and our daily life-situations we constantly use substantial arguments and analytical arguments are used not so frequently.

Apart from such logical crisis there also *other problems when rationality is restricted to deducibility.* For instance, a) The role of the person who formulates / receives the argument is insignificant; as the premises are already proven the conclusion is binding for all. A sort of intolerance towards others' views and dogmatism about one's own position emerge; b) Due to this intolerance, there does not seem to be any place for any fruitful dialogue with those who accept different perspectives; this may hamper the growth and development of the accepted theory itself. Similarly, being dogmatic about one's position may lead one to fanaticism and violence. There won't be any effort for self-change. c) There is no room for meaningful talk of values because values are not empirically verifiable and not logically demonstrable; and d) The rationality of action can be judged only with reference to their efficacy as a means of achieving some goals; no rational discussion about goals and objectives of an individual is possible, because the goals involve individuals' values and interests. Since these values are not 'empirically verifiable', nor 'logically demonstrable', there don't seem to be rational solutions to moral conflicts. Thus "all our practical moral life, [will end up] without any rational foundations".[16]

The notion of rationality must transcend deducibility and include elements of non-deductive reasoning as well. One may, however, still hold that deducibility as the ideal type of rationality, as it gives complete certainty. Unfortunately it is practically unattainable, except in a few areas like Mathematics and Logic. Thus due to a) *the importance of the non-deductive reasoning* (especially in tracing out the backing of the warrant, because the warrant itself need not be the result of deduction and the backing also need not be deductive and still it is accepted as rational!) and b) *the inevitability of substantial arguments in everyday life and science,* the notion of rationality has to go beyond deducibility. Our day-to-day life situations and empirical sciences mostly involve non-deductive type of reasoning, where complete certainty is, in principle, not available. So, it is better that we explore non-deductive type of reasoning though it lacks complete certainty.

Inductive Reasoning – A Positive Legacy from the Traditional Accounts of Rationality

I now analyze the concept of inductive reasoning, a positive legacy from the traditional account of rationality in science. This shows that not all the features of traditional accounts of rationality need to be dropped. I highlight the perennial problem of providing a rational justification for inductive reasoning. There have been many attempts to solve, even to dissolve, this problem, but they have not been quite successful. I make a brief presentation of 'the Bayesian theory of Probability', which is hitherto said to be the best attempt to solve the problem of induction. However, this theory too is not without problems. I end with the discussion that science is rational, in spite of inductive reasoning.

Inductive Reasoning – Essential, Yet Problematic!

Aristotle in his *Prior Analytics* and *Posterior Analytics* used both the inductive and deductive reasoning for scientific inquiry. From individual instances, indemonstrable first principles are inductively inferred, and from them by deduction scientific knowledge is obtained. In the modern times Francis Bacon, in his *Novum Organum* (1620), emphatically proposed that the chief method of scientific inquiry is induction. It seems, however, that there are proofs in the recorded history to show that humans used inductive reasoning even before Aristotle. For instance, an event narrated in the book of Judges of *the Holy Bible* (Chapter 6, Verses 36-40), written around 12th -11th century B.C. can be given as an example, where the method of agreement and difference of inductive reasoning is used. God chooses one Gideon to liberate the Israelites form the tyranny of the Midianites. He asks God for a proof that he will be able to deliver them. He keeps a fleece of wool out in the open ground during the night and asks God that only that fleece to be wet with dew while the ground around to be very dry. Next morning he finds exactly as he has asked. Next day, he asks God that the fleece to be dry and the ground around to be very wet and the following morning he finds it to be exactly that way. From these miraculous events he inferred that God would deliver the Israelites through him.[17]

Without involving an inductive inference from individual instances to generalizations, I think, the world can't be comprehended at all. Inductive reasoning is so common-sensical that no one doubts it in ordinary circumstances unless there is a specific reason to do so. Yet philosophers don't seem to be convinced of this. N.R. Hanson sounds rather strong in criticizing them for this. He argues that the actual problem with the inductive thinking "is not the traditional problem philosophers have supposed it to be. It is the problem of showing how 'The Inductive Principle'... can at once be synthetic (in that its negation is not demonstrably self-contradictory), yet necessary (in that its negation describes no intelligible state of affairs at all)"[18].

In spite of its wider utility ad practical success, it seems that there can't be any logical justification for the principle of induction. Hume argued that there was only psychological justification for the inductive reference. He put forth the problem of induction first in his *Treatise of Human Nature* (1739): "Let men be once fully persuaded of these two principles, that there is nothing in any object, considered in itself, which can afford us a reason for drawing a conclusion beyond it; and, that even after the observation of the frequent or constant conjunction of objects, we have no reason to draw any inference concerning any object beyond those of which we have had experience.[19]

It is widely assumed that science is the best example of rationality and that science is rational because of its method and that method is the method of induction. If induction is rejected it is like rejecting science itself. Therefore John Wettersten observes: "if inductivism cannot succeed, then there can be no adequate theory of rationality".[20] But *the crux of the problem* with induction is that to justify induction we need to involve inductive reasoning itself; it can be illustrated as follows: The principle of induction worked successfully on occasion n1, n2, n3....nX. Therefore the principle of induction always works. But the irony is that one can use the same inductive reasoning to show that it does not work by giving many examples where it does not work. For instance, one can argue that so far in the past 42 years my age has been below 42 and therefore, I will ever be below 42! Indurkhya gives another example: "1 is less than 1000, 2 is less than 1000..., 999 is less than 1000, therefore all numbers are less than 1000". So "one can inductively infer from them (the examples) that induction is not justified".[21] Thus inductive reasoning has been shown to be problematic in spite of its essential role in practical life and in natural sciences as well.

Solutions and their Limitations

To defend induction a number of avenues have been opened up in the past few decades and even now attempts are being made. But no solution seems to solve the problem of the inductive reasoning satisfactorily. As an alternative to inductive method Popper popularized the Hypothetico-deductive method (H-D method). But H-D method brought in its own problems: All the followers of the H-D method agree that some sort of background ideas, a set of hypotheses are necessary for science. For instance, Kuhn says scientific investigation presupposes a paradigm; according to Feyerabend it is a high-level background theory and Lakatos declares that scientific investigation presupposes a research programme. All these pre-suppositionists assume something to begin the scientific process with. But, as Shapere points out, don't tell us whence the hypothesis comes and that is the fundamental weakness of this method. Cohen and Nagel say that the antecedent hypothesis is just an assumption and for Popper it can be just out of an intuition, falling in the context of discovery. For Kuhn there is no positive reasoning to introduce a new paradigm, as the choice of the new paradigm is like a religious conversion experience. Thus the major difficulty of this method is that one does not have a clear idea regarding the 'whence' of the antecedent hypotheses.[22] Even with efforts to overcome these difficulties H-D method is still not satisfactory as a methodology of science.[23]

Further, David Stove (1986) claims that he has refuted Hume's problem of induction. He argues, by randomly choosing a fairly large number of samples at various time intervals and locations we can increase the confidence in the conclusions about the future phenomena that are inductively derived from these samples. But unfortunately that confidence depends on how we choose the samples. Samples are limited in space and time. It is nigh impossible to have samples from all the locations of the earth and it is unthinkable to think of samples of the future. Meanwhile Nature too has its say in changing certain parameters over the long period of duration. Bipin Indurkhya, however, objects to Stove's refutations and suggests a way out for the problem by making a distinction between *the cognitive* and *metaphysical notions of induction,* where the former can be solved but not the latter. To summarize Indurkhya's understanding, "... induction is a cognitive mechanism... The best we can do is to understand them fully, and make ourselves less vulnerable to the erroneous judgments caused by them... As far as the metaphysical dimension to the problem of justification of induction is concerned, it can no more be justified than we can justify the fact that we have two eyes".[24] One may take the uniformity of nature as the fundamental assumption for all the scientific activities and inductive reasoning is based on this assumption. This fact is said to be enough to explain the need and the validity of inductive reasoning. But this kind of explanation is not logically satisfactory as he tries to justify it in terms of an assumption.

As the efforts to solve the problem did not meet with much success, some philosophers made attempts *to dissolve* the problem rather than solving it. B. Russell (2002) for instance, seems to advice us that we live with the inductive reasoning since, any appeal to experience can neither prove not disprove inductive reasoning. Interestingly P. F. Strawson *dissolves* the whole problem of induction. The demand for justification of induction, he clarifies, arises due to a wrong expectation of getting a deductive support for an inductive reasoning. He goes even to the extent of calling it an absurdity to look for a deductive support for an inductive inference. If one is able to achieve that, then that induction ceases to be induction at all and turns out to be deduction! Therefore, "Inductive reasoning must be assessed, for soundness, by inductive standards".[25] He gives concession in asking for the inductive support for any particular belief. One could apply the inductive standards and find out whether that belief is well-supported or not. But if one demands for the justification of the very inductive reasoning itself, then one does not have any other set of standards to appeal to! It is as problematic and unanswerable as asking "Is the law legal?".[26] We may answer whether any given rule or practice is legal or not, but we don't have legal standards to appeal to, to answer the question whether the legal system as a whole itself is legal. Further, he makes a bold statement. Instead of finding out whether any inductive statement about the unobserved future has any support, he declares "any successful method of finding about the unobserved is necessarily justified by induction. This is an analytic proposition, the phrase 'successful method of finding things out which has no inductive support' is self-contradictory. Having or acquiring inductive support is a necessary condition of the success of a method".[27] Finally he says that the reason the process of induction works out very well is because our universe is such that it works well. Induction "is certainly a factual belief, not a necessary truth; a belief, one may say, about the constitution of the universe".[28]

Popper's Solution

In Popper's view, as discussed earlier in Chapter 1, the problems of demarcation and the problem of induction are the main problems of philosophy. When we solve the former, it becomes easier to solve the latter, for the latter is only an aspect of the first one. Popper's solution to induction has been criticized by many, like Salmon, Lakatos, Ayer, Putnam and so on. Alan Musgrave thinks that Popper's solution has been misunderstood more than been understood. Musgrave[29] tries to explain what that solution is, leaving the question of its acceptability aside. I would like to address the issue as I understand and make my comments here. Of course one must not fail to note that Popper includes dissolution or rejection of a problem too as a solution of the problem. All may not agree with this kind of solution though! Therefore, he seems to dissolve, not to solve, Hume's problem of induction. Popper finds problem with the traditional formulation of the inductive problem: "What is the justification for the belief that the future will resemble the past? What is the justification of so-called *inductive inferences?*" – He argues that this is a bad formulation as "there is simply no such logical entity as an inductive inference".[30]

There is a *crucial shortcoming of Falsificationism* apart from those that I discussed in Chapter II. Many important theories in science, and otherwise, are glaringly non-falsifiable. Grünbaum joins many others to show how 'deductive falsifiability' is too strong a demand for a hypothesis to be included in scientificity. It can be shown with an example as simple as this universal hypothesis: 'All men are mortal'. It implies that all men die at some time or the other, and no one lives for ever. But "To falsify this deductively by a test statement, we must produce at least one man who *never* dies. For no matter how long any Methusalah[31] lives, he may still die later on. Hence no basic observation statement can deductively falsify the old law that we are all mortal".[32] So universal claims like, 'All men are mortal' – down the ages has been confirmed whenever a man dies, but it has never been falsified! Thus, Falsification does seem to be *neither a solution nor the methodology of science,* and that in turn shows that the problem of induction too remains unsolved.

Now I proceed to another solution to the problem of induction, the Bayesian Theory, to see how far it succeeds in its claim.

Bayesian Theory of Reasoning – A Reasonable Solution!

Bayesian theory is an attempt to answer Hume's challenge of the validity of inductive reasoning in science. It was proposed by Thomas Bayes (1702-61), an English non-conformist clergyman, a renowned mathematician, a fellow of the Royal Society - the source of the great epoch-making theory of probability. His seminal work on probability was posthumously published in 1763, after two years of his death. *Richard Price discovered two unpublished essays among Bayes' papers which he forwarded to the Royal Society. Though it was not given importance immediately, it was independently discovered a few years later by the great French mathematician Laplace. He* proposed a theorem to justify the acceptance of the conclusions of probable arguments. Both in everyday life and in the world of science evaluating an uncertain claim in the light of available evidence is very crucial. The degree of certainty of the conclusion drawn by investigator is measured by the probability of the conclusion. He explains what he means by probability, in his *An Essay towards a problem in the doctrine of chances*, in *1763,* "a rational acceptance: The probability of any event is the ratio between the value at which an expectation depending on the happening of an event ought to be computed, and the value of the thing expected upon it's happening".[33]

As complete verification is not available, science has to be satisfied with degrees of certainty, which is a spectrum of probabilities, of which the state of being fully correct and fully wrong are the two ends. No scientific theory can escape the skeptic danger of being disproved in future, therefore the probabilistic character is indispensable for any scientific theory. Hence, in the words of Poincare, "all the sciences would only be unconscious applications of the calculus of probabilities. And if this calculus be condemned, then the whole of the sciences must also be condemned".[34]

Howson and Urbach[35] explain the Bayseian confirmation in simple terms. The relationship between the empirical data (evidence) and the hypothesis is crucial, because it helps in accepting or rejecting a hypothesis. A piece of evidence can be confirming or undermining or neutral to the hypothesis under the test. Let h be a hypothesis; e – a piece of evidence, P(h) – the prior probability of h, and P(h|e) – the posterior probability of h; e confirms h, if P(h|e) > P(h); and e disconfirms h, if P(h|e) < P(h); i.e. if the posterior probability is more than the prior probability, the evidence confirms the hypothesis and vice versa. When the confirmation is at maximum, P(e|h) = 1 and then h logically entails e; when the disconfirmation is at maximum, P(e|h) = 0 and then h is refuted. No further evidence will confirm it, unless the refuting evidence or part of background knowledge is revoked. In other words, according to the Bayesian theory of *inductive support* (i.e. the degree to which evidence e, supports h), is given by: S(h/e) = p(h/e) – p(h). If the inductive support is positive, e is confirming h, if negative, e is disconfirming h, and if zero, e is neutral. The authors, in fact, claim that the "Bayesian approach is the only one capable of representing faithfully the basic principles of scientific reasoning".[36]

There are some more advantages with Bayesian reasoning: a) It provides a solid foundation to the deterministic and statistical theories. It seems to satisfy our intuitive expectation too; b) The conclusion in the Bayesian approach is derived from the mean of the experimental sample alone, not the means of possible samples. Hence it is unaffected by the experimenter's arbitrary decision of stopping with some samples; c) The posterior distribution is very insensitive to the prior distribution and this insensitivity increases with sample size; d) Therefore two people starting from very different prior beliefs must converge in their posterior beliefs as evidence accumulates; thus the objection to Bayesianism as purely subjective is disproved; and e) The Bayesian approach is said to solve the Quine-Duhem problem, that is, the problem of distributing falsity between a theory and its auxiliary hypotheses, when a prediction is falsified.[37]

Due to the advantages mentioned above Bayesianism seems to be reasonable to many who prefer to rely on it. Salmon[38] lists several leading candidates, who are highly influenced by Bayesian reasoning. In the scheme of inductive logic, elaborated by *Carnap* (1950) in his *Logical Foundations of Probability,* we find a technical language which assigns a priori weights to all statements in that language, including all hypotheses. The concept of prior weights is similar to the Bayesian prior probability. Among the exponents of the frequency theory of probability *Reichenbach* (1949) explicitly invokes Bayes' theorem to talk about the probability of scientific hypotheses; *L. J. Savage* (1954) in his *The Foundations of Statistics*, which popularized Bayesian approach in the foundations of statistics, introduced the concept of 'personal probability', which is actually the subjective interpretation of probability. Prior probability is just a degree of prior belief in the hypothesis even before any convincing evidence is obtained; *Nelson Goodman* (1955), in his famous work *Facts, Fiction, Forecast*, clearly demonstrates that there is always more to confirmation than mere confirming instances. Though he has not made any obvious reference to Bayes theorem or its prior probability his solution to 'the new riddle of induction' can be read as an influence of Bayesian reasoning and finally *Popper* (1959) though he astutely denies the notion of confirmation and inductive reasoning, his well-cherished concept of *corroboration* in the theory of scientific methodology involves lots of factors similar to *the plausibility consideration* of Bayesian reasoning. The process of corroboration include features such as simplicity, content, and testability of hypotheses, and also the seriousness of the attempts made to falsify them by experiment.

There are, however, objections against the Bayesian theory of reasoning and the probabilists in general. Popper and Miller[39] have argued against the probabilists to show that the probabilistic calculus could in no way furnish logic of inductive inference. Further, science is never in need of any inductive principles and all the processes of the scientific inferences are only deductive and not inductive; and if at all there is any inductive support is available it is only for probabilistic counter-support (disconfirmation). They claim to have eradicated the inductive support of probability altogether in the scientific reasoning. Their position has created a considerable interest among the philosophers. Some, like Richard Jeffrey (1984), and I. J. Good (1984), have criticized it, while certain others, like Isaac Levi (1984), have come forward to make it better. (However, Ellery Eells argues that they have failed to show that all probabilistic support is purely deductive in nature. He shows that purely deductive support does not come in degrees; either the evidence fully guarantees the truth of the hypothesis (deductively implying it) or does not fully guarantee the truth of hypothesis (does not deductively imply it). So what Popper and Miller argue for is also implying inductive support.[40]

Bayesian models seem to be problematic in evaluating the probabilities, especially the prior probabilities. It involves lots of background information determining prior probabilities. This opens the door of subjectivism. There are attempts to answer this objection; by providing objective rules for calculating priors; e.g. if the realm of possibilities is finite and has m members, and if h is true in n of these, then the desired objective prior possibility (also called 'informationless prior probability) of h = n/m (like the dice throw, 1/6). However, when the possibility is *infinite*, uniform prior distribution can't be properly defined. There is arbitrariness implicit in the choice of the space of the equiprobable alternatives. The choice is bound to be arbitrary because it is supposed to be made in advance of any empirical information.

Bayes is blamed for making problematic assumptions in order to make his attempt successful: For instance, the unknown chance of an event occurring does not change over time or from place to place. (e.g. The chance of the sun rising in the East, on any particular day, is taken to be the same as its chance on any other day). In case this assumption is wrong then none of Bayes' calculations works out. No probable argument would generate any degree of certainty. Therefore Hume's deep-seated worries about probable reasoning seem to be still staring at us![41]

The *most basic objection* to the Bayesian approach is that it gives importance to subjective factors in the scientific appraisal, while objectivists like Popper, Lakatos, Fisher, Neyman, Pearson and many others, look for uncompromising objectivity in science. For instance, Lakatos claims that the states of mind (like one's beliefs, commitments, understanding, one's psychological approaches) don't affect the cognitive value of a scientific theory, which is actually "independent of the human mind, which creates it or understands it, its scientific value depends only on what objective support these conjectures have in facts".[42] Bayes does not seem to satisfy the demand for objectivity in the scientific theory. But how far objectivity is a reachable ideal in science is a difficult question to answer. So, it is true that there are strong opponents to Bayesian theorem, especially against its 'subjectivist' character. However, the admirers of Bayesian reasoning seem to suggest that the so-called 'objectivists' of scientific methodologies deliberately close their eyes towards the indispensable subjectivist elements in the scientific reasoning. That is why a leading proponent of Bayesianism, I.J. Good, argued persuasively that "the subjectivist (i.e. Bayesian) states his judgments, whereas the objectivist sweeps them under the carpet by calling assumptions knowledge, and he basks in the glorious objectivity of science".[43] Some element of subjectivity seems to be inevitable in any knowledge process, including defining the prior probabilities. Given the scenario, Howson is confident that one day all will become subjective Bayesians: "For long people rejected a purely regulatory role for deductive logic, as opposed to one offering substantive 'self-evident' premises'. Eventually they accepted it. Eventually we shall all be subjective Bayesians".[44] Therefore, in spite of some disadvantages, Bayesian solution seems to be a reasonable solution to inductive reasoning.

Science is Rational – In Spite of Induction

That there is no pure observation is almost evident to philosophers of science. Thereby, the very foundational claim of inductivism is shaken. This, in turn, sends out shock-waves to science, which is built upon inductivism! Scientific theories contain universal propositions. They cover infinite instances. All observations are theory-dependent. Even in very ordinary, commonsensical observational statements too, we assume some theories, without which those so-called direct statements will be impossible. So Popper asserts, "'Here is a glass of water' cannot be verified by any observational experience. The reason is that the universal terms which occur in their statement ('glass', 'water') are dispositional: they 'denote physical bodies which exhibit a certain *law-like behaviour*'".[45] However, in spite of all the difficulties with inductivism and the principle of falsificationism for him science is still rational and that rational element is assured by corroboration of theories. That is why he raises a question: "For if science does not advance rationally and critically, how can we hope that rational decisions will be made anywhere else?"[46]

Inductive Reasoning – A Species by Itself: In Inductive reasoning we need to leave a margin for any change in the future, for the simple reason that conclusiveness always eludes us. According to Mellor, "in inductive reasoning about Nature... conclusiveness always escapes us". Nevertheless, it need not make us nervous or skeptic towards the whole enterprise of inductive reasoning. For, he continues to record, "... although that (i.e. the possibility of rejecting the present inductive inferences) always could happen, we can still know that it won't: namely, by inductive inferences that have as much right to 'the complimentary title "reasoning"' as deductive ones, since they have as little chance in fact of going wrong".[47] Therefore, inductive reasoning seems to be *a species by itself.* All the efforts so far to solve the problem of inductive reasoning, are somehow to reduce induction to deduction. But inductive reasoning cannot be put in the deductive jackets. Now both scientists and philosophers seem to realize that it is a futile task to look for such a reduction. That is why Rudolf Carnap too reasons out, "It is generally acknowledged that induction is fundamentally different from deduction and that any prediction of a future event reached inductively on the basis of observed events can never have the certainty of a deductive conclusion".[48]

Inductive Risk – For Ever!: Hempel, who coined the phrase, "inductive risk," recognizes the risk only in accepting / rejecting a hypothesis, because no evidence can determine a hypothesis with certainty. Judgment of value lacks "all logical relevance to the proposed hypothesis since they can contribute neither to its support nor to its disconfirmation".[49] However, Heather Douglas says, inductive risk is present throughout science, not only in accepting / rejecting theories, but also in accepting *methodologies, data* and *interpretations*.[50] The traditional normative standard of science (i.e. 'value-free' account of science) is challenged now. Feminists (E.g. Rooney, 1992) have criticized the normative standard, by challenging the epistemic/non-epistemic distinction on which it rests. We have not been, at least so far, able to overcome the problem of induction, and at the same time inductive reasoning is so essential that we cannot get rid of it either. Highlighting the need for being satisfied with probable certainty of inductive reasoning, Poincare argues that though Newton's law has been upheld by numerous observations there is an element of chance also; one can never be sure that it will be always true in future as well; it is only highly improbable that the law will fail to work in future. Therefore, "From this point of view all the sciences would only be unconscious applications of the calculus of probabilities. And if this calculus be condemned, then the whole of the sciences must also be condemned".[51] He goes on further to demand that "Thus, in a multitude of circumstances *the physicist is often in the same position as the gambler who reckons up his chances*. Every time that he reasons by induction, he more or less consciously requires the calculus of probabilities".[52] Similarly, Bipin Indurkhya (1990) has suggested, as mentioned earlier, that the principle of induction has to be accepted as a brute fact that we have two eyes, about which we can't argue anything! As our eyes are equipped to see the light of only a certain wavelength and our ears to hear the sound of only a certain decibels, our mind has been wired by Nature (God!) that we reason out in an inductive way. Induction may be seen as a part of our neural wiring. No logical or deductive evidence is needed to support its working.

It seems knowledge cannot be grounded *only in reason* or *only in the senses*. The rationalists and the empiricists of modern age tend to replace one sort of authority (a sacred text or institution), with another (a human mental capacity). Both kinds of intellectual authoritarianism hold the mistaken opinion that knowledge can be full rooted in the sense experiences or reason alone, but such an approach is not conducive for science. However, one thing is very clear: There always seems to be a tension between empiricality and rationality (logic) in science. For, the empirical experience can't give universality, as it is always instantaneous, and on the other hand, rationality (logic) alone will not give any substance to our knowledge. If we were to rely only on empiricality even our ordinary life would be impossible. We need to transcend our immediate environment, to think about the past and to plan about future for which rationality is very essential.

Our knowledge includes tacit and personal components; "because all our knowledge is to some extent tacit, and no set of cognitive procedures can unambiguously grasp the meaning dependent upon empathic indwelling of the human subject".[53] The recognition of all these tacit and personal elements does not make knowledge relative but helps us become more critical and realistic in understanding the rationality in science. Human knowledge is always on the move. We can't be complacent with that, rather we need to subject that reasoning into criticism. Though science is the best available rational enterprise yet we must possess the critical attitude of Popper. It cautions us not to end up being naïve. An individual or a particular community, cannot be, and for Popper even *should not be*, totally rational and objective. However 'inter-subjective' testing enables us to address the subjective elements and biases to the extent of making science *(reasonably!)* rational. The insolubility of inductive reasoning, however, need not bother us, as long as that works out well, enabling us to have many scientific breakthroughs. At the same time, the mere pragmatic success of induction cannot serve as its justification. As Strawson clearly points out that to move, "...from the fact that it has worked hitherto to the conclusion that it will continue to work" *involves circularity.*[54]

With this background of thinking, induction, therefore, can't be lamented as a 'thorn in the flesh', nor ignored as signifying nothing. Hume has after all shown that inductive reasoning is fallible and our empirical claims don't have the certainty that one would usually expect science to have, but he "did not show that they (empirical beliefs) are unreasonable".[55] I am sure, therefore, *science can still be rational* (or reasonable?) *in spite of its reliance on induction.* Here I am even tempted to claim that science is rational not only in spite of, but even *because of* inductive reasoning. For, it is not enough for science to describe / explain the past and the present, but also to predict the future. Science is what it is today because of its predictive power. Induction makes the process of prediction rational (or reasonable!), by basing itself on the background information. Even if the predictions are deductively made from a universal law, one needs the assumption of the Principle of the Unity of Nature, which greatly involves inductive reasoning. The Bayesian approach, as we have seen, is one strong way to show how the inductive reasoning can be redeemed from skepticism and pessimism. If the predictive ability is denied, science loses its character. Thus, I am led to hold that inductive reasoning, in a way, is that which makes science rational (reasonable).

The inductive reasoning is so indispensable a part of our beliefs that we cannot give it up. Inductive reasoning, therefore, seems to be *a species by itself.* Given the nature of the world, where causality, stability and uniformity are present, it is not a mere chance or coincidence that things occur but the going together is necessitated by the very nature and capacities of things, and we have a reason for having assurance in inductive inference. Of course, this kind of pragmatic justification may not satisfy Hume, as his challenge was, not about its utility but logical justification. Induction may be seen as a part of our neural wiring. No logical or deductive evidence is needed to support its working.

Here it is clear that it is the community that has decided to give the definite validity to the inductive reasoning. Recognition of such non-epistemic values in science makes science fundamentally social, but not weak or irrational. Science is still the best enterprise in revealing nature to humanity. While admiring science humanity needs to be aware not to be carried away by the spirit of scientism, but to realize, as Newton did, that *we just play with a pebble or two on the seashore while the vast ocean of reality lies in front of us unexplored.* This realization will, no doubt, do a great deal of good not only to science but to human lives as well!

SECTION 2
THE NEED TO REVISE THE TRADITIONAL ACCOUNTS OF SCIENCE

Science is usually seen as an exemplar of rationality, so much so, as Plazowski puts it, "today rationality is not the ideal for science, on the contrary, science forms the ideal for rationality".[56] Its rationality is seen in terms of its progress, accumulation of more knowledge and obtaining a true picture of the world. It is even maintained that progress is available only in science. According to Popper, in science "errors are systematically criticized and fairly often, in time, corrected... in other fields there is change but rarely progress".[57] Rationality consists in the unbiased applications of the scientific method. However, as mentioned earlier in the Introduction, this rational model of science has been severely criticized by the second half of 20th century; with more knowledge about actual history of science, the rational model is shown to be highly wanting. The critiques of the rational model of science claim that the scientific changes and the theory choice can't be rationally explicable as there are lots of social, psychological and non-rational elements at work.

Due to such hurdles and critiques now there are efforts to revise the understanding of rationality. Here, in this second section, I present, in brief, a few significant alternative accounts to the traditional view of rationality.

Putnam's 'Integrated View of Rationality'

Putnam criticizes the major views of science, like the *verificationism* of the Logical Positivists and *anarchism* of

Feyerabend. Verification as the criterion of meaning is self-defeating because it is neither 'analytic' nor 'synthetic'. Incommensurability thesis of Kuhn and Feyerabend is also not acceptable, because if theories are totally incommensurable there is no way at all to meaningfully speak about them. It is ridiculous, for Putnam, "To tell us that Galileo had 'incommensurable' notions, and then to go on to describe them at length is totally incoherent".[58] He also rejects the *instrumentalistic* view of rationality, according to which only success matters. For the instrumentalists reason is equal to instrumental reason. It is true science is highly pragmatic and the success of meeting the ends is very important but we must not forget the importance about what ends we must look for. He rejects *majoritarianism* as well, for which rationality is the satisfaction of majority and the checkability by the public. It is possible for one to be rational even though one goes against the majority, as the expressions of rationality must also include "the ability to judge correctly in precisely those cases where one cannot hope to 'prove' things to the satisfaction of the majority".[59]

Putnam looks for *an integrated view of rationality.* Such a view will admit the interpenetration of facts and values. Criticizing the fact-value dichotomy, he argues that "every fact is value loaded and every one of our values loads some fact".[60] With an integrated approach to rationality one would easily realize that experience, reason and faith are also interconnected. One cannot speak of any one of them without any reference to the other two. For an adequate and refined notion of rationality we need to go beyond the traditional notions of objectivity, universality and necessity, without slipping into relativism. He sees, therefore, reason as both immanent and transcendent at the same time. The notion of reason presupposes a cultural context and vice versa: "The 'standards' accepted by a culture or a subculture, either explicitly or implicitly, cannot define what reason is, even in context, because they *presuppose* reason (reasonableness) for their interpretation. On the one hand, there is no notion of reasonableness at all *without* cultures, practices, procedures; on the other hand, the cultures… we inherit are not an algorithm to be slavishly followed. …Reason is, in this sense, both immanent (not to be found outside of concrete language games and institutions) and transcendent (a regulative idea that we use to criticize the conduct of *all* activities and institutions".[61]

Stephen Toulmin's Criticism about 'Over-reliance on Rationality'

Stephen Toulmin argues that the notion of classical rationality of the 17th century has done a great damage to humanity, as it has treated reasonableness as something distinct and separate from rationality. His work, *Return to Reason,* is a clarion call to humanity to come back to reason, from what he calls as *'over-reliance on rationality'*. For, somehow modernists have been obsessed with certainty, absolutes and stability. In the process of working for such ideals they have not paid due attention to reasonableness and this has deepened the wounds on human rationality. Among many other things, he also points out that scientific method can't be the absolute ideal of rationality, and science can't be taken to be absolutely value-free, which makes it objective. He maintains that the 17th century natural scientists had a dream of uniting the idea of rationality, necessity and certainty into a single mathematical package. The exact sciences had dreams for 'a Universal Method', 'a Perfect Language' (which would need no interpretation) and 'a Unitary System of Nature'; but unfortunately all these dreams had to face insurmountable attacks. Further, their efforts regrettably inflicted a deep wound on human reason as such, as their unification has left out the aspect of reasonableness. In order to heal that wound one has to re-establish the lost proper balance, as Toulmin attempts to do, "between Theory and Practice, Logic and Rhetoric, Rationality and Reasonableness"[62]; Further, "the future depends just as much on our ability to recapture the values of the sixteenth-century humanists and maintain the fragile balance between the refinement of our practical skills and the human interests they serve".[63]

The turn of 20th century has brought in lots of changes in the epistemological beliefs and metaphysical principles. Their thinking about 'thinking' has also been changed and they are forced to revise the very notion of rationality, which has been cherished for several centuries, at least from the times of Descartes. Along with the notion of rationality other aspects of necessity and certainty are also raised: "Intellectuals in the year 2000 – philosophers or social scientists, literary critics or economists – have inherited a family of problems about the idea of Rationality and its relations to those of necessity and certainty".[64] He argues that humans now live in unreasonable times where they are led to be unreasonable. He opines that rationality itself is the problem; they seem to cling on to it too blindly that they seem to defend their own unreasonableness.

We need to learn, Toulmin suggests, to live with uncertainties and ambiguities as in the pre-modern times. The post-modern intellectuals, while they are busy with the notion of rationality, seem to forget the 'complementary' idea of reasonableness and the balance between rationality and reasonableness is lost.[65] In the post-modern times people have rightly understood the need for inter-disciplinary approaches to gain knowledge. All the approaches increase in us our professional and intellectual capacities, but "the revival of more humane and reasonable ideas proceeds more slowly that it need do".[66]

The classical rationality involves theoretical, scientific, deductive account of rationality, while reasonableness has the narrative and rhetoric elements, which are essentially uncertain and contextualized. *Rationality is proof-oriented, while reasonableness is sensibility-oriented*! He believes in the power of reasonableness to restore the holistic welfare back to society. He earnestly appeals to the thinkers of our times: "Our first intellectual obligation is to abandon the Myth of stability that played so large a part in the Modern age: only thus can we heal the wounds inflicted on the Reason by the seventeenth-century obsession with Rationality, and give back to Reasonableness the equal treatment of which it as for so long deprived".[67]

Stephen Nathanson's 'Reasonable Form of Rationalism'

Humans as we are, endowed with a certain type of very valuable reasoning power that we can never settle with anything less than rationality. However, serious reflections upon our own reasoning capacity, the admission of the infinite life-situations force us, as

Stephen Nathanson seems to suggest, to have a re-think of our conception of rationality. Some seem to end up with highly negative conclusion of denying rationality altogether; for example, John Barth, as well as Philip Slater, seem to conclude that trying to be rational in the multidimensional human life-situation is a "stifling and alienating activity, that rationality is a destructive ideal"; this conclusion is, for Nathanson, far-fetched, 'shocking and implausible' and he strives for, what he calls, *reasonable form of rationalism.*[68] Reason held as the measure to demarcate between legitimate and illegitimate actions, expressions and decisions. But this kind of formalist view is under attack by the Postmodernists. The Post-structuralists, the Feminists, the De-constructivists also criticize the traditional notion of rationality for different reasons.

According to the traditional understanding the two important aspects that make a person rational are: the ability to have a 'rational deliberation' and 'the quality of the reasoning power'. However human life is a vast, complicated arena where sometimes it seems that *reasoning is irrational.* Nathanson lists out certain situations. Deliberation is irrational *when it is harmful:* certain situations demand a quick action, often to be taken trusting on one's instincts rather than spending on time in long thought (e.g. when caught in a burning building one can't spend a long time in deliberating all possible ways of escape. One has to act swiftly after a reasonable consideration of alternatives); *when it is unproductive*: when there is a range of choices available on a trivial matter. To go on contemplating all minute details of all the choices is actually unproductive; *when it is baseless*: if one is ignorant of the possible consequences of one's choice and if the result depends completely on matters of chance, deliberation will be fruitless; *where it is redundant*: deliberation is unproductive because already in the past one has carried on the deliberation in a similar situation and found a choice that has worked out well. Now again to deliberate would be unnecessary, unless the deliberation promises to lead to a better action, the action without deliberation is rational; *when it is inconsistent:* with other ideals like love, friends, human dignity, self-esteem etc. One needs to swiftly plunge into action without wasting time in deliberation or else one may endanger other great values, like spontaneously reaching out to a friend when she just slips on wet floor. In all these cases, the result of the actions provides the criterion of rationality. Acting in a given way depends on the awareness of the situation: "Part of being rational, then, involves being flexible and open to new information. New knowledge about one's situation may make deliberation necessary in cases where it would once have been pointless".[69]

Rationality of an action, some argue, can be determined by the result of the action; if the result is good, it is a rational action. But Nathanson argues that an action can be rational even if the result is bad. If someone gives a bottle of salt taking it to be sugar; I add it to coffee without realizing it; now the coffee is wasted and the result is obviously bad. But this does not make the action irrational. For, an action based on false belief need not be irrational, even if the result is unfortunate; it is enough that the agent is reasonable, with the available evidence at hand, to believe in the case, though it leads to an unexpected and harmful effect. "Our judgments about the rationality of action need to be based on the action's *foreseeable results*, in the light of evidence *available* to the *agent*".[70] Similarly, an action may lead to a good and desirable result, but it may be only a matter of chance that it turned out to be beneficial; this does not automatically make the agent rational.

McMohan's 'Collective Rationality'

We need to recognize the importance of the collective wisdom. The integrated view of rationality shows the need of *collective rationality* in science. Humans, highly social beings as we are, need to act cooperatively and reason collectively. Christopher McMahon (2001) proposes a *Principle of Collective Rationality* as a rational foundation of normal practices. Humans, social as we are, need to rely upon one another for a holistic welfare of the individuals as well as the society. Since no one is omniscient, we need to help one another mutually. The sincerity and the expertise will, no doubt, enrich every other person in the community.

The collective rationality assumes the utmost importance where there are controversies over border-line cases, which lack strict rules and regulations to guide the process. Then it is *quite rational* to rely on the collective wisdom of the experts, who are also sincere about their exercise. I insist upon their *expertise* and *sincerity*; if either one is lacking the project of arriving at a rational decision will naturally be jeopardized.[71] Borderline cases arise not only in the realm of moral philosophy but also in scientific researches. For example, i) as the contemporary science unfolds itself, it becomes amply clear that to accept something as observable in the case of unobservables we need the decision of the scientific community. ii) similarly even to justify the time and energy that is spent to decide upon the goals of science we need the consensus of the community. For instance, instrumentalists restrict science to the explanation and prediction of observational results, and realists insist that science should attempt to reveal the underlying causal structure of reality. Which one is to be accepted? What are the types of researches to be allowed to proceed, especially those which are very expensive and involve great risks? — All these questions need a collective effort to be settled. The study of the actual practice of science gives a mixed verdict – in some context, only the prediction and observational results are important (e.g. thermodynamics in the 19th century), and in other contexts it attempts to reveal the reality (molecular biology). As Ronald Giere puts it, "In any case, there is no longer much reason to think that these issues can fruitfully be pursued *a priori*. What we need are more empirical studies of science, at both the descriptive and theoretical levels".[72]

Jurgen Habermas' Notion of Consensus as a Form of Rationality

Habermas is one of the serious and constructive critics of rationality. He upholds rationality and he is against only glorifying it as a unique, abstract and ultimate principle that affects and guides every aspect of human existence. Since science is equated to rationality, to have the right picture of science one needs to look into the notion of rationality and vice versa. Here I make only a brief mention of his idea of consensus.

The Notion of Consensus

In the analyses of the notion of modernity and rationality, Habermas (*The Theory of Communicative Action*, 1981, and *The Philosophical Discourses of Modernity*, 1987) great importance to the notion of 'consensus'. He raises it to the level of shaping human rationality; it occupies the center stage in his analysis, because, as he claims, any understanding of rationality requires a prior understanding of consensus and vice versa. He proposes consensus 'as a form of rationality'. Consensus has the transcendental nature that influences the ideas of reason, rationality and reasonableness. He does not define what 'consensus' is, because the task of giving definition belongs to the realm of formal logic. He is more interested in exploring the 'whatness' and 'howness', rather than the 'thatness' of consensus.

The received notion of rationality, like Kantian or thc Posivitists' notion, is impracticable, though it is not wrong; it is not practical because it leaves out the important factor of consensus in the process of 'decision-making'. Consensus has both objective and subjective characteristics because the decision is made in the inter subject community. There is a kind of evolution in the arrival of the notion of rationality as we have it today. Human rationality is linked with their capacity for self-reflection, and this aspect is ignored by both positivists and metaphysicians. Without reflection there would not be any consensus; and this consensus is a consensus of human interests.

Consensus can be a priori or a posteriori: the former is shared intuitively by all in a community and this does not emerge from a discourse or analysis; while the latter is obtained by means of an analytical and empirical analyses, as a result of mutual trust and accord. Habermas treats the a priori consensus as transcendental consensus, and this has to be accepted by all the partners of the inter-locution. The acceptance of language presupposes intersubjective mutuality or reciprocity in understanding or shared knowledge, and mutual trust of one another.

Habermas elaborates the relationship between consensus and the fundamental human interests. These interests are seen under three categories: *technical, practical* and *metaphysical*. Among these interests it is more demanding to address the practical interests; for, empirical rationality or metaphysical reason are not adequate; neither any analytical or mechanical method, not speculative reason is sufficient to deal with the practical interests, because the practical dimension is very rich, vast and open. Each society is expected to construct its own special form of rationality. But since human societies are multidimensional, one has to find a common root, say language, to build social rationality. Language is the locus where the universality of consensus is expressed. Though Hempel does not go deep into the aspects of how language comes into existence, he projects language to be most expressive form of human nature.

An Appraisal of Habermas' Consensus

Habermas' notion of consensus is very attractive and pragmatic as it avoids the a-human and formal characters; it overcomes the rigidity and neutrality of science. It gives sufficient importance to the human factor in the decision-making-process. It gives equal importance to social reality and human idealism, avoiding any over-stress on society as the only reality. In spite of many such advantages and positive elements, Habermas' notion has some shortcomings, as Doan points out. Some of them are as follows:[73]

a) Habermas aims at providing absolute objectivity to consensus but he does not seem to succeed much in this ambitious project.

b) There is a little problem in his claim that consensus is the means and the aims of human praxis; all human practices are oriented towards consensus, and that a unity of theory and praxis is best seen in the act of consensus. He has not in fact settled the conflict between theory and praxis. He has not adequately explicated the inner relationship between theoretical and practical consensus. For example, the Kantian categorical imperative, 'you ought not to lie' does not mean 'you cannot lie' – one can lie and still claim that 'one ought not to lie' at the same time. Similarly, consensus cannot claim a transcendental status so long as it cannot resolve the contradiction between theory and praxis. Until this issue is reconciled it is premature to identify consensus with rationality and rationality with reason.

c) Whether consensus could be treated as rationality and whether rationality in the moral field could be equivalent to the scientific principles is still a moot point.

d) Even the most universal consensus cannot totally be identified with rationality as principle, or as Reason (Logos); i.e., the meaning of rationality is not simply identified with principles, at least in social sciences; and in turn, one has to take into consideration the dilemma that arises there: whether to take rationality as principle or only a form of pre-principle.

e) Habermas, in order to substantiate his thesis on consensus, has done a detailed investigation in various realms of language, psychology, psychoanalysis and so on. But he does not seem to have considered another important and vital area of inquiry, the analysis analyse the ontological state of human beings; he has not sufficiently dealt with the question of what or who a human person is.

Another basic and simple difficulty with the notion of consensus is that if consensus is based on the interests of the society, how to settle the conflicting interests when they exist, say the interests of socialism and capitatlism? Habermas has not provided the ideal conditions upon which consensus can be constructed. Consensus may not be always sufficient to decide upon rationality. Sometimes the claim of natural sciences may contradict the consensus of the community, like the Galileo case. History of science has plenty of such conflicting situations, with very serious consequences. What would be the guidelines and conditions to resolve such issues?

In Section Four of this Chapter, I am going to explore the aspect of 'Consensus' in a elaborate manner. I would argue that this consensus would be an essential component to understand the rational picture of science and, in fact, this consensus makes science *reasonable*, rather than rational. While I explore the notion of 'consensus' in my scheme of reasonableness in science, I would mention that in the name of consensus anything cannot go as there is a *hidden-hand of reality*, as we have the hidden-hand of the editor in a journal. Therefore, while appreciating all the advantageous elements of Habermas' notion of consensus, I would rather hesitate to raise it on a pedestal to make it a 'form of rationality' for the difficulties mentioned above.

Section 3
Reasonableness: An Integrated Account of Rationality in Science

Both the models, normative and descriptive, have their own problems. If logic and methodological rules are stressed too much the actual practice and the content of science seem to be at stake; on the other hand, if the latter are stressed, relativism looms large, threatening the very rational character of the scientific inquiry. After all the discussions of these models, the picture of science that emerges at the end of the day seems to be very gloomy. Both Verificationism and Falsificationism are problematic; the issue of theory-choice, the role of evidence in theory confirmation – all seem to be problematic. If so, is science only a house of an organized delusion and a systematic illusion? It need not be. For considering the enormous success of the scientists in the various fields of technology, as Henry Harris puts it, "it is difficult to escape the feeling that the scientist's view of the world must touch reality at least at some points".[74] True, science is fallible, but not all the scientific conclusions are bound to be replaced in future. For instance, in the fields of biology it would never be disproved that the blood of animals doesn't circulate. Though, as we have seen, most of the traditional aspects of rationality need to be modified, still science can be rational, or more correctly, reasonable. I attempt to develop an account of rationality, in terms of reasonableness, which transcends the constraints of traditional understanding of rationality. It will, I believe, look beyond mere logical consistency and justification.

Normally we tend to dwell at the two extremes of rationality of an argument – one consists of completely *irrational* ways like force or bribing, and the other is *purely rational* ways of logical deduction. We have been trained to think that 'logical deducibility' is *the* rationality; it is a burdensome legacy form Aristotle, Descartes and the like. We are made to gloriously forget that there can be, and in fact are, many layers in-between the two ends of the spectrum; some of them may be persuasion, reasonableness, use of common sense etc. When one attempts to critically look at rationality it does not amount to the total denial of rationality as such. At no point, reason can be done away with, because to condemn rationality involves a paradox or even self-contradiction because one has to assume some sort of reasoning to deny rationality! The notion of rationality is an interesting topic even today, as it attracts many debates; the defenders praise it to be the hallmark of human beings, while those the attackers see it as a source of varied ill-effects on human society. Traditionally rationality has been given a predominant role in human life; but down the centuries this view has been challenged by skepticism, romanticism, fideism, and irrationalism. All these critical approaches to rationality will take us to consider 'reasonable form of rationalism' seriously.

In this section, therefore, I elaborate on the notion of reasonableness to show how it is useful, and even necessary, to speak of reasonableness as part of the holistic understanding of rationality. I derive some inspiration from the works of Philip Kitcher (2000), Stephen Toulmin (2001) and the like to develop my account of reasonableness.

The Notion of Reasonableness

Reasonableness, perhaps like Rationality, poses many problems in having a clear definition. Various contexts of its usage seem to suggest many layers of meaning with nuanced differences. For instance, one way of understanding reasonableness is as follows. According to the online-free dictionary, the noun *reasonableness* refers to:[75] a) The state of having good sense and sound judgment; b) The quality of being prudent and sensible; c) The quality of being affordable (moderate in price); d) The quality of being moderate and avoiding extremes in expectations; and e) The quality of a claim or an argument being acceptable to a reasonable person. Similarly, the adjective *reasonable* refers to: a) The capacity of reasoning (a reasonable person); b) The state of being in accordance with reason or sound thinking (a reasonable solution); c) The state of being within the limits of common sense (arriving home at a reasonable hour); and d) Being not excessive or extreme, being fair (in prices, in emotions).

Deepti Gangavane echoes a similar understanding of reasonableness. Her analysis of the notion is as follows:[76]

a) Reasonableness is that feature of rationality which is not subsumed under the category of deducibility; it includes all those field dependent criteria, used for the evaluation of substantial arguments;

b) It is not that reasonableness is just the second best alternative for rationality, for the contexts where the strict deducibility is not possible. In fact, logic itself can be shown to be dependent on something else. For the norms of a system cannot be validated within the given system, they can be critically discussed only outside the framework of the given system;

c) Carnap distinguishes (in his article "Empiricism Semantics and Ontology", 1972), between the problems that are solved within the framework of a system and the problems about the framework itself. He makes a distinction between *internal questions* (which emerge from within a system, and answered within the system, either factually or logically) and *external questions* (which are about the existence or reality of the system). These external questions don't have cognitive content. There is no question of truth or falsity; it is about the acceptance of a certain kind of linguistic framework. But Carnap seems to belittle the significance of these questions as they lie beyond the limits of rationality. Deductive reasoning presupposes a very complex framework (as we pointed out in the previous section). Reasonableness can handle not only the internal questions but also the external ones, because it goes beyond the considerations of mere validity, but also other field-dependent criteria, which operate in that field;

d) Reasonableness is not bound by any particular criterion and therefore one can question the criterion / presupposition underlying its use; but it does not make the acceptance or rejection of the criterion arbitrary or subjective;

e) Reasonableness does not demand unconditional or uncritical surrender to the axioms / rules of a theory; one may interpret them differently within the boundaries of the theory; "Obviously disagreement and choice do not have to be characterized either as impossible or as necessarily irrational. There can be reasonable criticism and justification of different positions, and choice can be based on rational, i.e. reasonable grounds";[77]

f) It enables one to move differently, to the more radical levels of argumentation, which can't be given within the limits of rationality;

g) Unlike in the notion of restricted notion of the traditional rationality, in the wider perspective of reasonableness it is possible to compare the merits / demerits of the available theories;

h) The advantages and disadvantages of adopting alternatives can be critically considered; "On the background of such a comparative assessment of the alternatives, it is certainly possible to make a reasonable choice, if not a rational choice in the narrow sense of the term."[78];

i) The arguments used within the scope of reasonableness are all non-deductive. Therefore the conclusion is not entailed by the premises in a mechanical, deductive way. So one can accept the premises and deny the conclusion without landing into an explicit contradiction;

j) The non-compulsive nature of arguments gives an important role to the 'other', the receiver of the argument; the rational assent of the receiver becomes the real criterion on which the conclusion is said to be established. The assent of the receiver is not subjective / arbitrary, if it is based on reasons; or on the premises themselves if they are acceptable; or if the premises are not acceptable either the conclusion be rejected, of different grounds can be offered to ground conclusion;

k) Here it is necessary to listen to the new proposals, compare the reasons put forward and thus interaction is made possible; reasonableness includes concern for others; "reasonableness consists on the one hand, in grounding one's own claims in sound reasons and in being prepared to discuss these reasons with others, and on the other hand, in understanding the claims made by others, acknowledging the reasons on which they are based and also in being prepared to recognize the limitations of one's own approach in the light of the views of the others and to change them, if necessary;"[79]

l) Reasonableness as wider notion of rationality has a place for practical reason; *in scientific-instrumental rationality only empirical verifiability and logical demonstrability are taken to be important*, and therefore there is no room for values, morality, or aspirations. It helps to evaluate not only with respect to the utility and efficiency to attain certain ends, but also with respect to the normative standards of an individual or society; the goals can be evaluated on the basis of accepted value system of the society.

Actions, decisions, choices, even human beings and their attitudes are brought under the purview of rationality; here rationality is not a mere structural attribute to arguments, but it also involves "a concern and respect for the other, a willingness to participate in dialogues with the other etc."[80] As the modern world becomes more interactive and inter-disciplinary this dialogue becomes an essential dimension for science and scientists as well. Therefore science also needs the 'communicative virtues', like tolerance, patience, respect for pluralism, readiness to listen to others[81] and further 'virtues of reasonableness' like being objective, fallible, pragmatic and judicious.[82]

Gangavane speaks about the advantages of reasonableness from various other angles as well: At the *interpersonal* level, reasonableness appreciates and it is ready to listen to the positions, approaches, and systems of the other; At the *theoretical* level, it makes one very receptive towards the views of the others and not rejecting them just because different from one's own; At the p*ractical* domain it refers to the readiness to consider and respect a value system other than one's own, and so there is no room for dogmatism, intolerance, or reluctance and at the *personal* level, it enriches personality; it enables one to accept one's limitations, and this acceptance is necessary to go beyond the limitations.[83] *Thus reasonableness has more scope and depth than the restricted notion of rationality, going beyond mere deducibility.*

Confucian Notion of Reasonableness

Confucius tries to understand reasonableness in terms of human nature. Human nature is not so much determined as historically and socially constructed; it is known by virtue of various types of human activities, like solving problems, cherishing enjoyments, hoping for something in life, making contracts with nature, others and even with one self. Human nature is revealed in how she deals with her interests. So the forms of reasonableness can be found out in human activities and human relationship with regard to human interests. Reasonableness, therefore, is born and revised in history; it is incorporated in different forms of regulation such as customs, laws, rites, orders etc. It is the principle of social order and the essence that determines human nature. The fundamental features that make up the kernel of human reasonableness are grounded on human being themselves: "it is man that can make the way great, and not the way that can make man great".[84] Though Confucius invites us to be reasonable, he maintains a distinction between rationality and reasonableness. Reasonableness concerns itself mainly with human life-world and world-view, but it cannot be excluded from other forms of reason, i.e. rationality and reason as logos.

Since reasonableness is grounded on human beings, it changes according to human development at every stage. The new relationship of humans with nature, God and other fellow people require new principles, that is, new forms of Reasonableness. It accommodates priorities and degrees of importance in one's duties. Man's duty to humans in general is less than his duty to his parents; his love to a colleague is much less than his love for his best friend, and so on. In case he loves all people just like his wife, such an act is theoretically possible, and laudable (to Kant), but unreasonable to his wife and all members of his family. Actually, the so called 'caritas' or love for all equally, has been impossible for human beings, though, it may be the most ideal and divine law.

Various Forms of Reasonableness

Confucius captures the human forms of reasonableness in three aspects: anthropological, cosmological and theological orders. Doan makes a brief presentation of these forms of reasonableness:[85]

a) *Anthropological Order or dimension*: i) Reasonableness is the principle guiding or helping man to live up to human nature. It is also the measure (which is born in and from human consensus) to judge the degree of humanness and to differentiate man from beasts etc. Actually, this kind of reasonableness is formed after human nature itself; ii) Reasonableness expresses a common form or measure accepted by humankind or at least by the majority of a society. That means, what appears to be unacceptable could not be reasonable (though it may be rational); iii) Reasonableness means sometimes a dynamic force which can transform man into a better one (superior); iv) Reasonableness expresses a *medium* and a *means* in which and with which man lives harmoniously. Human co-existence, co-living are built on such a medium and with such a means; v) Reasonableness sometimes reveals the inner feeling which makes man at ease with himself, which he shares with others; and finally, vi) a new form of reasonableness born from the spirit of legalism. This reasonableness bears some resemblance to the Kantian rationality. Here, reasonableness means rational act conforming to the required rules.

b) *Cosmological Order or Dimension:* i) Reasonableness expresses the human act of conforming to external natural rules; ii) Reasonableness means a certain principle without which one cannot understand or attain the Cosmic Tao; and iii) Reasonableness makes man harmonious with nature. (Note that Confucius is always against any kind of domination, especially the kind of domination practiced by the rationality and capitalist).

c) *Theological Order or Dimension:* i) Reasonableness is seen in the act of respecting, obeying and following the Mandate of Heaven; ii) Reasonableness is seen in the human action of self-learning and correcting to become more noble and holy.

Characteristics and Functions of Reasonableness

Confucius enumerates the characteristics and functions of reasonableness as follows:[86]

a) It is quasi-universal, in the sense that it is not global but more regional. It is necessary but in such and such a world;

b) It is applicable or practicable. i.e., it closely connects with our concrete life;

c) It is acceptable; its acceptability can be tested by the members of a community;

d) It is open to change. Due to lack of absolute necessity and universality, because of its practicability and acceptability, and because of its inner existence in the human world, reasonableness is up to any possible change and transformation; for instance, a Chinese who lives in Western countries has to adopt to Western customs, and such an act is reasonable; if not he may be ridiculed as to being 'irrational';

e) It is dialectically self-developing; due to its flexibility, and by virtue of human self-reflection and self-correction, reasonableness is a self-changing principle form; and

f) It is self-conscious. Reasonableness reflects both individual and social consciousness; it reflects an harmonious way of solving the conflicts of individuals, between individual and individual, between individual and community, and between different communities;

g) It is changeable and adaptable.

Other Forms of Reasonableness

Three more logical consequences of the first three forms:[87]

a) Reasonableness expresses both the medium and the means in and with which human beings co-exist. Harmonious peaceful ideal society is the *medium* where we discover the expressive form of reasonableness; as the *means*, reasonableness takes the functions of an instrument. There is a difference between technical instrument and human instrument; the former is fully neutral, unchangeable, while human instrument varies;

b) Reasonableness expresses human feeling. To be sad with the people in distress, to be joyful with those in happiness, are what Confucius calls sympathy. One who lacks this sympathy will act unreasonably; great emphasis is given to sincerity; it is expressed in the forms of trusting, loving, understanding; Confucius compares it to Heaven;

c) Reasonableness is seen in the act of obeying or following the rules and laws; All rules, including golden rules, are purposively serving man; they are not beyond or superior to man. *The most important law is humanity itself.* So when we say that reasonablenes means acting in accordance with laws, it just means that in terms of humanity: "only the man of humanity knows how to love people and hate people"(*The Analects* 4:3);and "if you set your mind on humanity, you will be free from evil;"(*The Analects*, 4:4).[88]

Confucian notion of reasonableness is very deep and profound as he relates it with human nature. Only an ideal and cordial society can be the medium and means where reasonableness can be realized.

He makes it clear that being reasonable obviously includes obeying rules and regulations; but it has its priority clear to have humanity as the most important and fundamental law. One who is clear and convinced about this principle will never go wrong in any manner, which would affect himself or others or nature or God. In fact, this is what I am going to show that this reasonableness must be taken seriously in science as well, for the simple reason science is for humans and not vice versa.

Some Features of Reasonableness in Science

Reasonableness, I think, can be thought of at different levels – at the level of an individual person, or a thing / a phenomenon, or a claim or a belief or a method. Similarly the adjective 'reasonable' can also be applied to variety of entities, like one's capacity, attitude, arguments, explanations, beliefs, actions and so on. *Reasonableness in science* also can be seen at various levels. I like to look at some of the features of reasonableness in science under three categories, though the distinction need not be taken as water-tight compartments[89]. These are: *Reasonableness at the Personal Level of the Scientists* – Imagination, the Agent, Judgment and Intuition; *Reasonableness in Scientific Methodology* - Rejection of Zero-tolerance, An Enriched Notion of Objectivity, The Need for Skeptical Attitude, and The Need for Common Sense; and *Reasonableness in Scientific Practice* – The Consensus of the Scientific Community and Embracing Pragmatism. These features, as I attempt to show, constitute the notion of reasonableness in science. These features are, I argue, so important and constitutive of science that it is more meaningful to speak of *Reasonableness in Science* rather than Rationality in Science.[90]

Reasonableness at the Personal Level of Scientists

Imagination: Not only is it that imagination plays an important role in the understanding of rationality but also many authors, like Ruth Burne, consider *imagination itself to be rational.* Moving far away from the earlier position that imagination has nothing to do with rationality, and vice versa, they hold that the guiding principles of both the imaginative and rational thoughts are the same and they show that imaginative thoughts are more rational than what the cognitive scientists have so far thought, because "human rationality depends on the imagination of possibilities and the set of principles that guide the possibilities people think about when they reason also guide their imaginative thoughts".[91] It is our basic experience that our ability to imagine counterfactually has helped us, individually and as a society, to make big strides in different domains of life. Therefore a revised account of rationality (reasonableness) needs to include imagination as one of its important components. When scientists arrive at conclusions it is not just out of mindless repetition of the experiments or observations, rather, lots of imaginative explorations are involved in arriving at them. That is why, Einstein rightly declares, "Imagination is more important than knowledge".[92]

The Role of the Agent: A comprehensive understanding of rationality, I am convinced, must involve in a very essential way the *agent*. The role of the agent in the understanding of science is clearly brought out by Henry Harris. A scientist explores a hypothesis in his head first whether to proceed further or not. "Each scientist has his own mental store of facts, theories and associations, and a private set of value judgments about the relative importance of the different elements in that mental store"[93]. Some scientists do end up being more fortunate than the others, not just as a matter of luck, but the imaginative world of the former has a closer relationship to reality than that of the other. The understanding of rationality has to involve the context as well. It can't be merely a rule-oriented activity. Rules are made for humans and not vice versa.[94] When we give due importance to the agent and the context in which she acts, rules take the back burner. The classical model of rationality was guided by philosophical overtones and considerations. In the classical model the central emphasis is on the logical relations between evidence and belief; and the role of the agent is minimized or ignored; whereas the alternative model "will take the agent to be basic, and the way in which an agent deals with evidence in arriving at a belief to be determinative of the rationality of that belief for her".[95] This, in fact, may be a relativization of rational belief to individuals, but that this is *not the same* as relativization of the notion of rationality itself.

Judgment: The ability to make a judgment is central to our notion of rationality. However, our human capacity and potentialities are so very rich that they can't be limited by rules. It is true, established rules, methods and algorithms are essential to prevent going off the track of rationality, but at the same time it must not cripple human abilities. It should be rooted and yet must be able to reach out to new horizons; both the rootedness and the ability to reach out are the two eyes for human society; one without the other will result in a dangerously distorted and stagnant society. This is applicable to the notion of rationality also. To stress on the one at the expense of the other is our human tendency. To be completely free from the community is in fact destruction for humanity; it is like the liberation from the soil is not the real freedom for the tree, but its destruction.

Harold Brown maintains that an account of rationality necessarily acknowledges the critical role of *judgment*. He points out three important characteristics of judgment:[96] a) judgments are not made by following rules; b) judgments are fallible; and c) individuals, who have a sound knowledge of appropriate and relevant information, must make judgments. There are situations where we lack sufficient rules to determine our decision; it all depends on one's expertise to make decisions; when I lack expertise, I need to consult the experts, and it is rational to do that. How one exercises one's judgment is very important. The role of judgment in science is involved at various levels: in choosing a hypothesis for further investigation, as all the hypotheses may not be taken up for the test; in deciding upon the exact pursuit of the chosen hypothesis; in devising testable consequences; in choosing among many alternatives available principles of logic or the accepted scientific practice and above all in attributing the falsity to any of the hypotheses or the auxiliary hypotheses, as there is no logical and error-free way of doing it. Therefore, contrary to the demand of the traditional account there is no one rigorous method for science. According to Brown, "the accomplishments of science depend on the ability to exercise judgment, an ability that scientists develop as they learn and practice their craft".[97] We have already seen in Chapter I, while dealing with Kuhn's contributions, how 'judgment' plays an important role in 'theory-choice'.

The dispositional account of judgment is prior to 'logicality', because the chief characteristic of a reasonable person is the ability to judge, to distinguish between situations where a rational calculation in the narrow sense is called for *and* when it is not. She knows when not to figure out a thing in a particular rational way and when it is not reasonable to apply *analysis of logic, the strict rules of evidence, and the critique of informal fallacies* to a situation. The ability to judge enables the agent to have prudence and moderation even in the exercise of reason itself.[98] Further, one would generally understand that "rationality is a goal-oriented action"[99] and "the worthiness of goals can't be specified without some reference to the agent".[100] Goals involve an individual or a community; and this in turn involves a particular perspective. This perspective is not always explicit, but presupposed. Thus it is clear that an action is rational from a particular perspective. The *judgment of rationality* arises within purposes and contexts. When one claims to be rational, it is quite natural for us to ask which perspective one is speaking from. For, "there is no single perspective that is adequate for understanding the entire range of our judgment of rationality".[101] Therefore the notion of reasonableness in science has to have a significant place for judgment.

Intuition: Science has hitherto been seen as an enterprise strictly of reason and logic, where one normally does not find any respectable role for intuition. But now more studies in History and Philosophy of Science seem to suggest that intuition too has an important role in the scientific enterprise at various levels: in proposing hypotheses, in conceiving, executing and interpreting a scientific experiment and learning how to learn from our interactions with the nature around us. We do find, as I have elaborated elsewhere,[102] many scientists, especially among the quantum physicists, who seem to be convinced of the role of intuition and mysticism in science.

It is perhaps impossible to define exactly what intuition is, though we seem to have a first-hand experience of what it is. Richard Rorty identifies four important levels of meaning of intuition: a) Intuition simply means a hunch; a true belief but not justified; there is no process of inference involved in arriving at it; b) it means a piece of immediate knowledge of the truth of proposition. It is immediate because no inference is involved here too. Philosophers seem to show interest in this type of intuition because it claims to give knowledge, but the process of arriving at the truth is not known. No inference is given to justify; c) it also means immediate knowledge of a concept. Here one has the knowledge of a concept but unable to define it exactly and d) it can be seen as non-propositional knowledge of an entity. It includes universals and insensible particulars, which are essential requirements to have intuitive knowledge of some analytical truths. As a sub section, it may also refer to mystical or inexpressible intuitions of certain entities.[103] Science seems to show an interest in all these four types of intuition. Scientists do get some rare and valuable ideas just as a flash of lightning in their minds. For instance, Frederick Kekule is said to have stumbled upon the idea of the molecular structure of benzene in a dream – six carbon and six hydrogen atoms are positioned in a ring shape, as though a snake tries to sink its teeth into its own tail. Similarly, Dmitry Mendeleyev is said to have got the idea of the Periodic Table of the elements in a dream!

Modern science seems to realize very strongly that the traditional epistemologies, rationalism or empiricism, or even the combination of both, are not sufficient, to understand the reality around us. Science needs to transcend these two in order to include other valuable modes of knowing, like intuition, aesthetic perspectives and so on. As Richard Tarnas puts it: "We need to move beyond the very narrow empiricism and rationalism that were characteristic of the Enlightenment and still dominate mainstream science today. We need to draw on ... *the wider epistemologies of the heart.* We need ways of knowing that integrate the imagination, the intuition, the aesthetic sensibility, the revelatory or epiphanic capacity, the capacity for kinaesthetic knowing, and the capacity for loving".[104]

More reflections on the role of 'Intuition' are given in Chapter IV.

Reasonableness in Scientific Methodology

Rejection of Zero-Tolerance: Rule-oriented notion of rationality leads one to *zero-tolerance*, which generally implies that the rules should be followed no matter what. But an integrated view of rationality rejects it and invokes reasonable tolerance. Zero-tolerance is fine, but when we realize that human situations are indefinitely complex, the rule-makers could not have, humanly speaking, anticipated all types of happenings. A situation might arise where human concerns demand deliberate breaking of those rules. I can give a real-life example: A church in a village gathered thousands of devotees on the occasion of the annual festival. They were praying and making offerings in the church. As a part of the ritual, many people held lighted candles in their hands while praying. Due to the carelessness of a little boy, a young lady's sari caught fire. In an utter panic she was screaming; hundreds standing around were so stunned by that gory scene that they could not do anything to save that lady. My priest-friend who was just outside the church, on hearing the screams of the desperate lady, entered the church. Realizing the seriousness of the situation he managed to rush in, and pulled away the burning sari. She became unconscious due to that trauma. He removed the cloth from the sacred altar and covered her body with burn injuries.

If one holds the view of *zero-tolerance rationality*, one would not approve anyone (forcefully) undressing a woman in public. Society would not tolerate any man, especially a priest who has taken the vow of celibacy at that, to strip a woman, to expose her… or to use a piece of cloth used in sacred worship to clothe someone – but all these were blatantly ignored by that priest in the process of saving the lady's life. It was the most rational thing to do at that moment. The people not only overlooked his breaking those rules of society and religion, but also acclaimed him to be a brave, wise and virtuous man! In fact, it was the *reasonableness* of his action that was appreciated by them!

An Enriched Notion of Objectivity: One of the important demands of classical rationality is pure objectivity. History of science shows that pure objectivity is not possible; going a step further we say it is useless as well. Human beings, experimenters as they are, can't be purely objective. 'Pure objectivity' would imply complete detachment and this would discourage 'our motivation to seek truth, and rational action'. We always choose from the infinite data or options available, any choice would always involve some specific interest and purpose. An adequate ideal of rationality cannot focus on the results of an action in isolation; the agent's perspective must also be taken seriously in assessing actions and beliefs; it will have to acknowledge plurality of values and interests that are important to human life rather than limiting itself to cognitive ideals.

Objectivity may mean many things, as Elizabeth Lloyd points out: Objectivity means detached, disinterested, unbiased, impersonal; or publicly observable, available, accessible; or existing independently of the observers; or something really existing, really real.[105] Contrary to the common understanding, objectivity does not automatically rule out value orientations. Values are very important to both natural and social scientists with equal force. "Objectivity means the clear and unambiguous formulation of solutions proposed to emerging problems and their critical discussion"[106] and it does not necessarily mean that value considerations have not gone behind the process of arriving at those solutions. For Popper, objectivity is possible even at the level of community, for "Objectivity of science is not a matter for the individual scientist but rather the social result of mutual criticism".[107]

Nicholas C. Burbules[108] proposes to derive a notion of objectivity that is enriched by tolerance. He thus translates this notion from a formal decontextualized language and shows it to be concerned with personal character, practical context, and communicative relations. To be objective, one needs to be basically tolerant. This aspect of tolerance is not emphasized by the classical rationality. Being objective demands that one must not rush to conclusion; one must have the ability and patience to listen to the alternative ways and to recognize one's own biases, one's limits in appreciating fully the viewpoints of the others. A sense of pluralistic sentiment is needed; tolerance implies an openness to listen to all points of view, to consider the merits of each, without being prejudiced that one's own views are the best. To value tolerance and pluralism are methodological and moral edicts. Two extremes of people, where both fail to be reasonable, are: a) highly intelligent people, who have no self-critical attitude, to hear, consider alternatives; b) persons who uncritically accept everything. Therefore such a revised notion of objectivity is needed in science to make it reasonable, if not rational in the traditional sense of the terms.

The Need for Sensible Skepticism: Skepticism is an essential dimension of rationality in general. Rationality in science is no exception either. *Scientific thinking is critical thinking* and as Steven Schafersman[109] points out, it essentially involves three elements: a) empirical evidence, b) logical reasoning and c) skeptical attitude. Empirical evidence[110] is something repeatable and can be available to others as well. Others too can check the conclusions arrived at based on that evidence, for corrections or improvement or even rejection. This fallible nature is the undeniable characteristic of the empirical sciences. So, skeptical attitude is essential in science because of such tentativeness of beliefs. It demands openness to check whether one is deceived by self or by others. Skepticism does not buy any argument that easily and if they take a stand it is only provisional. It abhors authoritarianism and orthodoxy, subjecting everything to a severe scrutiny. It is a part of rationality in science. Without this, one may be blinded not to see the truth and be strengthened in one's own ignorance.

Nevertheless consistent skepticism is not possible; it is only a myth. Riniold and Nisbet argue that *no one is consistently skeptic but only selectively skeptic*. Skeptic attitude shows that we all want to be objective and unbiased. But "we all have limitations and built-in biases that hinder our ability to apply the methods of skepticism objectively and consistently".[111] Even hard core skeptics are prone to beliefs that are non-skeptically formed. Three biases are predominant among the skeptics: a) *Confirmation bias* – It refers to the tendency to look for evidence that is consistent with our beliefs; we look for supporting evidence with more vigour than the discrediting evidence; b) *Biased assimilation* – It refers to the bias in assimilating information into our belief system. We seek information that supports our belief system and we also assimilate that very fast and comfortably; and c) *Belief perseverance* – It is the tendency not to give up a long-cherished belief, though there is pressing evidence against them.[112]

Even a staunch skeptic does not always want to check all his beliefs thoroughly. A certain block of beliefs are kept aside from the skeptical inquiry. This selective skepticism is not only in practice, but even in principle. Selective skepticism is not just due to the lack of time and expertise to be consistent skepticism "but our minds have a variety of built-in biases that directly hinder nonselective skepticism".[113] For example, Einstein, a great mathematical genius, who revolutionized the way humans looked at the universe, was voted as one of the ten outstanding skeptics of the 20th century by the Fellows and Scientists (*Skeptical Inquirer*, Jan/Feb 2000). However even he did not seem to be equally skeptic or objective in his outlook outside science. He had a great desire and belief in political liberty. Yet he refused to criticize the Soviet Government and even justified the murders and creation of slaves' labour camps.[114] All these reflections go to show that *it is reasonable to be skeptic; and it is not only unreasonable but also impossible to be absolutely skeptic at all times, in all matters.*

The Need for Common Sense: A comprehensive account of rationality has to realize the role and importance of *common sense*. Rationality in science, in my opinion, can't be incompatible with common sense. Of course, it should not be confined to mere common sense but it should never negate or violate it either.[115] Common sense involves many assumptions. These assumptions, as Henry Harris shows, are constituents of rationality in science. He gives a list of such assumptions[116]: Scientists intuitively accept that there is a real world out there, existing independently of them; that there is a coherent structure in it which makes the interactions possible; that the world is knowable and the interactions reveal the nature of the world; that there is a steady progress in this knowledge and finally, that there is a distinction between facts and theories, though it is not often easy to say what a fact is. One may argue that these assumptions are so fundamental and undisputable that even the proponents of the rational model of science would also agree with them. However I still see the need of highlighting common sense as a feature of reasonableness so that the rationalists may be aware that not everything in science can have logical or rational justification. Many philosophers, like Howard Sankey, even hold that science starts with common sense and goes beyond that. Science never eliminates it, but illuminates it. One need not expect science to prove or approve common sense, as it has already been tested and approved by innumerable instances of day-to-day life. In fact, we would not have been what we are today but for the assumptions and beliefs given by common sense.[117] In order to have a holistic view of science and its rationality one needs, in my opinion, to be aware of the role of common sense.

Reasonableness in Scientific Practice

Embracing Pragmatism: Pragmatism is essential for science. It refers to an attitude of tolerating uncertainty, imperfection, and incompleteness at the existential conditions of human thought, value, and action; and yet it demands for persistence and flexibility in confronting such difficulties. The spirit of pragmatism enables us to be sensitive to the needs of context, variety of human needs and purposes. In some situations conventional responses fail and doubts prevail. Then it is reasonable to acquire appropriate skills and to look for experiences with similar situations and success-rate. The role of society, where failure is tolerated easily as a part of growth, is very important and the co-operation, constructive suggestions from others matter a lot.[118]

Further, the consideration of pragmatic factors gains importance in many areas of science, specifically in the context of 'theory choice'. The scientists seem to choose a particular theory when they believe it to be true or at least true to a certain extent. This epistemic concern is very important in theory choice. But given the challenges that come from the camp of anti-realists it is problematic to settle with the claim of corresponding-theory of truth. Theories are often accepted because they are 'workable' and there are chances of their making the existing theories better; here the choice does not involve any belief in the 'true' character of the theory. True, as empiricists claim, in many cases the theories can be chosen on the basis of observational statements whose truth-value can be inter-subjectively determined, there are indeed theories that are completely theoretical, as they themselves are not observable, nor are they able to give any deductively inferable consequences. These theories seem to be 'trans-empirical' and can never be established by data, as for instance, "No amount of evidence about tracks in cloud chambers and the like can deductively establish that those tracks are produced by 'trans-observational' electrons".[119] Along with the epistemic aspect, there are many pragmatic aspects like simplicity, better scope, more problem-solving ability, unity, and even aesthetic considerations. "Accepting in *this* sense involves a commitment" and in the case of commitments it is more appropriate to see whether they "are vindicated or not, rather than true or not".[120] Thus to embrace pragmatism makes science reasonable.

Consensus in the Scientific Community: Research frontiers in modern science come across serious situations where the experiments or empirical observations don't seem to arrive at unambiguous conclusions. Certain experiments need the consensus of the community to reach some meaningful and useful conclusions, without which those experiments are incomplete and unproductive. So we are forced to include the aspect of consensus in the holistic understanding of rationality (reasonableness) in science. Of all these features discussed above I tend to think the aspect of consensus plays a really very important role in modern science. So I elaborate the notion of consensus, with the help of a case study of Gravitational Wave Experiments, later in Section 4.

Reasonableness and Other Domains

In the recent times reasonableness seems to be given more importance in every knowledge seeking enterprise, especially in dialogues with others. Dialoguing with others is one of the best ways of clarifying and correcting, acquiring and transmitting one's knowledge. As the modern world becomes more interactive and inter-disciplinary, this dialogue becomes an essential dimension for science and scientists as well. I like to mention here a few such domains where reasonableness is gaining importance.

Reasonableness in Judicial Systems

The need to be reasonable is stressed among the judges. For instance, as Douglas A. Berman explains, after the *Sentencing Reform Act*, the federal judges in the USA are required to give away sentences which are both reasoned and reasonable. Thus they are reminded not just to follow the guideline-specified facts and calculation in order and to mechanically impose on sentences. It is true unnecessary sentence disparities are to be avoided. Though the legal systems want uniformity in punishing the same sort of crimes at different places, yet 'perfect sentencing uniformity' is an *illusory and non-achievable goal* and since humans are involved in it we need to make the "sentencing system humane and respectful to all the persons it affects".[121] As long as humans are unique and no two incidents (or crimes) can be exactly similar in every respect, there will be some sort of *uniqueness* attached to every case. It is desirable that every case is studied in itself before passing the verdict on it. Rules and laws need to be seen in their context. So it is rational to go by laws, but it is reasonable to be sensitive to the demands made by different situations and persons involved in interpreting and applying them.

Reasonableness in Environmental Protection

Ecological concern is yet another sphere where reasonableness is given great importance these days. There are authors[122] who explicate the need and relevance of reasonableness in the environmental pursuits as well. That would probably strike a balance between the ecological conservation and the developers of ultramodern world. No one denies the need to develop places and enhance the comforts for people. But in the name of developmental activities we incur great harm to the environment at large. We sadly fail in our serious duty of handing the planet over to the future generations, if not in an enriched manner, at least as good as we inherited from our ancestors. We are only the caretakers and not the owners of this planet. The need and the urgency to be ecologically concerned is very seriously felt today more than ever before. Various biotechnological advancements threaten not only the identity and the dignity of human beings, but also the very fabrics of ecology. The intrinsic value of nature seems to be very badly ignored, as the modern researches and technologies seem to be focused only on the economic considerations. For instance, those who are in favour of *Genetically Modified Food* (GMF) claim that this will eradicate hunger in the world. But we need to realize that the world, on the whole, produces food-grains much more than what is necessary for the whole population of the world, but still millions all over the world die of starvation every year. So it is not because of lack of food-grains, but because of many other serious reasons, like the unequal distribution of the resources and the opportunities, the landless poor farmers, exploitation of various sorts and so on, that we sadly encounter hunger deaths in the world today. *In fact as responsible citizens of the world and the members of the human family, we need to be ashamed if even one person, in any corner of the world happens to die of hunger.* Genetically modified crops cause irreversible damage to the land and the earth's biosphere. For instance, the toxic materials used in the plants to kill the pests or to eradicate weeds spoil the land in course of time. Further, the pest gets used to the toxic materials and they are not deterred by the pesticides, as it has been recently shown that the worms get immune to the bt-toxin. So the uncertainty of the full consequences and the risks of biotechnology far outweigh the benefits. Humanity needs to be reasonable in all its undertakings in the realm of biotechnology.

Reasonableness in Feminist Approach to Rationality

There are serious attempts to see rationality in a comprehensive manner, which would include a woman's perspective too. Feminist thinkers are quite vehement in criticizing the traditional view of rationality which they claim to be gender biased. For instance, Wendey Kohli points out that reason and reasoning need to be addressed historically within the context of gender differences; for centuries down history masculinity and reason have been treated as synonyms, making feminine inferior, and even subservient to masculine rationality. 'Feminine Notion of Rationality' is a vast area in itself that recently attracts more and more sympathizers.[123] Feminist authors look for a view of rationality which is neutral to gender differences. The society, which has been very easily biased to the thinking and interests of men, very easily stabilizes a notion of rationality, which is considered not only not neutral but also even against women.

The traditional understanding of rationality seems to have not paid attention to non-cognitive elements of thoughts and feeling. Formal, impersonal ways of expression in social interactions have sidelined women and children. Men and women, as the recent studies on human psychology and behaviourism seem to show, are certainly complementary to one another, in more than one sense of the term. The feminity in men and the masculinity in women need to be roped in to have a holisitic and healthy lives in a family and in the society. The rational mind of men and the intuitive mind of women need to be in collaboration with each other for the betterment of humanity. Very important decisions and policies are, by and large, finalized and executed by men. Women are systematically ignored in this whole process. If they are also taken into confidence and their views are also taken seriously one can certainly expect a better 'human touch' in all those policies. For instance, it is a painful irony to see in India that while there is a surplus of food-grain production (about 60 million tons), there are thousands of deaths of starvation every day. Due to a lack of an efficient way of storage, about 20% of it is eaten away by rats or destroyed by rains. The glaring mistake lies in the callous attitudes of beaurocrats and the politicians. The government machinery does not seem to be bothered about the starving people when such an enormous amount of food-grains are in excess. If women are involved more and more in the administration and decision-making process, I strongly believe, that with their intuitive and maternal touch will they not easily allow the wastage and the starvation deaths. With their in-born quality of dividing the stuff among the children lovingly and equally, they will certainly bring down, if not totally eradicate, the hunger deaths on the streets. Thus, the substance of our moral beliefs, social values and so on can be much more enriched, comprehensive, and holistic with the inclusion of women and children. Similarly the notion of rationality will be more human and enriched if it involves emotions, empathy and so on. I am convinced that when women who form half of the population are not properly recognized and encouraged to contribute to humanity in any opinion, society cannot have a holistic growth and development.

Reasonableness in Critical Thinking

'Critical thinking' is equated to rationality. As Howard Gabennesch argues, there is no pure, objective, and a-historical definition of critical thinking. It involves skills, worldviews and values; and a possible definition would be: "Critical thinking is the use of rational skills, worldviews and values to get as close as possible to the truth".[124] The *reasoning skills* are necessary but not sufficient for critical thinking; they need to be taught. Often we are not aware of our own worldviews in which we live, move and have our being; it is like the fish that is not aware of the water in which it lives. Unless these *worldviews* are taken seriously we are often ignorant of our ignorance; we mistake pieces of truth to be whole truth; we don't see the difference between something to be partly true and part of truth; we don't realize that partial truths are often more misleading than full lies. The awareness of worldviews teaches us that self-deception can be more problematic than deception by others. *Values* include the temperaments like the ability to say 'I don't know', even at the risk of appearing unintelligent, a sense of more integrity, humility and tolerance of uncertainty.[125] A reasonable ideal of rationality will have to attend to the results of action; the agent's perspective to be taken seriously in assessing actions and beliefs; will have to acknowledge plurality of values, interests and values and ideals that are important to human life rather than limiting itself to cognitive ideals.

Humans seem to be ill at ease with irrationality and want to be rational in all their undertakings so as to achieve their goals, whatever it might be. However since we are led, not only by rules of logic, but also by emotions and sensitivities, the attainment of rationality will ever be a matter of degree. As Russell points out, "Complete rationality is no doubt an unattainable ideal"[126], we need to be satisfied with being reasonable.

SECTION 4
CONSENSUS IN THE SCIENTIFIC COMMUNITY

In this section, I explain how the need for consensus among the scientific community is felt more, in my opinion, in the contemporary scientific research frontiers. For example, I take the field of cosmology and notion of observation to reiterate the need and the role of consensus. With the help of an analysis of the Gravitational Wave experiments I try to argue that it is reasonable to include the consensus of the scientific community in the holistic view of rationality in science.

The Need for Consensus in Science

Even by the beginning of the 20th century only the theoretical calculations and speculations were given more importance than the actual practice and content of science. For instance, Einstein was so convinced of his theoretical calculations, that when his student was excited about the successful confirmation of his theory of relativity by the English expedition in 1919, the unmoved Einstein, who had been always sure of the success of his theory, is said to have quipped, if the experiment were to disprove his theory, "Then I should have been sorry for the good Lord, for the theory is correct".[127] But the notion of rationality becomes bankrupt without due importance given to the actual practice of science over the years. Granted, science is rational because it is methodological; it is aiming at a goal. In speaking about goals, according to Hugh Lacey,[128] values, both epistemic and non-epistemic, enter. Once these values are there, it is inevitable to bring in society or the consensus there[129]. It was only with Kuhn's *Structure*, that the philosophers of science came to realize the need for considering the actual practice and experiments of science. The strict demand for testability for something to be rational is just one of the several elements that construct the edifice of rationality in science. In accepting a hypothesis many criteria are followed, among which there can be non-rational ones too. For instance, Quine and Ullian give the following as the desiderata:[130] a) conservatism, i.e. not to question the existing beliefs much; b) generality; c) simplicity; d) refutability and e) modesty. This list obviously includes non-rational desiderata like conservatism, simplicity and modesty.

As Harold Brown shows, a comprehensive understanding of rationality in science requires the social element: "for a belief based on judgment to be a rational one, it must be submitted to the community of those who share the relevant expertise for evaluation against their own judgments".[131] Thus he proposes a *consensus model of rationality*. He also says that in a revolutionary period the rules of normal science break down; then decision making is a socially mediated process, and not rule-dominated process. The social aspect of science seems to make science more complete and realistic. Science, being one of the forms of our social enterprise, science and its rationality cannot but be dissected from the social and cultural network.

Scientists and philosophers of science have so far spoken about the demarcation between science and non-science. Now, I think, the time has come to speak of such a demarcation within science itself, like *empirically based science* and *speculative science*. This is very true in the modern researches in fields like Astronomy and Cosmology. Speculation is very essential in these areas of science. Till relevant observations are made to confirm or disprove the theories, the community comes together to accept speculation as something meaningful, useful and rational; not only that, it is the community that is to decide what observation is. That is why, they even sometimes going against observation for something they hold to be rational.

Consensus in Cosmology

In Cosmology, the theory of Big Bang can be taken for example. Till today the general theory of relativity seems to be a well-accepted conceptual framework of the model of the universe as a whole and the Big Bang is the so far accepted cosmological model for the origin of the universe. Till the 20th century from the time of ancient Greek philosophy, the universe has been seen to be permanent, steady and stable; but Einstein argues with his general theory of relativity (1915) how the universe has an actual beginning at finite time (with the Big Bang) and will be destroyed at a specific time in future. He is not able to envisage a universe that lasts forever, with no beginning, and no end. So he introduces an ad-hoc term cosmological Constant A, to avoid such a dynamic picture of the universe.

But by 1929, Edwin Hubble empirically discovered the ever-expansion of the universe. He even came up with the theory, relating distance and velocity of the expansion. Einstein gave up his cosmological constant A, but Hubble himself was not at ease with the changing, expanding universe. So he spoke of 'apparent velocities' and tried for some other alternative explanations, which would preserve the static picture of the universe. By the late 1940s, different versions of *Steady State Theory* came up, as alternatives to *the Big Bang* theory. But the discovery of microwave background radiation by 1965, the accurate calculation of actual abundance of hydrogen and helium seems to support empirically the Big bang theory. Though Steady State theory was philosophically and aesthetically pleasing to many, they were forced to drop it in favour of the Big Bang theory.

Lots of speculative and theoretical constructs are used in the Big Bang. Many constructs are adopted for easiness of computation, like Friedmann solutions (with zero pressure and zero cosmological constant), Hawking-Penrose singularity theorems that actually indicate the big bang proper, thermal history of the universe from 10^{-10}seconds after the big bang till now, including the passage from a radiation to a matter-dominated age, chemical history of the universe, including the account of primordial nucleosynthesis, and so on.[132]

Though the general relativity can't give the complete answers to all the questions (like it is not able to satisfactorily explain the big bang singularity), so far it is the best available theory to deal adequately with three large empirical facts: the expansion of the universe (empirically discovered by Hubble in 1929), the abundance of chemical elements (75% hydrogen and helium 25% in the galaxies, the expected amount of lithium produced in the big bang, all were proved with empirical evidence in the past 2 or 3 decades) and the cosmic background radiation (discovered by Penzias and Wilson in 1965). Speculations even include the existence of many universes, causally and informationally isolated from ours. They all seem to be ineffable, beyond any science or empirical checking, as they are spatially, temporally and other-dimensionally multiple. Just to suit the currently available theories, the value of the Hubble Constant is proposed to be 30, though there are strong empirical evidences that indicate the value to be 65. Thus consensus is seen to have an important role in Cosmology.

Consensus in Observation

An elaborate discussion on the notion of observation is in order, I believe, because of its importance in the modern science. One of the burning issues in Philosophy of Science is also the concept of Observation. Many questions in Philosophy of Science can be either traced back to or derived from the issue of observation. Unlike many other philosophical issues, I think, this one has a very direct relevance to the breakthroughs of modern science. Hence modern science, especially the discoveries and advancements in the contemporary Particle Physics and Astronomy, have kept up the debates alive. Such debates seem to force us to change our very idea of observation.

In natural sciences, being empirical as they are, observation plays a very important role. However, almost all seem to agree now that the traditional understanding of observation has to be revised. The tendency to give ultimate importance to observation is questioned on two accounts: At times, an observation has to be ignored in favour of a theory (and this fact has a lot of meta-scientific importance!) and there is no 'pure' observation. Secondly, if observations themselves are theory-laden, the inductive reasoning (upon which the logical empiricists base their demand for an absolutist notion of rationality in science) is questionable, because induction already assumes on the purely observational, un-interpreted premises.

Further, this theory-ladenness is often seen as a threat to the objectivity and rationality of the process of testability of theories, and in turn a threat to science in general.[133]

I now take up an experiment from the modern Astrophysics, that is, *Gravitational Wave (GW) Experiments.*[134] It is an attempt to see whether Shapere's conditions for something to be an observation (or observable) are fulfilled in this experiment as well; or to see whether there is any need to modify or elaborate Shapere's conditions in the light of revelations of the GW experiments. With this analysis I also try to explicate how *the role of the scientific community* is very prominent in deciding on important concepts in science, and in this case, what an observation is. I argue that *a holistic understanding of rationality* has to essentially take the consensus of the scientific community into consideration. (More on the issue of observation in Chapter IV).

Gravitational Wave (GW) Experiments – A Case Study

What is a Gravitational Wave?

Gravitational wave is now understood to be a fluctuation in the curvature of space-time which propagates as a wave. GWs are similar to sound waves. As air can vibrate with the sound waves (like when a thing is dropped the sound waves are created and they travel through the air breaking the calm), whenever any (massive) object moves in the fabric of space-time it produces vibrations, known as gravitational waves. The existence of GWs was predicted by Einstein's theory of the general relativity. Einstein proposed that matter pulled on the fabric of space-time around it. That warping was called gravity. He also said the movement of matter, such as the rotation of a planet, would send out ripples across space and those ripples are known as gravitational waves. "Ripples in space-time are waves of gravity emitted by the movement of massive bodies, such as black holes or binary systems, in space. The movement of these massive bodies disturbs the 'fabric of space-time' around them, sending ripples of gravitational waves radiating outward"[135] and these ripples propagate at the speed of light. These GWs are too weak to be measured.[136] Measurable amounts of gravitational radiation are generated only when stars collide or explode.

The Need for Gravitational Wave Experiments

Many serious efforts are undertaken at the expense of billions of dollars to study the gravitational waves, in order to unravel the mysteries of the universe. Gravity is a fundamental interaction of nature. Detecting GWs is a key problem in modern physics. It has attracted a great deal of attention from both the theorists and the experimentalists. GWs experiments affect the theoretical foundations of modern physics. Gravitational waves, unlike electromagnetic radiation, are capable of giving information about astronomical sources. For example, a blackhole does not give out any observable radiation as a star does, but gravitational waves can be emitted when an object falls into a black hole or when two blackholes merge together. Unlike electromagnetic waves, GWs are not much scattered. They are produced in compact, dense, high-velocity regions of matter in the universe. It is expected that they carry information about the very early stages of the universe and its formation and that may lead to a great revolution in the understanding of cosmology. GWs can propagate while the universe was still opaque to light, they can reveal information about the structure of the universe. The universe became transparent to light by 380,000 years after the Big Bang. So one can't use the electromagnetic radiation to directly study the universe before that stage. GWs on the other hand, propagate directly to our detectors from the very beginning of time itself, carrying information about cosmic events on all scales throughout cosmic history.[137] The information that we would get from detecting and studying about GWs is unique; it may not be obtained otherwise; it can also give new revelations regarding the ideas about the existence, nature and behavior of black holes.[138] Experiments would enable us to detect GWs from *cosmic inflation* a period of accelerated expansion that began immediately after the Big Bang during which the volume of the universe increased by a factor up to 10^{80} times, in a tiny fraction of a second.[139]

Thus though it is the most challenging task to observe GWs, its success will have lots of benefits for both astronomy and fundamental physics. All these advantages justify the enormous amount of money and energy that the scientists invest in the experiments to observe GWs.

The Basic Mechanism of Observing GW

There are *many sources* for GWs and each source needs different methods of analysis and mathematical calculations. Each source has its own advantages and disadvantages in the process of detecting GWs. Similarly there are *various detectors* as well; the three broad categories of them are Bar Detectors (Weber's types), Ground-based Detectors and Space-based detectors. The precision and sensitivity in detecting GWs are steadily increasing. Each detector of course has its own limits and limitations, where the latter one is an improvement of the previous ones; thus ground-based detectors, like LIGO, have an edge over the bar detectors and the space-based ones, like LISA, perform better than the LIGO.

The basic mechanism behind the GW experiment can be given as follows:

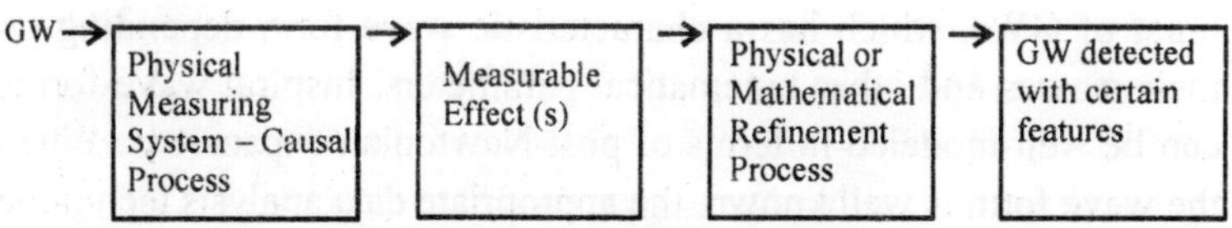

A Gravity Wave is sent through a carefully-designed physical apparatus, where GW produces some sort of effects as a result of some causal processes. These effects, which may be extremely minuscule in nature, are measured with appropriate devices. These devices produce some sort of pictorial, graphic or audible results, which are subjected to strenuous physical, mathematical or statistical refinement processes. Using different techniques of analyses the scientists come out with certain features which tell us (or prove!) that GW is observed.

Types of GW Sources

Several types of GW sources (which could be directly observed by Earth-based detectors):[140] a) Burst Sources – like binary systems consisting of neutron stars and/or blackholes in their in-spiral phase or merger phase; supernovae explosions; their signals last for a time much shorter (from milli-seconds to a few minutes) than the typical observation time; b) Stochastic background of radiation, either of primordial or astrophysical origin, and c) Continuous wave sources; e.g. rapidly rotating non-asymmetric neutron stars – where a weak sinusoidal signal is continuously emitted.

Of all these sources, *Burst sources* seem to be very convenient for observing GWs as they emit powerful bursts of GWs. A binary star system, the most common and the easily understandable source, consists of two stars, orbiting around their center of gravity, but slowly and steadily losing their energy and spiraling into one another. Some of the orbital energy that is lost is emitted as gravitational radiation, as suggested by the theory of general relativity. In the beginning, the period of rotation of such a system is quite large (may be hours or even days) and the GW frequency is too weak to be detected by the ground detectors. When the stars start spiraling the orbital period is reduced and there is a direct proportional increase in the frequency of GWs. Just before they merge completely, perhaps a few tens of seconds or so, they emit a very powerful burst of GWs, which has a characteristic wave form depending on their masses and other kinematical parameters. Inspiral wave-forms can be well-modeled in terms of post-Newtonian expansions. When the wave form is well known, the appropriate data analysis technique is 'matched filtering'. "It is the optimal strategy under the circumstances: it yields the maximum signal-to-noise ratio (SNR) among all linear filters, independent of the noise characteristics and secondly it is optimal in the Neyman-Pearson sense – if the noise is additive and Gaussian, it yields maximum detection probability for a given false-alarm rate".[141] The signal always comes as family of signals and not as a single one. Inspiraling binaries have been considered highly promising sources, not only because of the enormous GW energy they emit, but also because they are such 'clean' systems; towards the last stages of coalescence a gravitational wave is emitted, which has regular waveform, growing amplitude and frequency. The inspiral waveform can be calculated rather accurately, which is adequate for optimal signal extraction techniques such as matched filtering.

Types of Detectors of GWs

There are three types of detectors: Resonant Bars, Interferometric detectors (LIGO) and Laser Interferometric Space Antenna (LISA). The present scenario in the GW experiments has moved beyond Resonant Bars and LISA is yet to be launched (2011).

Bar Detectors (Resonant Bars): Resonant bars were first designed by Weber in the late 1950s. These detectors basically use the physical property of resonance.[142] Every object, including human bodies, has what is known as 'natural frequency'. If waves of any kind (electromagnetic, sound, or gravitational) pass through an object at its natural frequency, that object will vibrate more strongly at that frequency than any other frequency. Resonant detectors are constructed to vibrate strongly at specific natural frequencies that are in the predicted range for GW.[143] They make use of the change of distance between two test masses caused by GW in order to measure the strength of the wave. The detector is shaped as a cylindrical bar with heavy mass. When a GW passes through the bar perpendicular to its longititudinal direction, a change in spatial distance between the masses at the ends of the bar is measured. This change is felt as an applied force to the ends of the bar over the duration of the wave (an impulse). The force applied to the ends of the bar sets up a vibration throughout the bar, which will oscillate for some time after the wave has passed. The oscillation can be measured by accelerometer and the strength of the impulse, and thus the wave, can be ascertained.[144]

Using this basic principle, Weber hung a massive bar (as heavy as a ton or so) in a vacuum chamber and protected it from all known disturbing sources. If any wave passes through it there will be vibrations on the bar. He compared the residual vibrations in the bar with the vibrations in similar bars separated by thousands of kilometers. When the separated bars vibrated in coincidence it was shown to be evidence for the existence of GWs.[145] The sensitivity of the bar detectors has been steadily improving: in 1970, the sensitivity of the bar was h ~ 10^{-16} (h= gravitational wave strain amplitude) by the later 1990s, it was h ~ 10^{-18} and it could be further improved soon to 10^{-21}. More sensitivity is obtained by cooling the bar to low temperature (close to absolute zero) by reducing the thermal noise. This method was popular in 1960s and 70s. It is said to be the *first generation* of detectors. By 1975 it was argued that Weber was wrong in giving the vibrations of the bars as the evidence for GW existence. *The second generation* is similar to Weber's bars but they are cooled to liquid helium temperatures and this method was dominant till 1990s. *The third generation* detectors were using the interferometry technique, based on the famous Michelson-Morley experiment of 1887. This technique was proved to be more sensitive to the radiation and to "see" the pulses of radiation, not just their energy.

Laser Interferometer Gravitational-Wave Observatory (LIGO): Weber bars are not sensitive enough to detect anything but very powerful gravitational waves and therefore another technique, called Interferometers, is used. In the last decade several laser interferometric detector projects have been initiated: *LIGO* (Laser Interferometer Gravitational-Wave Observatory) in the USA (with two full length and one half length detectors), *GEO600* in Germany (with arm length of 600 meters), *TAMA300* (with arm length of 300 meters) in Japan, *VIRGO* in Italy-France (with arm length of three km on each side), and *AIGO* in Austria. "The goal of these instruments is to catch these waves in flesh and unravel the physics of the universe encoded in them".[146] Efforts are on to increase the sensitivity of the detectors. Currently LIGO is the most sensitive detector, which has crossed well below the mark of h ~ 10^{-21}.

Interferometric detectors to detect GW, use the same technique and the set up of Michelson-Morley's experiment. By the early 1970s, Robert Forward and others developed the idea of using interferometric detectors, which would make detailed measurements of GWs possible, including their waveform, their direction and the type and origin of the source. Later, Rainer Weiss, an American scientist, improvised further elaborations on the interferometer to use freely suspended masses at the ends of the arms of the apparatus to ensure a better noise resolution and the reduction of material constraints.

By 1992, 4-kilometer-long interferometric detectors were constructed, and collectively came to be known as the 'Laser Interferometric Gravitational Waves' (LIGO) project; after more than a decade great strides have been taken in the collection and the interpretation of the data from the interferometers. A greater sensitivity is achieved by effectively increasing the length of the path of the laser beam, which is reflected very many times through each arm. Ground-based interferometers, like LIGO, can detect even very minute motion – a small fraction of the width of an atom at the distance of four kilometers. Each arm has the mechanism, called Fabry-Perot cavities, to do the job of reflecting the beams. Two successful runs were made in August 2002, and February to April 2003. In collaboration with other observatories around the globe LIGO has been collecting data from 2005.

The instruments are sensitive in the bandwidth from few Hz (~ 10 Hz) to a few kHz, having the best sensitivity around 500 Hz. However the lower limits of a few Hz of the bandwidth of the ground-based detectors is a serious limitation to observing GW events. With the help of general relativity theory (applying to astronomy) it is learnt that the most predictable and most powerful sources of GW (such as super massive blackholes) emit GWs below 10 mHz; due to seismic activity, atmospheric disturbances, it is difficult to get sensitivity at low frequencies below 10 Hz. So the next option is to build a detector in space. Another technique to increase sensitivity of the GW antenna is to cool it near absolute zero. For, at this temperature the background noise is reduced very much.

LIGO's strategy is to split a laser beam down two long perpendicular tubes and look for differences caused by a gravity wave passing through the beams. In the plains of eastern Washington State and in a Louisiana forest lie two peculiar L-shaped buildings. Their skinny arms stretch 4 kilometers across the land, and inside each arm, a flashlight-sized laser beam bounces back and forth. The facilities, called LIGO, are feeling for tremors in the fabric of Space-time, called gravitational waves.

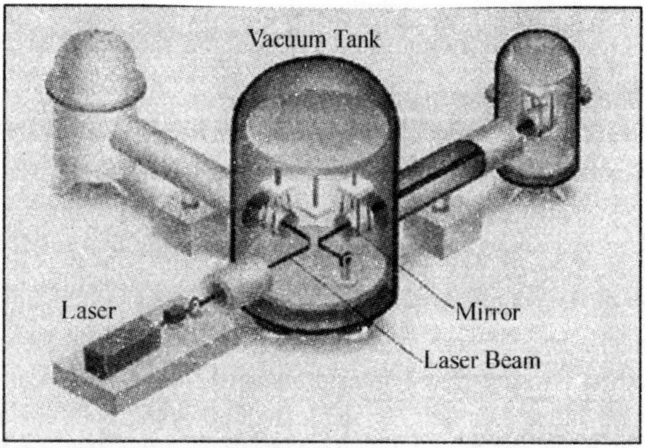

Schematic view of Laser Interferometrer Gravitational Wave Detector

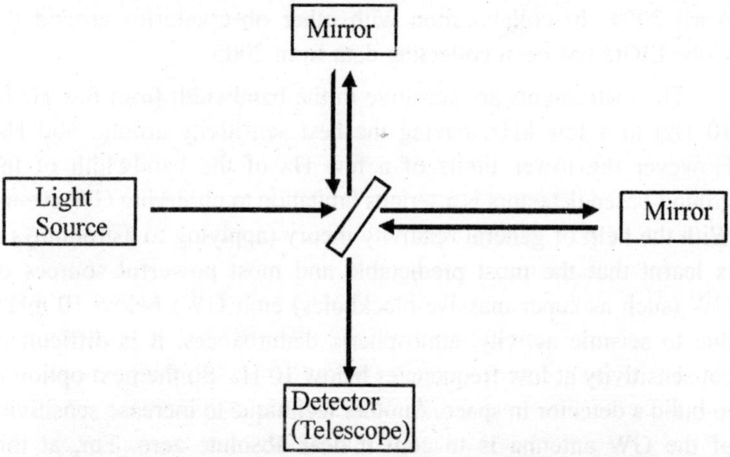

(The unnamed box at the center is the ***Half-silvered Mirror***).

The principle used in the laser interferometric detectors[147] is *the principle of an invariant speed of light* to calculate the change in relative spatial distance of two axes produced by a passing gravitational wave.

Interferometers pass laser light from a single source through a 90 degree beamsplitter; the light then travels in phase through two perpendicular arms, reflecting off of masses freely suspended at the end of the arms and recombining at the center again. As a GW passes through an interferometer the distances between the test masses and the centre of the apparatus vary; this changes the relative distance traveled by the two halves of the original laser beam. When the two halves recombine in the centre of the apparatus, their waveform will be out of phase with one another by an amount proportional to how far the mass distances changed. By destructive interference, the recombined laser beam will be less bright in proportion to how out of phase the two halves of the original beam are upon arrival. This can be calculated by a photodetector. The speed of light is invariant in any reference frame, the light travel time from the center of the apparatus to the end of the arm and back at any given time can be assumed to be $t = L/C + L/C = 2L/C$. The technology and knowledge of the Interferometric detectors need many other technologies and advancements as their background information: the technology of lasers, laser-optics, vacuum systems, seismic isolation, electronics and so on. However with all these advancements one has to have efficient data analysis to extract the GW signals from the background unwanted noise.

The following picture gives an idea of the long arm (2.5 miles) of the detectors:[148]

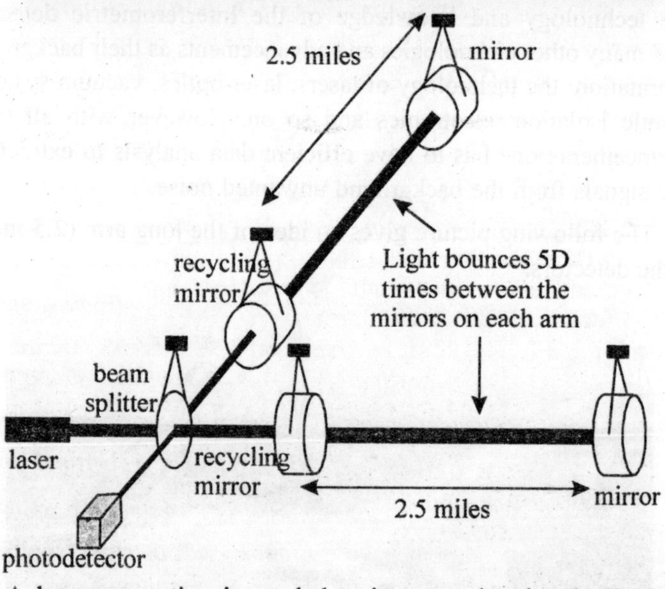

A large separation is needed to increase the signal. Therefore freely suspended test masses are placed at a very long distance separation. Though Weber himself had this idea, in the mid 1960s, of the interferometers he could not develop the configuration of such detectors for the lack of growth in laser and mirror technologies. Presently the designs range up to 3-4 km of armlengths, with sensitivities of h ~ 10^{-22}. They are used to obtain good isolation as low as 10 Hz. (e.g. LIGO) (Interferometers with armlengths range of 10 – 40 meters are able to provide sensitivities up to 10^{-20}).

Hulse and Taylor received the 1993 Nobel Prize for proving the existence of GW, basing on the observations of the binary pulsar PSR 1913 + 16, where they showed decay in the orbit, as exactly predicted by Einstein's general relativity: "These observations matched with the predictions of the theory of general relativity to better than 0.5%. This was the first strong indirect evidence which conclusively established the existence of GW".[149] But it was only an indirect and not a direct evidence of GW.

Shortcomings with Interferometric Detectors: Various noise sources infect the data received in the interferometric detectors. *At low frequencies* there is seismic noise and the data has to be segregated from the noise. (The seismic isolation acts as a filter, attenuating high frequencies but low frequencies get through). This results in 'noise well' at low frequencies around 10 Hz. *At mid-frequencies* (up to a few hundred Hz) the thermal noise is important and is due to the thermal excitations in the test masses, the mirrors and the seismic suspensions. *At high frequencies* the shot noise (which is due to the quantum nature of light) dominates. From photon counting statistics and the uncertainty principle we learn that long arm lengths, high laser power and extremely well-controlled laser stability are very important for reaching the required sensitivity (because the phase fluctuation is inversely proportional to the square of root of the mean number of photons arriving during a period of the wave). It is proposed to have all the five (may be including the sixth AIGO) together in a network, as it has more advantages than having a single detector; namely[150], a) we can improve on the confidence in detection by coincidence analysis, b) ascertain the location of GW source, and c) obtain polarization. Efforts are on to prepare detectors with aluminium cylinders (which are much larger than Weber's) which would weigh 360 kg and measure 3.5 m long, and would be at least 50 times more sensitive.

At lower frequencies ground-detectors are noisy, and therefore detectors in space are preferred. Ground-based detectors can operate at frequency ranges of GW of ~ 10 Hz to a few kHz, but not below

10 Hz. This 10 Hz is a natural limit because it is not feasible to enhance the arm-lengths on ground. Further it is not possible to avoid the gravity gradient noise below the level of 10Hz. The lower limit of a few Hz of the bandwidth of the ground-based detectors is a big limitation to observing GW events. This limit is due to our inability to avoid gravity gradient noise occurring due to seismic activity, atmospheric disturbances and also because of the technological limitations on long arm lengths due to earth's curvature. Thus the next option left is to build *detectors in space*. The ground-based detectors are not sensitive below this level of 10 Hz. They operate in the high-frequency band ~ $10 - 10^4$ Hz, while the space based ones operate in the low-frequency band $10^{-4} - 10^{-1}$ Hz.

Laser Interferometric Space Antenna (LISA): Till now, the radio signal detectors, on their way to various planetary targets, have been used to record GWs also. Only LISA has been specially designed with advanced system of laser interferometry and the finest measuring mechanisms, to detect GWs. Presently the European Space Agency (ESA) and the NASA (of USA) undertakes LISA (Laser Interferometric Space Antenna) aiming to observe low frequency gravitational radiation and it is to be completed by 2011. All these great works will lead us to have some glimpses of what is happening at the heart of the Milky Way. Burst sources are the most suitable sources for detection by LISA.

a) The Mechanism of LISA:[151] The goal of LISA is to detect and study low frequency astrophysical GWs in the band 10^{-4} Hz to 1 Hz. It could observe the sources like galactic binaries and coalescences, stochastic GW background from the early universe. These space detectors complement the ground-based ones. Since both have similar energy sensitivities their different observing frequency bands will provide crucial spectral information about the source. This is as important as complementing the optical and radio observations from the ground with observations in space at sub millimeter, infra red, ultra violet, X-ray and gamma-ray frequencies.[152]

LISA is a combination of three spacecrafts. It has three feely falling satellites, which maintain an equilateral triangle formation while orbiting the sun. It will be launched from a single Delta-IV rocket and inserted into an earth-escape trajectory. After about 13 months of its launch, the three spacecrafts will then separate from the rocket and from each other. Once separated, each spacecraft will be guided by an individual propulsion module to its own independent orbit around the sun. Each spacecraft's orbit will then revolve under gravitational forces only – not maintained by any onboard propulsion system – for the expected duration of the mission, about five years. All three spacecrafts need a large distance of separation among them; it is about three million miles (about 5 million km) from one another, which is about 1.3 million times greater than that of LIGO observatory arms. Each satellite will always maintain their orientation relative to one another and fire a laser at its companions. Each satellite will judge the relative distance shifts of the laser by using the interference pattern of the incoming lasers. Each spacecraft orbit is slightly elliptical and tilted a bit with respect to the other individual orbits and to the plane of the Earth's orbit, and thus it will maintain the triangle formation of LISA. The selected orbits will reduce the changes in the distances between spacecraft. Since laser interferometry measurements are more difficult to make if the distances between pairs of spacecrafts are not nearly equal, the maintenance of the distance among the spacecrafts is very crucial. Further, the distance between spacecraft was carefully chosen to allow LISA to observe in the frequency band of the most interesting sources. The centre of LISA's triangle will follow Earth's orbit around the sun, trailing 20 degrees behind, maintaining – like the Earth – an average distance of 1 AU (Astronomical Unit – the average distance between the earth and the Sun) from the Sun.

LISA's orbit is kept at 20 degrees behind that of the Earth to minimize the effects of the Earth's gravity and to tackle the solar disturbances. Similarly in order to reduce the effect of internal disturbances the spacecraft is controlled to follow the test masses with an accuracy of 10 nanometer, or about 1/100th of the wavelength of light. The spacecraft structure will be a sort of shield to protect the test masses; The maintenance of the distance between test masses floating freely is monitored by LISA's gravitational reference sensor because such distance is very crucial to avoid further disturbances. With all these measures LISA is expected to detect passing GW in low-frequency bands which were not hitherto available.

They form an equilateral triangle with sides of 5 X 10^6 km inclined at an angle of 60æ% with the earth's orbital. Each spacecraft houses two lasers (1 watt Yag) and each laser is phase locked to its companion and to the incoming light from the distant space-craft. The space-craft rotate in a circle drawn through the vertices of the triangle and the LISA constellation as a whole revolves around the Sun. **[See the diagram below]**[153]

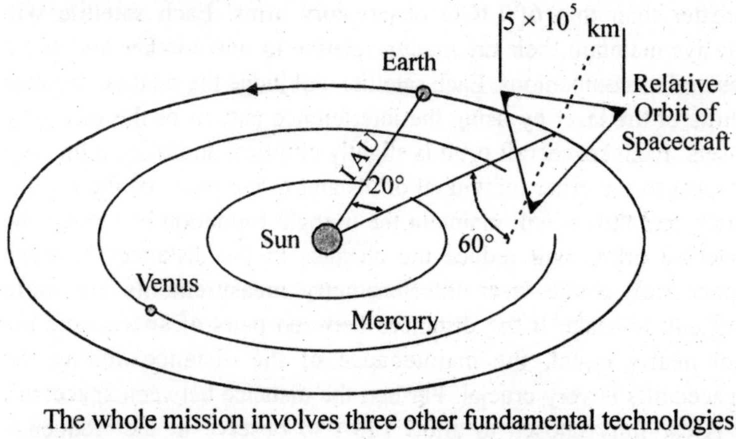

The whole mission involves three other fundamental technologies: *gravitational reference sensors, micronewton thrusters* and *laser interferometry.* All these are needed to fight against various disturbances that hinder the GW detection; "Because LISA detects gravitational waves by measuring the change in distance between freely floating test masses, sources of both external and internal disturbance need to be eliminated or damped to extremely low levels – much lower than what is needed by other missions. By minimizing such disturbances, motions that would imitate or mask the effect of traditional waves are less likely to occur".[154] LISA's position is carefully fixed so as to minimize the effects on the spacecraft of changes in the Earth's gravitational field and the need to be close enough to the Earth for easy communication. Moreover it is less expensive to construct the spacecraft and science instruments necessary to operate under the temperature at 1 AU. The triangle will rotate as the spacecraft orbit the Sun and this rotation is helpful in determining the direction of the sources of observed GWs.

b) The Main Problem for LISA: LISA has to eliminate the laser frequency noise found in the data. The noise in the data is due to i) frequency fluctuations of the lasers used in transmission and reception – laser frequency noise, ii) fluctuations due to non-inertial motion of the spacecraft, iii) beam-pointing fluctuations and shot noise. Eliminating such laser frequency noise in interferometers is very essential to obtain the required sensitivity. The laser noise found in the data received is much higher than the other noises. Since it is impossible to maintain equal distances between spacecrafts, laser noise elimination needs to be achieved by suitably combining the six beams with suitable time delays. Such processes have been proved to be feasible. The actual process can be easily understood in terms of properly defined time-delay operators that act on the one-way Doppler measurements. A rigorous formalism has been given involving the algebra of the time-delay operators, based on the theory of rings and modules and computational commutative algebra. This is the mathematical structure used in the time-delay interferometry in LISA. This methodology is suitable for a stationary LISA and efforts are on to extend it to a moving LISA with changing arm-lengths as well.[155] LISA's data analysis differs from the data-analysis for ground-based detectors in a very important way – i.e. because the observations will be at low frequencies, the computational cost, data storage is not an important issue here. Due to the fact that GW multitude scales inversely with the frequency for a given energy, the GW sources tend to be relatively strong at low frequencies. This in turn gives rise to a high signal-to-noise (SNR) for several GW sources. Maximized SNR in different scenarios, optimized directional sensitivity are some of the advantages with the LISA detectors.

c) LISA's Proposed Achievements: LISA will enable us to learn more about gravity, the source and the propagation of GW. It will probe the energy content at the very beginning of the universe, just after one second after the Big Bang. It will test the predictions of Einstein's theories to see whether they are correct, by looking into the unpredictable signals coming from a compact star spiraling into a black hole and by checking the structure of space and time around the black hole. It will check the correctness of Einstein's prediction that GW and light waves travel at the same speed, by comparing and contrasting the light and GW produced by the same source. Till now all our knowledge about gravity is obtained from the experiments within our solar system and we are led to believe that gravity is a universal force linking all forms of mass and energy in the universe. Some powerful sources emit GW at very low frequencies, as low as less than 10 mHz. Through LISA, by observing the low-frequency GW, one expects to learn about the universe beyond our solar system. [156]

Thus in the effort of GW detection the data analysis is very important. The data analysis differs according to the sources of GWs. The weaker the signal the more sophisticated method and strategy is required. Technical codes are developed based on the algorithms and they are being tested and adapted to the real data from the detectors. Regarding LISA getting rid of the laser frequency is a big challenge in order to achieve the goal sensitivity. No doubt, our theoretical understanding of GW is steadily increasing. When we know more about their sources the easier it becomes to learn about GWs.

Better detectors with greater sensitivity are being designed to study and learn about the hitherto hidden mysteries of the universe. Many scientists are so optimistic about the whole scenario that they declare "There is no doubt that the quest will be over, the only question is, when!"[157] Nevertheless all is not well. There are certain problems that seem to be invincible as we are going to see now.

Difficulties in Detecting and Interpreting Gravity Waves

GW has a specific feature of periodicity; it has a particular periodic fashion. Whereas unwanted noises from other various sources do not have any such periodicity; even if there is some periodicity it gets cancelled over a long period of time. In order to study the GWs we need to extract the periodic nature from the jumble of noisy data. Competing causal process works against the detection of GW, and thus giving false impression of the detection of GWs. We need to attenuate the effects of these competing causal processes. Following are some of the serious difficulties:

a) GWs are too weak to be measured, even if they are from powerful sources. Even the powerful sources like supernovae emit GWs which produces only minute changes (on the order of one part of 10^{-21} of an atom, and the diameter of an atom is in the order of 10^{-8} cm or 0.1nm; the minute changes are in the order of almost one thousandth of the 'diameter' of proton) in the apparatus. The strongest GWs that we can expect to observe on earth would be those that are generated by very far off and the earliest event in which a huge amount of energy moved very violently (like the collision of two neuron stars, the collision of two super massive black holes). These waves would create respective extremely miniature changes in distance everywhere on Earth, but it is highly burdensome to detect even the strongest GWs. To observe such changes from the earth our instrument will have to be competent to detect changes approximately 10^{11} times smaller than the size of a hydrogen atom. Thus the detective science of GWs indeed involves very precise and minute operations along with highly sensitive experimental instruments. Therefore some scientists even today doubt the existence of GWs.

b) Highly advanced algorithms and much complicated methods of statistical analyses are needed to study the data.

c) GWs have to compete against other causal processes, like background noise. For example, other sources of noise are: "The movement of the earth below the detector can produce random oscillations in the system, which are easily detected. Noise can also come from the detectors themselves; the Brownian motion of the particles in the detector creates random fluctuations that influence the result of the test. Even the pressure from the laser light in an interferometer can move the test masses by a detectable amount, changing the phase shift of the light".[158]

d) The GWs that are produced by our day-to-day objects, are extremely weak; the waves become significant only in systems, where the 'matter' is very massive, moving at significant fractions of the speed of light. To create GWs from such objects in our laboratories seems to be impossible now and at least in the near future. The appropriate sources to produce GWs that are strong and significant enough for detection are the astrophysical phenomena; those huge bodies moving at massive speed in the dark space up there can produce the GWs.

e) Above all *the crux of the problem* is: if all these extra noises are removed often the whole experimental set-up becomes inefficient to receive any signal at all. Therefore one has to trace out the patterns of GWs within the random noise of the received data. A mathematical function called a power-spectrum can be used to differentiate a periodic signal buried in a vast amount of random noise data recorded by the detector.

In spite of all these problems today's scientific community claims that GWs have been observed. This detection process and the results do always involve probability. The detection of GW, for that matter any entity, is basically a ratio between the *false alarm* and detection probability. For every millisecond there are some data recorded. For inspiraling binary sources, it is expected to have 40 detections in a year, and the false alarm is one in 40. So it is highly probable to observe GWs.

Elaboration of the Essential Requirements (Aspects) for Observation: I elaborate on the requirements for observation based on some of the insights that emerge from the case study of GWs experiments. These conditions are in addition to those three of Shapere, as mentioned above.

Causality in Observation: The principle of causality has to be accepted to a great extent to speak of observation meaningfully. For in most of the cases causal procedures are involved, and the effects are traced back to the causes. (For instance, in the GW experiments, after all, only the vibrations in the bars are observed, and this is attributed to the detection of GWs).

Selection of Certain Dimensions: Observation reveals information only regarding the dimensions under which we make the experiment. The apparatus of the experiment, and even the mathematical algorithms that are employed in the data analysis, play a role in restricting the dimensions of observation. For instance, in GW experiments, the calculations even with the use of highly advanced complicated algorithm, the digits involved run up to 10^{24}. Thus the data analysis has to be limited by the computing power available. Similarly, as per the predictions, there are 2 x 10^5 pulsars and there are about 10^8 neutron stars in our galaxy. So far we are looking for GWs from the sources less than one percent. We cannot have a 'blind' search for GWs from infinite directions, involving infinite time-periods. Thus the features of the present observation cannot be extended to the yet unknown or unobserved regions.[159]

The Reliability of the Background Information: An observation has to rely on the necessary background information (that has gone into the conception / execution / interpretation of the cxperiment), numerable mathematical techniques, statistical analyses and different calculus has to be assumed. The acceptability of the theories that are used in the making of the experimental set-up has to be assumed.[160] Without this the observation can't be put into any meaningful use. Observations are made with the help of instruments and every instrument is designed on the basis of certain principles and theories. They are assumed to be reliable and observations presuppose prior acceptance of those theories. In physics 'observation' is not that one makes, but one calculates, and advanced calculation at that, basing on complicated physical theories. For instance, in Wilson Cloud Chamber we just see a ray of light but we claim we observe a subatomic particle traveling through the chamber. Even a simple act of measuring the height of an object, say of the

Mount Everest as 29,002 feet, involves lots of theories behind; it is the height above sea level; and what is the sea level? "It is an arbitrary point selected for convenience as an average between high and low tides, which indeed varies from time to time during the year. So the height of the mountain is an arbitrary measure that depends upon a statistical smoothing out of figures measuring high tides and low tides".[161]

Observation and Its Interpretation: To be a meaningful observation it has to be interpreted, but it does not interpret itself. So observations always need an individual or a group to interpret them. In this process, the minds and beliefs of the interpreter(s) play an essential role. The very set-up established to make an observation is based on background information, convenience of the testable claims and the direction of the possible solutions. (Any question, for that matter, would not arise if the questioner does not know, at least vaguely, the possible directions from where the solution would come about!). It is quite natural for science to give more importance to the observable rather than the unobservable, but science must realize that *the demarcation of observable and unobservable is the result of a consensus of the scientific community.* To quote van Fraassen, "It is not irrational to commit oneself only to search for theories that are empirically adequate, ones whose models fit the observable phenomena, while recognizing which counts as an observable phenomenon is a function of what the epistemic community is (that *observable* is *observable-to-us*)".[162]

The Interaction between the Object and the Receptor: Observation involves an interaction between the object and the receptor. That interaction has to be such that some information is transmitted to the receptor, and at the end to a human observer. One needs further analyses to locate the elements of background knowledge in understanding the information coming from the object, and the auxiliary theories needed to interpret that information to make it sensible to the scientific community. The researches in Modern science seem to be so highly complicated that it is better to acknowledge the specificity in the notion of the understanding of observation to a given case at hand. From the interaction we infer some results. All the observations are in a way the results of some inference or the other down the line. The dividing line (of epistemic interests) between inferential or non-inferential observation "is drawn in terms of the distinction between that which we have specific reason to doubt (but which we are nevertheless still be able to use to a certain extent and for certain epistemic purposes) and that upon which we can build confidently".[163]

The Role of the Theoretical Context: To decide what an observation is, the theoretical context plays a very crucial role. Almost every time, the abilities of the scientific community and the context of the research, "will decide what should count as the right kind of evidence, and right kind of observation from which the evidence is constructed with the help of theoretical auxiliary assumptions".[164] For instance, when Galileo made telescopic observations he could not support them with adequate theories of optics. The opponents, not only the then religious authorities but also his own colleagues in science, were right in asking for a theory to justify the extension of the terrestrial theories to celestial spheres. Those advanced theories of optics were unfortunately not available for Galileo and they were brought out by Newton much later. This shows adequate theories were required to accept those telescopic observations as valid. The understanding of observation and the characteristics and the conditions of observation are not fixed once and for all. They are likely to be modified as science proceeds. More studies in the history of science show that the community decides what an unproblematic observation or data is. It may happen that what one community decides as evidence may be rejected by another community, and within a community the decision may change as time goes by.

Observation Cannot be Limited Only to Sense-Perceptual: Empiricists insist that all our knowledge must be shown to be ultimately based on sense perception. But unfortunately our sense perception covers only a limited section in the whole range of Electro Magnetic spectrum. (The EM spectrum consists of a wide range of rays and with vastly differing wavelengths. It ranges from short wavelength gamma rays through x-rays, Ultra Violet (UV) rays, visible light, Infra Red (IR) rays to long wavelength radio waves. Gamma rays (with extremely high frequency and wavelength as short as a billionth of wavelength of the visible light) are at the one end and at the other end are the radio waves, with trillion times higher wavelength of the visible light. "The eye", therefore, "comes to be regarded as a particular sort of electromagnetic receptor, capable of 'detecting' electromagnetic waves of the 'blue' to 'red' wavelengths, there being other sorts of receptors capable of detecting other ranges of that spectrum. This generalized notion of a receptor or detector thus includes the eye as one type".[165] Further, visibility or observability cannot be made a criterion to demarcate between objects of science as real or unreal. The summary of Maxwell's ideas presented by Hacking[166] can be given a scheme like the following: Seeing in vacuum à seeing in atmosphere à seeing through light microscope à seeing using a scanning electron microscope à and so on. The spectrum of observability can go on in future! "We can also come to observe", so Maxwell prophesizes, "other kinds of entities which were once theoretical".[167]

The Use of Language is an Essential Dimension of Observation: Without background information observation is not possible; and without communicating it to others there is not much of use in observations; therefore a language is an essential part of observation in making it meaningful. One may say observing an object may involve theories, but seeing a shape, 'I see a round shape', seems to be totally free of any theory; it is purely a sense-datum, private to the observer. But any observation, especially in science, must be intersubjectively open to inspection and it must be a part of the world. If it is not sharable then observation is not of much use in science. If it is totally possessed by an individual observer it may be incorrigible, but if it is available to more than one observer, it is possible that the other can see it differently, and so it is corrigible. So we have to settle with either theory-ladeness of observation or solipsism.[168]

To Detect the Existence of Something Some Sort of Interaction Has to be There: For, very often it is only the interactions that are measurable and not the thing (or the event/

process) in question as such. Scientists have often to work out some new algorithms of mathematical or statistical analysis and supplying imaginary constants in order to detect the existence of something.

The Notion of Observation Cannot Base Itself any Strict Demarcation between Observation and Non-Observation: Such demarcation is not eternal, not given *a priori*. It is a part of consensus. Even if any such distinction is granted it does not contribute anything significantly to the domain of science. That distinction will have nothing to do with the actual existence or non-existence of the theoretical entities. "Our drawing of the observational-theoretical line at any given point is," astutely concludes Maxwell, "an accident and a function of our physiological makeup, our current state of knowledge, and the instruments we happen to have available and, therefore, that it has no ontological significance whatever...".[169] The demarcation therefore is hazy and indefinite. It all depends on the technological advancements of the given time and not on any ontological grounds. For instance, the 'crobes', a theoretical term used to refer to the disease-causing micro-organisms, was thought to be too small to be observed. But with the invention of compound microscope those crobes were 'observed' in minute details and they were termed as *microbes*. Therefore to be observable, "there are no a priori or philosophical criteria for separating the observable from the unobservable" and the fact that "we can talk about the *possibility* of observing electrons without committing logical or conceptual blunders", shows that "any (nonlogical) term is a possible candidate for an observation term".[170] (One must leave the ends gloriously open, without defining what is observation and non-observation. Tomorrow it might change. How can one determine the ability of the future science to cross the limits and the limitations of the world of science?).

To Make an Observation an Artificial Order or Focus is Imposed on the Data: The world around us is always replete with physical data. But what data to select? Where to observe? and so on are determined with the questions that the observer in her mind. If one wants to study the cause(s) of light emission by certain fungi, one has to decide upon the relevant data to be collected. 'Is the time of the day relevant; the ambient temperature; the cycle of the moon; the presence of dog urine? In order to do the collection of data, it is already decided what causes can be comfortably put away from consideration; and this implies an order imposed by the observer on the data.[171]

The Notion of Observation Has to Necessarily Include an Element of Inference: Inference can't be ruled out in the process of any observation. Even to decide that a bright speck in the dark sky is a star, lots of background information is involved; as long as there is no specific reason to doubt, inference has to be accepted. All the observations are in a way the results of some inference or the other down the line. The dividing line (of epistemic interests) between inferential or non-inferential observation "is drawn in terms of the distinction between that which we have specific reason to doubt (but which we are nevertheless still be able to use to a certain extent and for certain epistemic purposes) and that upon which we can build confidently".[172]

Consensus in the Community: There must be a consensus among the scientists to consider something to be an observation (not only in using that information as a piece of evidence!). The consensus, even at the level of observation, is necessary because no one has any other mode of knowing the reality as it is! For instance, there is always a consensus among the scientists in general and about what an observation is in particular. As we saw, for example, in the GW experiments, a decision is made as to what is to be considered as a GW: "A gravity wave would be represented as, perhaps, a particularly high peak, but a *decision has to be made* (stress mine) about the threshold above which a peak counts as a gravity wave rather than unwanted 'noise'. However high the threshold it must be expected that occasionally a peak due entirely to noise would rise above it".[173] Therefore, consensus among the scientists over what is observation is as important as the role played by observation itself, for, "observation and formal techniques are not sufficient for dictating these decisions".[174] In turn, this observation is necessary but not sufficient for the whole of scientific process.

For, in choosing a theory and evaluating a hypothesis, scientists need more than strict rules and results of observation. As Harold Brown has it, "the process of evaluating a hypothesis requires multiple decisions that must be made by individual scientists or by an organized scientific community without benefit of formal rules... The reliability of science depends on the reliability of these decisions".[175]

From all these deliberations and considerations, one can easily see a powerful insight emerging quite emphatically. As commonly agreed upon, observation and experiments are very important pillars upon which the edifice of scientific rationality is built. But the experiment to observe gravitational waves has shown how the decision of scientific community is involved to decide what is, and is not, an observation. It emerges from the community. There is no *a priori* criterion for rationality, or a-historical epistemological standards, to be imposed upon science from outside the enterprise of science. Thus justification in science is essentially social and this is what is reflected by Kuhn as well.[176] Therefore, one has to consider the role of consensus of the scientific community in order to have a holistic view of rationality (reasonableness) in science.

Weber's Failure to Meet the Conditions of Observation: *Shapere's three conditions* for something to be observed / observable are: "a) information is received by appropriate receptor; b) that information is transmitted directly, without any interference, to the receptor from the entity"[177], and c) "The information is transformed by appropriate devices into humanly-accessible information which is (eventually) perceived (and used appropriately as information) by a human being"[178]. From the analysis of the GW experiments and the Requirements for making observation, I arrive at *the fourth condition*: d) The consensus of the scientific community is also required to consider something to be observation or not[179].

On applying these four conditions to Weber's experiments one may check as to which of these conditions his experiments could not satisfy.

Inappropriate Apparatus: Going against the first condition, the information was not received by an appropriate apparatus. The

Weber's apparatus was also at the rudimentary stage. That is, the detectors today are 1000 million times more sensitive, and in turn can detect GWs flux 1000 million times smaller than what Weber's bar could do.[180] Though Weber could subsequently improve the efficiency of his apparatus he could not reach the present day's efficiency.

Inefficient Interpretation of the Information: Contrary to the 3rd condition the results of Weber's experiments could not be efficiently transformed into humanly-accessible information. It may be due to the mechanical errors (not having efficient amplifiers, improper materials used in manufacturing the bars etc.) or the mathematical errors (like the variations used in signal processing techniques, the required complicated algorithms). Further, there was a lack of coherence with the well-established theories of that time. Weber found GW radiation too great to be accountable by the then-contemporary cosmological theories. The interpretation of GW by Weber seems to land us in a contradiction:[181] at the centre of our galaxy there are bursts of gravitational radiation that originate. A significant amount of matter needs to be destroyed to account for the energy of the gravitational radiation. Weber proposes a particular amount of matter that needs to be destroyed every year to account for the total amount of energy emitted due to radiation. But if his prediction were to be true, our galaxy would have already burned out!

Lack of Consensus: In the early 1970s Weber could detect above-threshold peaks simultaneously on two or more detectors separated by 1000 miles or more. This was convincing to some, but not to some others. For instance, a scientist wrote to Collins: "The chances of three detectors or four detectors going off together is very remote", while others found that the reason for simultaneous peaks were not strong enough to vouchsafe the existence of GW. The coincidence might be created by electronic manipulations or might occur by chance; for example, "(the bars) did not have independent electronics at all. ... There were some very important common contents to both signals. I said... no wonder you see coincidences. So all in all I wrote the whole thing off again".[182] These conflicting opinions clearly show the lack of the consensus among the scientists. In the following discussion I show how the consensus in the community is required to overcome the 'Experimenters' Regress'.

'Consensus' and the Experimenters' Regress

On the one hand experiment seems to be giving science the secure foundation, yet on the other, experimental results are obviously revisable, and sometimes even rejected. Though the experiment is objective (i.e. its repeatability) and observations are made with adequate care, yet the set-up itself becomes insufficient to answer the questions at hand. For, the experimental set-up itself poses its own restrictions in getting the results for the questions posed to nature. [183] Any experiment answers a specific question. What that specific question is, is decided by the practical and theoretical situation. The relevance and interpretation of experimental results always depend on the theoretical context. The experimental results are constantly revised, or sometimes even rejected, for various reasons: may be, sufficient precaution had not been taken to avoid possible sources of interference, or the measurements may demand new methods of detection, or the problem at hand needs a different experiment or the questions they intend to answer become irrelevant in the given context.[184] Therefore experiments don't seem to provide us with a secure foundation.

Further, in most of the experiments in science, especially in the modern science, 'Experimenters' Regress' assumes a greater importance. When the aim of an experiment is to determine the very existence of something, it becomes all the more crucial. The issue is: *a correct result of an experiment is one that occurs when the experimental apparatus is functioning properly* (and it uses proper method), *but we check proper function of the apparatus by whether or not the experiment returns the correct result.*[185] If the results are positive one can conclude that entity/field/force in question exists; if negative, it does not exist. But there can be methodological or mechanical or mathematical errors and because of which, the said entity is detected, or not detected. There seems to be no other way of checking it out independent of all these procedures.

In the case of GW experiments too, some thought Weber *could not detect* GW because of his ineffective methods or defective apparatus, or lack of perseverance or even due to bad luck, while others thought he *detected* GW precisely because of his faulty methods. To find the existence of GW, we need to build a good detector and do the experiments; but one cannot be sure whether one has built a good detector until one obtains the correct outcome. But one will not know what the correct outcome is until... and so on *ad infinitum.*[186] To break this circle the suitable range of outcomes is known by various theoretical calculations at the beginning of the experiments. Only in cases where such range is not known prior to the experiments, some other means of defining the quality of an experiment have to be designed and "the criterion must be independent of the output of the experiment itself".[187] Therefore it becomes obvious, as H. M. Collins argues, "Some 'non-scientific' tactics *must* be employed because the resources of experiment alone are insufficient".[188]

One of the best ways of breaking this experimenters' regress is to evolve a 'consensus among the scientists' over the criteria to decide on the range and characteristics of the expected outcome, and in the case of GW, whether the GW exist or not. This kind of 'closure' or 'resolving' mechanism does not count as 'scientific' in the usual sense of the term. However, without falling back upon them, it is unimaginable that controversial science would work. Very often neither the public nor the students of science are made aware of this face of science. It is very unfamiliar. However, it is necessary to be aware of this face of science as well in order to have a full and correct picture of science.[189]

Further, even the policies of a given government, the politicians' willingness to look beyond mere vote-bank, their educational backgrounds, and convictions, religious and otherwise – all these factors are important factors in allocating funds for scientific researches. Often massive lobbying, letter-writing and e-mail campaigns from the scientific community are needed to push their projects up for the required sanctions. Very often due to the lack of funds, rather the lack of willingness to allot such funds, the progress of many vital projects are hampered. For instance, LISA is to be launched by 2011, for which the preparatory works have been undertaken in a large scale. But now NASA is brought under heavy financial restriction such that it is highly doubtful whether it will carry on with the LISA project; about $545 million reduction is made in its budget from the Fiscal Year 2007. There is also a big hesitation on the part of the governments when it comes to the question of investing in non-defense scientific researches. Often they are unwilling to actually spend the money though they declare the big amount in the budgets. For example, after many years of persistent appeal and campaign for more federal science money from the Department of Energy's Office, the general scientific community and many other private industries, the government of George Bush allocated, in March 2006, rather a big money in the budget of the Fiscal Year 2007; but even after one full year the proposal was not only on paper.[190]

Many authors do criticize Collins for his allegedly hasty conclusion, which makes science merely a social construct. I highlight a few important ones.

Alan Chalmers argues that though experiment can't be an objective testing ground for scientific theories, yet science need not become totally relative. For, experimental results are determined by the nature of the physical world rather than by the theories entertained by the experimenters. In the case of GW experiments too, it was not that the stand on the existence of GW had already been taken and then the experiments were done; if that were the case, Collins would be right in accusing the experiments to be circular, but it was not the case. Granted, the experiments have many assumptions but not the ones that lead to circularity; there is some way of pitching against the real world out there. *The variability in the judgments need not be identified with the variability and uncertainty of scientific knowledge itself.* In fact, "what renders a scientific claim acceptable or utilizable is the extent to which it offers objective opportunities for future research or practical application... given the existing theoretical and technological resources".[191] Therefore, though the production of experimental results are not straightforward and yet one need not resort to an extreme situation, where social, political and extra-scientific factors are necessary for appreciating the epistemology of these situations.

Alan Franklin[192] also rejects Collins' conclusion (i.e. only social factors decide the outcome of scientific results!), because many scientific reasons can be given as to why the scientific community decided, by 1975, that Weber did not detect GW. It was not purely for non-scientific reasons that they decided. For instance, Weber used non-linear algorithm, while others used linear algorithm. Afterwards some others used non-linear algorithm too, yet they failed to detect GW; so the problem was with the apparatus and, as Weber himself agreed, the method of calculations. Thus based on these, it was reasonable to decide against Weber and there was no need for any social factors. Thus Franklin concludes: "Collin's claim that the experimenters' regress played a role [in the gravity wave experiments] is wrong. He conflates the difficulty of getting an experiment to work with the problem of demonstrating that it is working properly".[193]

Kenneth Ford and others criticize Collin's claim[194] that 'scientific theories rest on social agreement'. It is only a partial truth; for Collins seems to suggest that there is nothing of physical reality that affects scientific theories. There is certainly reality, "a hidden-hand editor" which is always there "to destroy errors, and science is unique among human activities in its sensitivity to these errors".[195] So, *experimenters' regress may be an important issue to be addressed but that does not cut the scientist completely away from reality.* In a way, science always remains open to correction and improvement. This does not make science the mere expression of collective whims and fancies of the scientific community. Scientific theories may be ever approximate but there are some in-built mechanisms in scientific methodologies for self-correction. The final outcome of science is under the grip of 'hidden-hand editor' of reality. "If Collins really believed what he says (i.e. the natural world has a small or non-existential role in the construction of scientific knowledge), he would not fly on aircraft for fear of a sudden change of (social) opinion about the possibility of flight.

However, aviators need not watch the opinion polls, because some truths rise above social opinion".[196]

Ian Hacking[197] also echoes the same idea when he agrees that experimental results do depend on theory, but experiments do involve actually acting upon the world out there, not just on theorizing about it. *Though experiment set-up depends on theory, once it is set up, the world decides the outcome of the experiment. Given the nature of the world the results of the experiment will be one way or the other.* Since the nature and the functioning of the world decide the experimental outcome, we can overcome (unreasonable!) skeptical relativism, though science can still be fallible.

Consensus – Making Rationality Social, Yet not 'Irrational'

The general accusation against Collins is that he lands us in relativism because he demands that sociologists need to explain the decisions of the scientific community using only social factors. But when we closely analyze Collin's position, he himself accepts rationality of the scientific practice. Collins' reply to the criticism of Franklin is quite revealing; for he says, "I never said scientists' actions were unreasonable".[198] Collins also always holds scientific knowledge in very high esteem. The social character of science need not minimize or remove the importance or the worth of scientific knowledge. He declares, "The social context of science is not thought of as polluting since all human activity is social; for those who work in SSK (the Sociology of Scientific Knowledge), scientific truth remains the paradigm of truth in respect of the natural world".[199] Reasons do play a role but that is not sufficient and therefore extra-scientific factors are also needed for the decision. Further, extra-scientific factors are also reasonable in some sense. Thus, Collins shows that *reasonableness is a social category, not derived from Physics.*[200]

Iseda argues that this way of understanding social factors as reasonable is possible, only when we understand 'social' in a new sense. Therefore, Collins is not a relativist of the traditional sense. Latour and Woolgar (1979) and Knorr-Cetina (1983) understand Bloor's 'social factors' in a new sense; that is, science is a social activity, because the scientific community with its own structures, norms, decision-processes creates scientific knowledge. In this sense science can be fully a rational activity; thus the shift in the meaning of 'social' makes it very rational. But not all sociologists are happy with this idea.[201]

Social constructivists of (scientific) knowledge claim that knowledge is merely a social construct. But many have argued against this; Stephen Cole (1992) raises two fundamental objections: a) the proponents of social constructivism of scientific knowledge have not been able to establish any linkage between social processes and particular outcomes of scientific research (e.g. $E= mc^2$). It may be granted that the activities of scientists are affected by social variables, but the constructivists have not clearly showed how these variables can have an impact on the final outcome of knowledge claims; and b) Constructivists fail to give convincing arguments why some ideas win over the others in the competition for the collective acceptance by the scientific community.

From these entire considerations one may conclude that *science is indeed a social and collective activity. Therefore, any account of rationality that underlies science and is sustained in science should have some institutional basis, which is analyzable in sociological terms.* However, social factors are not all that is there to decide the rationality of science, as there is a 'hidden-hand editor' of reality. *There is a reality out there, no matter how limited our understanding of it is, to influence the scientific theories.* For instance, if the whole humanity decides on one fine morning that hereafter in the electromagnetic spectrum the visible range of our human eyes will be higher or lower than the present range (0.4 to 0.7micro metre), will we be able to 'see' the infra red or the ultraviolet regions, with unaided eyes? Can we change the reality just by our social consensus? If at all one agrees that there is something special and peculiar about scientific knowledge it means that science has a better way of controlling nature than any other human enterprise and sociologists of scientific knowledge need to take up this issue to find out the appropriate sociological explanation to this.[202] *Thus, the consensus among the scientific community and the social factors that influence science, make the notion of rationality (reasonableness!) definitely social, but certainly not irrational, nor a-rational.*

Finally, the picture of science that emerges from such analyses is clearly fallible and anti-foundationalist, and yet he is epistemically optimistic. Just because foundationalism is rejected and because nothing is accepted as *a priori*, it is not the case that 'anything goes' in science. The insistence of going beyond the constraints of traditional understanding of science, and the recognition of social and historical elements in science must not lead one to think that rationality is denied altogether, for instance, as Feyerabend did, for whom as all methods fail, and "What remains (in science) are aesthetic judgments, judgments of taste, metaphysical prejudices, religious desires, in short, *what remains are our subjective wishes*...".[203]

Though the present knowledge may be questioned or even rejected in future, yet this mere possibility of doubt, as Shapere has astutely argued, is not a reason to doubt them now. Thus, Shapere argues against skepticism, relativism and any other threat to a sane, balanced, reasonable view of science and its progress. Like Shapere, Putnam too upholds the rational picture of science, in spite of all shortcomings. The limitations of science need not lead to a gloomy picture of science; for it "comes up with truths and approximate truths far more than any trial and error procedure could be expected to do".[204]

When one considers the various features of Reasonableness seriously, science can be understood as a very reasonable enterprise. Perhaps, more awareness of the social elements and the actual practice of science seem to suggest that *as long as those features of reasonableness,* that we discussed in Section 3, *are part and parcel of science, it seems to be not of much use to speak of Rationality* (in the traditional sense) *in science, but only of Reasonableness in science.* Envisaging rationality in science in terms of reasonableness, will have, in my opinion, some serious implications for both the contemporary science and society. In the next Chapter we discuss some of such implications.

Notes & References

1. Taylor, 1982, 105.
2. Amsterdamski, 1982, 3-17.
3. Putnam, 1979, 113-140.
4. Popper, 1994, 40.
5. Ronald Giere, 1989, 380.
6. Lewis A. Coser, 1996, 217-218.
7. Kitcher, 2000, 243 & 252.
8. Iibid, 261.
9. Plazowski, 1995, 183.
10. Dennett, 1987, 33.
11. Scott, 1995, 214.
12. Gangavane, 1995, 62-92.
13. Carnap, 1937, 1-40
14. Gangavane, 1995, 66.
15. Toulmin, 1964,163.
16. Gangavane, 1995, 84.
17. Schmidt, 1992, 11.
18. Hanson, 1971, 248.
19. As quoted by Indurkhya, 1990, 95-114.
20. John Wettersten, 1994, 719.
21. Indurkhya, 1990, 108.
22. Shapere, "On the Introduction of New Ideas in Sickness", in Jarret Leplin (ed.), 1995.
23. Park, 2004.
24. Indurkhya, 1990, 112.
25. P. F. Strawson, 1970, 591.
26. Ibid., 593.
27. Ibid., 594.

28. P.F. Strawson, "Dissolving the Problem of Induction", 1970, 595.
29. Alan Musgrave "Saving Science from Scepticism, 1989, 297-323.
30. K. Popper, *The Philosophy of Karl Popper*, Paul Arthur Schilpp, ed., 1974, 1015.
31. Methusalah is a person mentioned in the Holy Bible. He was the eighth member of Sethite line from Adam to Noah; the son of Enoch, the father of Lamech and the grandfather of Noah. (See: the Book of Genesis in the Holy Bible, 5:21-27 and the first book of Chronicles 1:3). He lived for 969 years, the oldest person in the Holy Bible.
32. Grunbaum, 1976, 227.
33. As quoted by Gower, 1997, 96.
34. Poincare, 1905, 186.
35. Howson and Urbach, 1993, 117-8.
36. Ibid., 112.
37. For more discussion on this last aspect, see: Michael Strevens 2001 and 2005; Branden Fitelson and Andrew Waterman, 2005.
38. Salmon, 2002, 397-99.
39. Popper and Miller, 1984, 434.
40. Popper and Miller, 1988, 115-6.
41. For more details, Salmon, 2002; and http://www.cs.ubc.ca/~murphyk/Bayes/bayesrule.html.
42. Lakatos, 1978, 1.
43. *http://www.cs.ubc.ca/~murphyk/Bayes/bayesrule.html.*
44. Howson, 1991, 82.
45. Popper, 1963, 387.
46. Popper, 1992, xxiii.
47. Mellor, 1989, 7.
48. Carnap, 1970, 473.

49. Hempel, 1965, 91-92
50. Douglas, 2000, 561.
51. As quoted in Howson and Peter Urbach, 1993, 9
52. *Ibid.*, 1993, 10; emphasis mine.
53. Zycinski, 1995, 253.
54. Strawson, 1970, 594-5.
55. Okasha, 2001, 327.
56. Plazowski, 1995, 180.
57. Popper, 1963, 216.
58. Putnam, 1981, 115.
59. Ibid., 179.
60. Putnam 1981a, 201.
61. Putnam, 1983, 234.
62. Stephen Toulmin, 2001, 13.
63. Ibid., 82.
64. Ibid, 1.
65. Ibid, 2.
66. Ibid., 138.
67. Ibid., 214.
68. Barth, 1969; Slater, 1974; as quoted by Stephen Nathanson, *The Ideal of Rationality – A Defense, within Reason*, 1994, xii.
69. Stephen Nathanson, 1994, 48.
70. Ibid., 1994, 49.
71. Christopher McMahon effectively discusses similar issues, in his book, *Collective Rationality and Collective Reasoning.* For Its précis, see: *Philosophical Studies – An International Journal for Philosophy in the Analytic Tradition*, Vol. 116, No. 2, Nov 2003, 153-157.
72. Ronald Giere, "Scientific Rationality as Instrumental Rationality", 1989, 377-384, 383.

73. Tran van Doan, 2001, 177-8.
74. Henry Harris, 1981, 38.
75. See: http://www.thefree-dictionary.com/ reasonableness
76. Deepti Gangavane, *Dialogues of Reasonableness* 1995, 85-90.
77. Ibid., 88.
78. Ibid., 1995, 89.
79. Ibid., 1995, 90.
80. Ibid., 1995, 91.
81. Nicholas C. Burbules and Suzanne Rice, "Dialogue across Differences: Continuing the Conversation", 1992, 411.
82. Nicholas C. Burbules, "Rethinking Rationality: The Importance of Being Reasonable", *Philosophy of Education*, Audery Thompson, ed., 1993.
83. Deepti Gangavane, 1995, 90-92.
84. (*The Analects*, 15:28). Chan Wing-tsit, *A Source Bok in Chinese Philosophy*, 1969, as quoted by Doan, 2001, 220.
85. Doan, 2001, 220-1
86. Ibid., 222-235.
87. Ibid., 235-7
88. As quoted by Doan, 2001, 237.
89. In the previous Chapters I have dealt with the issue of rationality in science mostly at the Methodological level, as most of the philosophers of science tend to look at the issue of rationality in science from the viewpoint of scientific methodologies. But since I think that reasonableness from other angles is also important for science, I try to look at reasonableness in science from these angles, which are of course not exhaustive.
90. However, one can still use the term of 'Rationality' in science, provided that account of rationality includes all these features of reasonableness explicated here!
91. Burne, 2005, 215.

92. http://www.quotationspage.com/quotes/Albert_Einstein/
93. Henry Harris "Rationality in Science", in A. F. Heath, 1981, 48.
94. Jesus Christ also reiterates the same, though in a different context, more than once, that the *Sabbath* is made for humans, and not humans for the *Sabbath*. The Jews were strict about following the religious rule of the *Sabbath*, which is observing the holy day without doing any work. On one such day, a man with a withered hand came to Jesus for miraculous healing. With the legalistic mentality, the Jewish religious leaders objected to Jesus' healing that man because they took the act of healing as doing some 'work', which was to be avoided on the *Sabbath* day. Jesus clarifies to them, saying that doing something good out of a genuine human concern cannot be considered as a violation of the rule of the *Sabbath* and he unflinchingly declares, "The *Sabbath* was made for humankind and not humankind for the *Sabbath*". (See: The Gospels of Mathew 12:9-14 and Mark 2:23-28, in the Holy Bible).
95. Harold I. Brown, *Rationality,* 1988, 185.
96. *Ibid*., 139.
97. Brwon, 2001, 201-2. Please Note: Brown treats the ability to judge as a cognitive skill. It is learnt like any other physical skill. But one might object that cognitive skill can't be taken on par with the physical skill. Moreover, to make a judgment one needs to operate on some sort of rules and principles, or background knowledge. This shows that judgment presupposes some meta-level rules to operate. Be that as it may, but my point is an account of rationality in science cannot ignore this important aspect of judgment from the purview of science. When one speaks of reasonableness in science, in my opinion, has certainly an important place for judgment.
98. Nicholas C. Burbules, http://www.ed.uiun.edu/EPS/PES-Yearbook/93_docs/BURBULES.HTM.
99. Foley, 1991, 366.
100. Kitcher, 2000, 245.

101. Foley, 1991, 369.

102. Jayard, 2004.

103. Rorty, 1967. http://www.intuition-in-service.org/main/intuition/quotations/creativity.html#tag.

105. Lloyd, 1995, 353.

106. Bazhenov, 2002, 21.

107. Popper, Kvantovaia Teoriia I raskol v fizike (Moscos, 1998), p. 17. As quoted by Bazhenov, 2002, 22.

108. Nicholas C. Burbules, "Rethinking Rationality: On Learning to be Reasonable", in http://www.ed.uiun.edu/EPS/PES-Yearbook/93_docs/BURBULES.HTM

109. Schafersman, "An Introduction to Science: Scientific Thinking and the Scientific Method", 1994; in http://www.freeinquiry.com/intro-to-sci.html.

110. Schafersman contrasts Empirical evidence with **hearsay evidence, testimonial evidence, circumstantial evidence, revelatory evidence, spectral evidence** (by ghosts, spirits etc.), **emotional evidence** and so on. All these unlike empirical evidence, are non-repeatable, non-objective and often not available to others for verification. (see: http://www.freeinquiry.com/intro-to-sci.html).

111. Riniolo and Lee Nisbet, 2007, 50.

112. *Ibid.*, 50-51

113. *Ibid.*

114. *Ibid.*, 51-52.

115. Einstein seems to go even further that one can be rational even when one transgresses common sense, in fact, he argues that it is essential for science: "If we should not sin against common sense, the progress, generally, would be impossible". As quoted by Plazowski, 1995, 193.

116. Harris, 1981, 45-48.

117. Sankey, 2008.

118. Nicholas C. Burbules, "Rethinking Rationality: On Learning to be Reasonable", in: http://www.ed.uiun.edu/EPS/PES-Yearbook/93_docs/BURBULES.HTM.

119. John Worrall, "Pragmatic Factors in Theory Acceptance", in Newton-smith, ed., 2001, 349-357, 351.

120. Ibid., 350.

121. Douglas A. Berman, "Reasoning through Reasonableness" 115 YALE L. J. POCKET PART 142 (2006), in http://www.thepocketpart.org/2006/07/berman.html

122. For example: Robert V. Bartlett and Walter F. Baber, "From Rationality to Reasonableness in Environmental Administration: Moving beyond Proverbs", 1999, 55-67.

123. See for example: Marie Fleming "Women's Place in Communicative Reason," in *Women and Reason*, Elizabeth Harvey and Kathleen Okruhlik, ed., 1992, 245-62; Elizabeth Lloyd, "Objectivity and the Double Standard for Feminist Epistemologies," 1995, 351-381. Some more issues in this regard are discussed in Chapter IV.

124. Howard Gabennesch, "Critical Thinking – What Is It Good for? (In Fact, What Is It?)", 2006, 36-41, 38.

125. Ibid., 39-40.

126. Bertrand Russell, *Sceptical Essays*, 1988, 41. First print in 1928.

127. This is taken from: G. Holton, "Mach, Einstein and the Search for Reality," 1977, 653.

128. Lacey, 1999.

129. Going a step further, some see even moral issues are also in science. For instance, according to Polanyi, "The choice between us is not simply an intellectual affair, or the matter of taste. It is a moral decision.... Argument cannot determine such a fundamental moral decision". [Popper and David Miller, 1984, 36].

130. As narrated in Quine and J. S. Ullian, 1970, 42-53.

131. Brown, 1988, 187.

132. Jesús Mosterín, "Observation, Construction and Speculation in Cosmology", 2000, 219-229, 222.

133. Wisdom, 1987, 71-72.

134. I like to acknowledge the guidance that I received from Prof. S. Dhruandhar, International University Centre of Astrophysics and Astronomy (IUCAA), Pune. The discussions that I had with him clarified some of the technicalities of this very important scientific experiment and gave me a lot of insights about the deeper issues of the same.

135. See: http://lisa.jpl.nasa.gov/SCIENCE/Science-goals.html

136. When a wave propagates through a massive body it will change the dimensions of that massive body by 10-21 meters! This, obviously, is too small to measure! [See: Resonant Gravitational Wave Detector, in http://www.macalester.edu/astronomy/research/chrissy/Links/resonance.html].

137. Hogan, 2007, 23.

138. Drewer, 1979, 11.

139. Inflation is the best model to explain kick-starting cosmic explanation for making the universe big, producing the fluctuations in space-time that seeded galaxy formation. Yet we know very little about the physics of this extremely brief chapter of history. [See: Hogan, 2007, 20-26].

140. Dhurandhar, 2003, 5.

141. Dhurandhar, 2004, 719.

142. Resonance Gravitational Wave detectors use theories, as background information, from many other domains of physics, like *thermodynamics* (to cool solids to near-absolute-zero temperatures), *electricity and magnetism* (to convert vibrations into electrical signals), *mechanics* (to design filters to remove external vibrations), *harmonic motion* (for the concept of resonance that is fundamental to these detectors) and so on. [See: Resonant Gravitational Wave Detector, in http://

www.macalester.edu/astronomy/research/chrissy/Links/resonance.html].

143. Resonant Gravitational Wave Detector, in http://www.macalester.edu/astronomy/research/chrissy/Links/resonance.html

144. Resonant mass (Bar) Detectors, http://web.syr.edu/~dmalling/hisotry.html;

145. For further details: Dhurandhar, 2004, 717-730; Also, See: (No author), 1972.

146. Dhurandhar, 2003, 3.

147. "Gravitational Wave Detectors", See: http://web.syr.edu/~dmalling/detectros.html

148. http://www.google.co.in/search?q=LIGO+detectors+pictures&hl=en&rlz=1W1ADFA_en&prmd=ivns&tbm=isch&tbo=u&source=univ&sa=X&ei=jxujTYmbOdGJrAf6uKn9Ag&ved=0CBkQsAQ&biw=1345&bih=567

149. Dhurandhar, 2003, 3.

150. S. V. Dhurandhar, "The Quest for Gravitational Waves", *Khagol*, 56, October 2003, 4

151. See: http://lisa.jpl.nasa.gov/STRATEGY/getThree.html; and file:///C:/Users/dell/AppData/Local/Microsoft/Windows/Temporary%20Internet%20Files/Content.IE5/ZG400TJ3/article%5B1%5D.xhtml

152. S. V. Dhurandhar, "The Quest for Gravitational Waves", *Khagol*, 56, October 2003, 4

153. http://www.google.co.in/search?q=lisa+gravitational+wave+detector+pictures&hl=en&rlz=1W1ADFA_en&prmd=ivns&tbm=isch&tbo=u&source=univ&sa=X&ei=58KaTav_B47NrQfpmsj1Bg&ved=0CBkQsAQ&biw=1345&bih=567.

154. See: http://lisa.jpl.nasa.gov/TECHNOLOGY/Challenges.html. [*Examples for external disturbances*: the pressure from the light of the Sun and its very small variations; the variable solar magnetic field, and distortion of the LISA array by the gravitational effects of the Earth and Moon. *Examples for*

internal disturbances: the interaction of the electrical field generated by the spacecraft computer acting on the test masses, effects from residual gas pressure near the test masses, and thermal radiation by the electrodes used to measure the spacecraft position].

155. For a brief discussion, see: S. V. Dhruandhar, "Data Analysis Techniques for Gravitational Wave Observations", 2004, 717-730, 728.

156. For further exploration on these topics the following sources are helpful: Websites of the GEO600 Project; LIGO Scientific Collaboration; LIGO Caltech Laboratory; LIGO Livingston Observatory; LIGO Hanford Observatory; LISA-NASA Projects site; the TAMA Project, and the VIRGO Project; and the scholarly works of A.A. Michael and E.W. Morley, "The Original Michelson Morley Interferometer Description," in *American Journal of Science* 34 (1887); J. Weber, *Observation of the Thermal Fluctuations of Gravitational Wave* Detector, 1966; P. Saulson, *Fundamentals of Interferometric Gravitational Wave Detector,* 1994, and *Physics of Gravitational Wave Detection: Resonant and Interferometric Detectors*, 1998.

157. S. V. Dhurandhar, "Gravitational Wave Detection: Current Status, Future Prospects", 1995, 549-557, 557.

158. "Gravitational Wave Detectors", in http://web.syr.edu./~dmalling/detectors.html

159. For detailed exposition of various difficulties involved in the data analysis; See: Dhurandhar, 2000, 545-558.

160. For some more elaboration on the notion of observation and the possible consequences of the revised understanding of observation in the world of philosophy / science / common people, please see my article: Jayard, 2007.

161. Wisdom, 1987, 85.

162. van Fraassen, 1999, 44.

163. Shapere, "The Concept of Observation in Science and Philosophy", 1982, 517.

164. Basu, 2003, 366.

165. Dudley Shapere, "The Concept of Observation in Science and Philosophy", 505.

166. Ian Hacking, *Representing and Intervening – Introductory Topics in the Philosophy of Natural Science*, 1983, 170.

167. Grover Maxwell, 'The Ontological Status of Theoretical Entities', 1998, 372.

168. J. O. Wisdom, *Challengeability in Modern Science*, 1987, 86.

169. Grover Maxwell, 'The Ontological Status of Theoretical Entities', 1998, 372.

170. Ibid., 369.

171. "The Postmodern Critique of Science", in http://www.xenos.org/classes/papers/pomosci.htm

172. Dudley Shapere, "The Concept of Observation in Science and Philosophy," 1982, 485-525, 517.

173. Collins and Trevor Pinch, 1998, 94.

174. Brown, 2001, 195.

175. *Ibid.*, 2001, 194.

176. Kuhn, 1970b, in Lakatos and A. Musgrave (eds.), 1970, 237.

177. Shapere, "Rationalism and Empiricism: A New Perspective", 1988, 299 – 312, 308.

178. Shapere, "The Concept of Observation in Science and Philosophy", 1982, 509.

179. 'The Consensus' aspect of observation, I think, deserves to be a separate condition by itself. Shapere's concept of 'background information' or Kuhn's notion of 'normal science' might seem to include this aspect as well. However, its role and importance are so vital, especially in the modern science that it needs to be elevated to the status of being a distinct condition for observation. Though, Shapere has elaborated the role and the need of background information in science, he has not included it as one of the conditions for observation. He thinks, probably,

that it is too obvious or too fundamental to make a mention of it in those conditions. But unfortunately we need to highlight what is too basic or essential, as we seem to have the tendency to ignore what is right under our nose!

180. Collins, and Trevor Pinch, 1998, 94.

181. (No author), 1972, 506.

182. Collins, and Trevor Pinch, 1998, 96.

183. Chalmers explicates this point with two examples from science: Heinrich Hertz' experiments with radio waves, from 1886 to 1888; and another one are about the measurements of molecular weight in the 19th century. See: Chalmers, 1990, 61-65.

184. Chalmers, 1990, 64.

185. See: http://www.galilean-library.org/blog/?p=105

186. *The Golem*, 98.

187. Ibid.

188. Collins, 1985, 143.

189. *The Golem*, 106-107.

190. Dawson, 2007, 27-2

191. Chalmers, 1990, 77.

192. Franklin, 1994, 471-473.

193. Franklin, 1998, 162.

194. See: Collins, 1994.

195. Kenneth Ford, 1995. See: http://findarticles.com/p/articles/mi_qa3651/is_199501/ai_n8722369. Article accessed on 14 Nov, 2007.

196. *Ibid.*

197. Hacking, 1983.

198. Collins, 1994, 501.

199. Collins, "The Science Wars". See: http://www.cf.ac.uk/socsi/contactsandpeople/harrycollins/science-wars.html

200. Collins, 1994, 503.

201. Alan Gross similarly argues to show that rhetoric, which is usually understood in the relativistic sense and having no epistemic role in science, also has a very significant role in science. This he does with a shift in the very meaning of rhetoric and shows that 'rationality' still has a central place. [See: Alan Gross, 1990)]. Even many rationalist philosophers accept divergent rationalities; for instance Putnam [1983, 184-204] and Cherniak [1986)].

202. Tetsuji Iseda, "Scientific Rationality and the 'Even Stronger Program'", in http://www.human.nagoya-u.ac.jp/~idesa/works/stronger/html.

203. Feyerabend, 1984, 285.

204. Putnam, 1979, 138.

Chapter 4

The Implications of Reasonableness for Contemporary Science and Society

INTRODUCTION

In the previous three Chapters so far, the difficulties of both the rational and non-rational models of theories in science have been presented and the need to look for 'reasonableness', rather than rationality (in the traditional sense of the term), has been strongly argued for. Now in the last Chapter I intend to discuss some of the possible implications, some of which may be very obvious and others quite subtle, of taking reasonableness seriously, both for the contemporary science and society. Some of the implications for science elaborated here in the first Section include: the need for a prominent place to be given for values, social and non-cognitive elements in science, the urgency to focus on the gender issues in science, the unavailability of 'pure' observation in science, the ever elusive nature of a total comprehension of reality, the need to learn how to learn, the requirement for 'reasonable realism' and the mandate to transcend rationality (science) for human fullness. Similarly, in the second Section, the implications of reasonableness for the contemporary society have been analysed. The implications include: the need for a holistic thrust in education, the relevance of the concept of '*intellution*', freedom from the clutches of scientism, the need to learn with unsolved mysteries and limits of science, and finally, the possibility of a meaningful integration of science and spirituality.

SECTION 1
IMPLICTIONS OF REASONABLENESS FOR CONTEMPORARY SCIENCE

When the notion of reasonableness is taken seriously one has to be ready for a drastic change in approaching several issues in science. The very outlook of science has to be changed. Science can no more be the reservoir of the all valid knowledge; it cannot claim to be the custodian of rationality and objectivity and much less can it project itself to be the final arbiter of truths. Science also has to find its way amidst struggles and uncertainties and it has to *learn how to learn*, because it is high time to realize that the methods used in science are not final and absolute. The vast and complex nature of reality out there demands that science has to modify its claims of both realism and anti-realism. More importantly science is forced to realize that human fullness, though this concept cannot be measured in unambiguous parameters, is found only when humanity is prepared to go beyond science and its (strict) rational demands. Ten such implications are put forward for our reflection.

Prominent Place for Values in Science

As the issue of value-judgment in science is very crucial I think it deserves a lengthy discussion. When enterprise of science is reasonable it will certainly give due importance, not only to the cognitive, but also non-cognitive values. It will treat epistemic and non-epistemic values equally, as science is for humans and humans need both these values equally, a reasonable approach to science will certainly treat all these on par. The need for value-considerations in science is predominantly the product of the scientific revolution of the 17th century. For the Greeks, the pioneers of the Western thought and culture, there was no significant distinction between science (knowledge) and values (good), science and philosophy, objective and subjective, and factual (descriptive) and normative (evaluative). The problem arises with the mechanical view of nature (or Newtonian science). The epistemological and methodological revolution in science and philosophy, brought out by Descartes, insisted upon a sharp distinction among the objective and subjective dimensions in the knowledge-seeking enterprise. Since science is said to be rational and objective, factual and experimental, values are beyond the purview of science.

Value, etymologically (*Valere*, in Latin), means 'to be strong', or 'to be of worth'. There is no unambiguous and precise definition for Values or Valuation. Various disciplines of Humanities have varied usages of values. Generally, value refers to the 'worth of a thing' and valuation is the 'estimation of that worth'. From the ancient times of Plato very many questions are discussed under this heading: good, end, right, obligation, virtue, moral judgment, aesthetic judgment, the beautiful, truth, validity. All these are related to what *ought to be*, not what was, is or will be. Some values are taken to be the end of life, known as *final values*, at which all humans do or should aim at. Aristotle saw happiness (*eudemonia*) as a value that all human beings seek, though the components of happiness differ for different types of individuals; however in every case development of one's rational capacities is to be included in the self-fulfilment of happiness. Aquinas accepted his analysis though he gave a transcendental extension of the meaning of the term. Pleasure was suggested as the final value by Aristippus (435-356 BC), the founder of Cyrenaic school. Hedonism as a theory of value tells us pleasure alone is valuable. Rational people act for the best and to act so is to maximize the pleasure experienced either by oneself or by others. Hedonists do value knowledge, virtue, love, beauty and other goods, but "the value of all these things is derivative, [and] they owe whatever value they possess to their tendency to produce pleasure".[1] Epicurus held happiness to be the end. In the recent times, Jeremy Bentham, James Mill, John Stuart Mill and Sigmund Freud held pleasure as the chief value of human life. For Stoicism, founded in Athens by Zeno of Citium, in 108 BC, tranquillity of mind (*apathia*) as the final value is to be sought. *Agape* or selfless love is the chief value of human life according to Christianity. Following Socrates and Augustine, Spinoza stressed wisdom and self-knowledge' to be the highest value. Going along the same lines, neo-Hegelians held self-realization or self-fulfilment as the final value in life. Jean-Paul Sartre (1905-1980) takes authenticity to be the final value. Many stressed social values as central; for Auguste Comte (1798-1857), order and progress was the central value of humanity. Albert Camus (1913-1960) suggested that human solidarity as the most important value for humans. Albert Schweitzer (1875-1965) stressed on 'reverence for life' as final value and the key to ethics. God is the supreme value for William Temple.[2]

As the notions of science and values differ, there are obviously different conceptions of the role of *Values in Science*. The issue of determining the role of values in science has been passionately discussed for a very long time. It spreads from one extreme of holding science to be completely value-neutral to the other, whereby one holds science value-laden, both in moral and non-moral sense of the term. One may grant the need for the so-called scientific values, like justification, coherence, truth, etc, but not the moral or social values.

The need and relevance of values in science will be explicated depending upon the understanding of what science is. If the traditional image of science and scientists is dogmatically held, that science is objective and rational to the core, then it is obvious that there is no respectable place for values in science. As science is objective and experimental, it would not entertain any moral consideration or personal consideration, as they are subjective; as science is rational and progressive, there is less room for any significant consideration for emotional or historical factors either. Similarly, if science is understood only from the methodological principles and the actual practices of science and personal dimensions of scientists are kept away, then science is taken to be busy with facts of the material reality out there. One way to understand *science* is to see it as "a profession in which *individuals* cooperate together in order to advance human knowledge, eliminate ignorance, and solve *practical problems*".[3] I like to stress on the two aspects of the definition: it is concerned with *individuals*, which automatically implies that goals, desires, interests, motivations and values are involved; also *practical problems* show how humans have to be taken seriously, their day-to-day concerns, aspirations, hopes and well-being play an important role in the scientific enterprise. It is, therefore, obvious that science can't just be an ideal or theoretical enterprise.

Though our intuition seems to caution the scientists to be morally responsible for their researches, yet there are many thinkers[4] who demand for a clear demarcation between moral responsibility and scientific research. They insist that the scientists can under no circumstance be made morally responsible for their theoretical researches in the scientific fields. The traditionally cherished reverential attitude towards mother earth and nature was diluted with the arrival of modern science. Knowingly or unknowingly great minds like Galileo hastened the death of such respectful attitude. For instance, when Galileo tried to convince the world that nature had been written in the language of mathematics and when Newton stressed that nature can be analysed and studied in full details as we do with a machine, humanity began the of objectifying nature.

But on the other hand the threats of nuclear bombs, the alarming increase of global temperature,[5] the degradation of ecology at an distressing rate, the umpteen numbers of issues arising out of biomedical researches have clearly taught us that scientists cannot be exonerated from their moral responsibilities towards the total welfare of humanity. Science and scientists are always an integral part of the human society, affecting and being affected by it. Their scientific activities cannot be seen in isolation as though they take place in a vacuum. Therefore we realize that the distinction between the theoretical pure research and the applied research is only an artificial fabrication.

But those[6] who argue for value-neutrality of science claim that science is not in need of any support from the political or metaphysical values, neither for its existence nor for its sustenance. Trying to fit the scientific findings to our value systems would be stifling science itself, therefore one must be ready to revise one's idea of value systems or the applications of such values in the light of new discoveries in science. In case values conflict humanity must choose those that augment the growth of science because science is above any value considerations. Science must be independent of human hopes, aspirations and ideologies. Confusing values with science would not only produce good science, but also it would not help for any better reflections on value systems. The material world independent of humans is the object of study and therefore there is no need for any value consideration in science.

The proponents of value-neutrality in science further argue that science is busy with only *descriptive claims* (what *is* the case), not normative ones. For, *normative claims* (what *ought to be* the case), they argue, don't matter to science, and therefore the whole question of values becomes irrelevant to science. Value-neutrality of Science is usually grounded on *Fact-value dichotomy.* Science is said to deal with facts, not with values, as the latter lie outside the purview of science. Science can show only 'how', not 'why', and will tell us how to achieve those goals, not about whether it is worth the while to achieve them; the quality of the goals and their worthiness are the subject matter of humanities, not science as such. In other words, science is for what 'is', not what 'ought to be'; science can only describe, not prescribe.[7]

But many voices[8] are raised against such "neutrality" or "irrelevant" thesis. Value-neutrality is also being questioned now, as scientific discoveries seem to have landed us in harmful and inhuman situations. Postmodern trends attack this fact-value dichotomy. They believe that facts are constructed. It is traditionally held that values are mere projections of the mind, while factual statements can be confirmed by scientific methods. It is now argued that there does not seem to be any significant difference between the so-called scientific values, like justification, coherence, truth etc., and moral or aesthetic values. Hilary Putnam[9] insists that if values like goodness, kindness etc. are a bit suspect from a narrowly scientific point of view, they have a 'lot of companions in guilt', namely the scientific values, we make not only values, but also facts. Moral images are human creations, but they are not arbitrary. As the knives that we make do satisfy real needs and can certainly be better or worse, the values we create can also be better or worse. He points out, therefore, that the traditional distinction of subjective and objective is in total collapse. All values, including the cognitive ones, derive their authority from our idea of human flourishing and our idea of reason. But both facts and values are thoroughly interdependent. There is no logical gulf whatsoever between them. Truth is linked to reference and reference is related to the *reasonable* intentions of the speaker. We at once realize that 'reasonableness' is a value question, and therefore facts and values are inter-related. Fact-value dichotomy blurs our conception of our reality. There is always interpretation, several interpretations at that, of facts and values in every area of human life.

Fact-Value dichotomy can be reduced to the dichotomy of subjective and objective, but the dichotomy of subjective and objective steadily loses its grips in the contemporary science. Quantum physics seems to clearly show that the reality cannot be known in its objective sense, without somehow the subject involved in the very act of knowing. The investigation essentially involves the investigator and the investigated. *The notion of objectivity* is also no more an absolute and unambiguous commodity. Objectivity can be understood in terms of social aspects such as competition among the individual scientists or various schools of thought patterns; or various traditions promoted and published by various groups in their own journals and open debates in congresses; or the amount of freedom granted and tolerated by the State for free expression of thoughts. Objectivity in science is never guaranteed by individual scientists but the community ensures it by mutual criticism. Popper, therefore concludes, "Scientific objectivity is based solely upon that critical tradition which, despite all kinds of resistance, so often makes it possible to criticize a dominant dogma".[10] Truth is considered as a scientific value, but relevance, interests and significance of such truths in the social contexts are also equally important scientific values. The latter may be called as extra-scientific values, but they are also equally important. To remove these value-considerations from a scientist is to remove his/her humanity as well. One cannot suppress his/her value judgment unless one is ready to destroy him/her as a human being and as a scientist. Scientists are not seeking for merely objective truths, knowledge, but *truths that are interesting, which are useful and valuable to us, humanity as a whole*. Therefore, knowledge, truth, objectivity and so on are rooted in values and human purposes.

The priorities of the scientific researches are strongly criticized by various human sciences in the context of the enormous amount of money and energy spent on the weaponry of mass destruction, while millions of hunger-deaths take place every year all over the world; while they are contemplating putting human colonies on the moon, they need to be emphatically reminded about their duty of putting humans right, here on the earth; while they make so much jubilation over the discovery of traces of water or hydroxyl-molecules on the moon's surface, why don't they pay much attention to provide safe drinking water to the millions on the surface of the earth?... Why should we pay for clean water and air?... Why should one pay for basic quality education?... Is it not the natural right of the citizens to have access to such basic needs? Is it not the duty of the governments to take care of such fundamental needs of their people? Science does not develop in a void. Technological revolution is not a non-social concept. Therefore, "It would be highly irrational to overestimate the powers at the disposal of science and the scientists in the humanization of advances in science and technology. It is still less rational, in fact, irresponsible, to ignore the active socio-ethical humanistic stand taken by the scientific community".[11]

With all these considerations, it is now almost granted that as the epistemic values (objectivity, progress, verisimilitude…) are part of scientific reasoning, science needs non-epistemic values (i.e. social, ethical, and political) also in its reasoning, especially when the consequences of error include non-epistemic consequences. In those cases, without the consideration of such non-epistemic values the scientific reasoning is incomplete and flawed (it is even dangerous to humanity, other creatures and the whole of environment). There are some areas of science where making a wrong choice is said to have no impact on anything outside the research area (e.g. the research into the coherence properties of atom beams). Here scientists are said not to worry about non-epistemic values. But here also, in my opinion, it matters; for basing on the properties discovered, some form of technology may be developed in future; if the properties are wrong that may have impact on outside the research also. So science involves moral decision. This basic moral responsibility cannot be overlooked or diluted by mere arguments. If one takes science to be a reasonable enterprise then it would be easy to find serious and sincere considerations of values in it.

Gender Issues in Science

One may be shocked to know that even gender issues also play a significant role in the developments of science. Though science has been thought to be beyond all these biases, the recent arguments by the feminist philosophers of science startlingly reveal that science is indeed coloured by gender biases. Given the important and surprising nature of this issue, this topic deserves, I think, an extensive analysis.

Need for a Feminist Epistemology!

In the recent years, Feminists, Sociologists of Knowledge and Postmodernists criticize science from various angles. By the early 1970s feminism began critiquing science. The research on the gender bias in the fields of science is highly complex and multi-faceted; it includes issues like the research on the presence of women doing hard-core science, the ill-treatments done to them in the scientific circles, the gender partiality exhibited in the content, methodology and epistemology of science, the selection of the problem for scientific investigations, especially in the fields of biology and health sciences, the decisions and policies drafted by the males, and so on. Feminists point out that scientific researches are gender-biased, for instance medical researches use only the males of any species for their study.

Feminist accounts of science, not only reveal the gender bias in theories, practices and assumptions of science, but also most of them go further to show how this male prejudices shape the very notion of rationality, objectivity and scientificity. They substantiate their claims with case studies that there have been systematic omissions or distortions of women / gender issues, which have resulted in faulty conclusions. Case studies show that even contradicting claims are given in order to claim the superiority of the males, in terms of smartness, intelligence and physical robustness. To argue for the natural tendency for women to have a subordinate role in the society and family, as Fausto-Sterling maintains, some evidence is consistently not considered, some questions are never raised, some hypotheses are ignored and some experimental controls are never instituted (Kathleen Okruhlik, 2001). Feminist critiques point out that the gender bias of Darwin in his theory that evolutionary development exclusively belongs to males, the 'hunter-gatherers'. This androcentric bias serves as an auxiliary theory in other disciplines, like Anthropology, to justify the behaviours and social roles of the males. Similarly, Francis Bacon made use of gender metaphors to refer to nature. He saw nature as female, as a wild and uncontrollable female to be restrained and controlled. Feminists argue that a dualistic and detached pursuit of science to study nature should be removed in order to save women from male domination. Even the scientific revolution of the modern science is seen as the reflection of gender ideology, whereby the mechanistic world view looked at the earth and nature as something to be dominated and controlled by humans.

Women's experience and women's epistemology must also be taken into consideration in order to have a holistic picture of science. In a reconstructed feminist epistemology of science, according to Anne M. Clifford, "the scientist is not an impersonal authority standing above nature and human concerns, but a person whose thought and feelings, logical capacitates and institutions are all relevant and involved in the process of enquiry".[12] They argue that the gender does influence the investigations and the conclusions arrived thereupon; they argue that feminists can point out the bias in the masculinist science and feminist can even produce better science (Harding, 1986 and 1991).

All are not ready to buy the arguments of the feminists. Some don't see even the theoretical possibility of feminist critique. Feminists' convictions that women can produce better science and 'feminist epistemology' is privileged are challenged by many critiques.

Cassandra Pinnick and others (2003), for instance, do not find any evidence to support such convictions. Though the feminists may exaggerate, there must be, I think, some grains of truth in their claims. For, down the centuries rationality has been equated to masculinity and women are said to be emotional, or even eccentric, without any capacity for logical thinking and so on. Their reasoning capacities or abilities to make decisions have been downgraded down the year. The traditional understanding of rationality seems to have not paid attention to non-cognitive elements of thoughts and feelings. The rational mind of men and the intuitive mind of women need to be in collaboration with each other for the betterment of humanity; this distinction reveals the complementary nature, not superiority-inferiority divide, between men and women. What the world needs today, both in the society and in the circles of science, is an attitudinal change with regard to treating women. If women are also involved in the important decisions and policy-making we can certainly expect, as we are going to see shortly, a better 'human touch' in all those policies.

Rationality = Masculinity?

Many of the feminist thinkers argue that even reason and reasoning in science has been coloured by male chauvinism; for centuries down the history masculinity and reason have been treated as synonyms, making feminine inferior, and even subservient to masculine rationality. Rationality and feminism is a vast topic in itself, as it is attracting more and more sympathizers recently (Fleming, 1992 and Lloyd, 1995). The society, which has been biased to accept the thinking and interests of men, very easily stabilizes a notion of rationality, which is considered not only biased but also even against women. This biased approach will be destructive to the 'objectivity' of science, as it would encourage sectarian and sexist ideologies in the pursuit of science (Sandra Harding, 1986).

Carolyn Merchant[13] points out, among many other changes, how changes took place in the social and political perspectives of the people of Europe in the modern age. For instance, when the Copernican theory of the modern science (15th – 17th centuries) moved the earth to the periphery from the centre, as a by-product women were also moved to the periphery. The earth, hitherto seen as the central embodiment of the maternal instincts, was suddenly thrown far away into the dark space to be just one among the many planets, while the sun, the sign of masculinity, was brought to occupy the centre stage. This new picture of the cosmos affected the world of science and scientists too. Misuses and abuses of the natural resources by the greedy humanity lead to a great environmental disaster. The metaphor used to express this abuse of nature is: 'nature is raped'. Sandra Harding raises a thought-provoking question at this juncture: "As nature came to seem more like a woman whom it is appropriate to rape and torture rather than like a nurturing mother, did rape and torture come to seem a more natural relation of men to women?"[14] Similarly, "Women's bodies were simultaneously presented as objects of scientific curiosity and as objects (socially constructed) sexual desire".[15]

Women's Intuitive Power for Science and Society

It is a sad fact that the metaphors and symbols used in social and political spheres are very often insensitive to the feminine gender. In many societies most of their proverbs are infected with gender insensitivity. The arrogance of the male attitude comes out in many of the proverbs, created by men of course, going against the dignity of women; e.g. "Even if you trust a serpent, never a laughing-woman"; "The beauty of a woman is to fix her eyes on her toe while walking"; "Late to bed and early to rise is the hall-mark of a good wife" and many more. This infection has unfortunately caught up with the world of science too; the role of women in science clearly seems to be overshadowed by male chauvinism. The dominance of male does not allow women to occupy the key posts in the premier institutes of science, nor to lead major scientific researches. Women are forced to be content mostly with secretarial jobs. They don't seem to be consulted in the process of decision-making of any major issue, nor their suggestions, ideas and inspirations, which are generally believed to be richer due to their intuitive power. But an extensive involvement of women, who are said to be the repositories of greater intuitive power, in science would make an obvious difference in the progress of science.

Almost all the researches in the field of human psychology agree that the gender difference plays a significant role in the emotional and intuitive capacities. As women are more sensitive to non-verbal communications they can read the mind of the others better. That keeps them better-equipped in spotting lies. Women are shown to be faster and more accurate in understanding the emotional communications. The gender-gap "helps explain why both gender report that their friendships with women are more intimate, enjoyable and nurturing than are their friendships with men. When seeking understanding, both men and women usually turn to women",[16] affirms David G. Myers. This intuitive power of women does certainly make a difference in science too. Therefore it is clear that the involvement of the intuition, especially that of women, is never a distraction for science rather a strength and a necessity. If half of the humanity is ignored how can one expect a full growth in any human endeavour?

Women's contribution will certainly enrich society also. For instance, In India, every day hundreds of people are dying of starvation. But the most saddening fact is that our nation produces about 60 million tons of surplus food-grains; more than 20% of the storage is eaten away by rats and wasted due to inefficient ways of storing. It is purely because of the mis-management on the part of the government and the irresponsible higher authorities. For instance, recently the Supreme Court (14 May, 2011) instructed the State Governments to distribute 500, 000 tons of rice freely to the people at the below the poverty line, rather than wasting the grains due to lack of storage facilities. The Court painfully pointed out that it is not digestible to see starvation deaths and millions of children with malnutrition while we let the grains get rotten in the storage houses. *If only women were given such authority, I do believe, they would certainly not allow such starvation deaths to take place, as they are generally bestowed with the sensitive maternal care and they know how to save things, much better than those men in politics and administration.*

The world is paying a big price for sidelining women in all walks of life down the centuries. The subordination of women has resulted in the ecological crisis, technological haughtiness and value degradation. The whole world is now geared to eradicate poverty, environmental protection and a sustained economic growth. This is possible and will have a lasting effect only if women, the nearly half of the population, are really taken into confidence. Their contribution is very essential for any real and substantial development of humanity. As a symmetric vision requires both the eyes, *a holistic and well-rounded development of humanity,* whether in science or society, *has to have the full partnership of women, who are genuinely empowered, in all domains of life.*

Attitudinal Change in Everyone, Including Scientists!

Evelyn Fox Keller, both as a feminist and scientist, explores in her paper, "Feminism and Science", the liberal and radical criticisms of feminism against science. She opines that only when these aspects of feminist criticism are taken seriously, science can have a liberating experience, as this criticism would help science to be more objective. She painfully points out that in every branch of science some sort of gender bias seems to be prevalent. For instance, in Medicines, more time and energy is spent on the contraceptive techniques meant for women rather than on male contraceptives. Science shows more interest in developing contraceptives for females rather than males. For, society thinks that contraception is the duty of women, as men are supposed to enjoy sexual pleasure without any restriction or anxiety. Contraceptives were produced several centuries before; it is said, Aristotle recommended olive oil as a contraceptive (spermicide) for women and probably he did not feel the need of thinking about male-contraceptives. Compared to the verities of female-contraceptives, the male ones are very few. Recently researches have been undertaken for the male contraceptive pills by the Columbia University Medical Centre. The pills have been tested with mice and if it is successful, in a few years' time, such pills for men will hit the market. However some studies show that men are already uneasy about it. Though contraception is desired and discussed by both men and women, "when it finally boils down to using it, it becomes the woman's responsibility again", says Dr. Arokiasamy, who teaches at Mumbai's International Institute for Population Sciences, which conducted the survey.[17]

Unless the dominating mindset of the males is changed there won't be an easy acceptance of the idea of male-contraceptive pills. Similarly, problems that are specifically feminine don't seem to be given due consideration in developing medications. No due attention, for instance, has been given to the problem of menstrual cramps, which is a very common crisis for many women. Also when a research is done on animals, usually only the male species is chosen; though they give the reason that the female ones may have biological cycles (e.g. female rats have a four-day cycle), which would complicate the research processes, yet the "Implicit assumption is, of course, that the male rat represents the species".[18] Therefore Sandra Harding is forced to wonder, "... if this masculinism is dangerous and undesirable for men as well as women – are the intellectual, ethical, and political structures of science also dangerous and undesirable?"[19] Such deep-rooted ideologies and attitudes against women determine, in general, the growth and the direction of science. Scientists, being the products of a particular society in space and time, can also imbibe the spirit of male chauvinism. There needs to be a drastic attitudinal change in the society, including the scientists. Only when this is achieved all other steps to empower women will be fruitful. The male chauvinism is glaringly reflected in the behaviours and dress-codes that are stipulated for women.

We have been indoctrinated, explicitly or otherwise, to treat women as secondary citizens. It is so subtle that very often we are not even aware of it. Both the print and the electronic media repeatedly reinforce this distorted ideology into our minds. For instance, we may be familiar with the TV advertisement for a particular detergent powder, where a small boy and a girl are walking; the small girl stumbles and falls into muddy water; the small boy, her brother, lifts her up and beats the water, telling the girl soothingly, "don't worry; I have punished the water for causing you to fall". Many of us have seen this and enjoyed the nice and spontaneous acting of those cute kids. But do we stop to think for a moment about the wrong ideologies that are gloriously perpetuated, in and through this advertisement? One, it fails to educate the children that they need to be careful while walking on the road, and falsely puts the blame on the road. Two, it encourages the revengeful attitude in the young minds to punish the water for causing her to fall. In fact, it is her fault to have been careless while walking. (Unfortunately this is what we find with the elders as well; when a child happens to stumble and fall on the ground, even before lifting the child, first we beat the floor for the child to see that the revenge is carried out. The child is pacified, of course, negatively and wrongly!). And finally, we see the boy coming to the rescue of the girl... Why can't it be the other way, as the boy falling and the girl coming to his rescue? It won't be and it can't be. It is because that our Indian psyche has been systematically *polluted*, over the centuries, to think that only men can save women, as we see in most of our Indian movies, and not the other way about. So this small little advertisement reinforces the male-chauvinism in its own subtle, yet strong, way!

Further, because it is believed that women can't think for themselves, men take the upper hand and try to decide everything for them. Even in the matters which are purely personal for women, men seem to decide for them. Since it is believed beyond doubt, as the proverb, which is certainly perpetuated by the male chauvinism of course, goes "Women act faster without reflection and regret later", men don't even bother to consult them for anything. At home they are silenced by their men. Women are not allowed to attend the *panchyat* (traditionally powerful village-level governing bodies) local meetings in villages. Even if women come, they can remain only as spectators, having no place for raising their voices on any issue being discussed there. Even the women-presidents of certain *panchyats*, which are reserved for women, cannot act freely and they are highly controlled by the male members in the family or the village.

It is not that women are oppressed only in the rural areas, among the uneducated pockets of the society.

Education and job opportunities in the urban localities don't seem to change the scenario. For, studies shows, about 800 dowry-related deaths take place every year in the urban and the cosmopolitan cities in India. Only the forms of discrimination or oppression may change, but women are still suppressed, whether urban or rural areas. Describing her experience during the World Women's March, held in New York, on October 17, 2000, Mercy Alexander, who travelled from a tiny village in Kerala, India, said, "Earlier, I thought only the women of poor countries face problems of domestic violence and denial of freedom. However, when I saw firsthand, women from the US and other European countries, standing with us, demanding end to violence against women, and seeking justice, I realized, it's an international phenomenon, faced by all women of the world".[20] *It clearly indicates that the education level, riches, or position don't really matter, but only the basic attitude of men, and of women in some cases, towards the womenfolk.*

Unless the male-population of the world realizes the value of the other half (shall we say, the better half!) of the population, they will not have the courage and the wisdom to divert more funds for empowering women all over the world. For instance, South Asia requires an annual increase of 3% towards education. The total additional cost for universal education is $10 billion, or less than 0.03% of world GNP and less than 0.14% of the combined GNP of developing countries. While billions are spent on the mad arms race across the globe, far less is being spent on education by every nation, pushing the world even further more destruction and underdevelopment. The world must wake up to see the need of the hour and respond to them wisely.[21] Women's empowerment must be holistic. Just throwing a few possibilities of education or job opportunities open to them here and there is not enough. True, many Social Organizations and Women movements do take efforts to make women aware of the social inequalities and discriminations, but quite often they fail to strike at the root causes of such issues. The Women's Conference held in Beijing, in 1995, demands that laws relating to family planning, contraception, abortion, divorce must be looked at from the point of view of all aspects of women health and well-being. Women must come together to work for their empowerment. They must not wait for the males to help them in their empowerment. Women must realize that they are no longer the weaker sex; in fact, as someone put it in a lighter vein, *women are of the stronger sex because of the weakness that the stronger sex has towards the weaker sex*!

Indispensable Role for Social and Non-Cognitive Elements in Scientific Knowledge

As we discussed the notion of consensus in the previous Chapter, we realize that there are many other non-cognitive factors too come into the picture of science. They even seem to occupy the centre stage in the understanding of science. In a much wider sense, the very personalities of the individual scientists play a role in determining the development of science. For instance, the ability to take risks, the level of honesty, the willingness to share the findings, personal longing for power and position, fame and name… all these cannot be ignored from science. The world-view of the given communities, its aspirations, its religious and metaphysical ideologies also play a significant role in the development of science.

Similarly, one cannot assume that scientific experiments are performed in a purely dispassionate manner. There are several non-scientific elements involved in scientific experiments. We need to take them into account in order to trust or doubt the results of an experiment. For instance:[22] i) Faith in a scientist's experimental capabilities and honesty, based on a previous working partnership; ii) Personality, the value-system and intelligence of the scientists involved; iii) A scientist's reputation gained in running a huge lab; iv) Whether or not the scientist worked in industry or academia; v) A scientist's previous history of failures; vi) 'Inside Information'; vii) Scientists' style and presentation of results; viii) Scientists' 'psychological approach' to experiment; ix) The size and prestige of the scientist's university of origin; x) The scientist's degree of integration into various scientific networks; and xi) The scientist's nationality.

Sociology of scientific knowledge is gaining more momentum in this decade. The staunch proponents of the view that knowledge, especially scientific knowledge, is social, like Helen E. Longino, attempt to dissolve the very dichotomy of social and rational aspects of knowledge. It is wrong, according to Longino, to distinguish between cognitive rationality and sociality, which definitionally exclude one another. She "integrates the conceptual and normative concern of philosophers with the descriptive work of the sociologists and historians".[23] Sociologists of (scientific) knowledge question the very assumption that there are asocial (non-social) mental ideas, which are generally taken to be 'pure' in contrast with the 'applied' or 'mixed' ideas. They deny such 'pure' ideas and even if any, those ideas also have social and material foundations. They claim to show that concepts like self, mind, knowledge and so on are all social constructs. Mathematical knowledge, which is traditionally set apart as general or abstract, so much so that it is beyond the purview of sociological, cultural and historical influences, is also examined to see whether one can talk about it in sociological terms. Sal Restivo[24] asserts that, it is not only meaningful to talk about pure mathematical knowledge in terms of sociology but also it is the *only* way to talk about it meaningfully. Science has, by and large, been influenced by the historical, sociological, cultural and psychological factors of the scientists involved.

In the recent years, Feminists, Sociologists of Knowledge, Postmodernists and the like, criticize science from various angles. They analyze the wide range of issues of science, like, scientific discoveries, justifications, its relations with values and so on. They pose very meaningful challenges to the scientists and the proponents of scientism, so much so the sympathizers of scientism call those critiques as 'anti-science'. Feminists point out, as we saw earlier, that scientific researches are gender-biased, for instance medical researches use only the males of any species for their study. Sociologists of knowledge show that one's values, religious or otherwise, personal idiosyncrasies, human foibles all matter a lot in shaping up the actual practices of science. Therefore science is not special in any way. Science is only a social construct, which is neither more nor less objective than stories, novels, and folktales. They claim that there is no rational basis to consider science superior to other disciplines and they accuse that "modern science is rooted in sexism, racism, euro-centrism, male bias and dominance, and is in fact 'privileged' for reasons that have little, if anything, to do with the greater rationality or evidential support of science over myths".[25] As we have already seen science is indeed a human and social enterprise, wherein all possible human and social factors greatly influence the content and the growth of science. Religious authorities and political supremacy can also influence the way science proceeds: e.g. the Catholic Church banned Copernican theory in 1616; similarly Mendelian genetics was declared pseudo-scientific by the Soviet Communist Party in 1949. One tends to think that if there had been a whole-hearted support and encouragement from all relevant religious and political sources science would have surely made much more progress in a much shorter time.

Further, once it is established beyond doubt that science is fundamentally a social enterprise, and then the social and ethical responsibilities of the scientists also come into the picture. Science has to operate for the total welfare of the humanity, and such responsibilities cannot be excluded from serious considerations. Science must never forget its ethical and social roots. It is criticized for what it has done, in the recent past, towards the destruction of humanity, in terms of nuclear and biological weapons and so on. So much so, Feyerabend (1975) appeals to the world community to liberate society from the clutches of science as once it struggled to liberate itself from religion. In a similar vein Michael Dummet (1981) wishes that Physics had stopped by 1900 so that the World would not have encountered such deadly Wars. *We must strive hard to remove human errors from science but it would be a grave error to remove humans themselves from science.*

Social and Moral Responsibilities of Scientists

Alfredo Dinis[26] identifies at least four reasons for the apparent lack of responsibility of scientists: a) the increasing objectification and mathematization of the world led to dichotomy of facts, the domain of science, and values, the realm of ethics; b) in spite of the romantic thinkers and the contemporary move for ecological concerns, major industries still see nature primarily as a source of profit; c) as a sense of privacy and individuality has increased in the social life the very concept of being responsible for something or for someone is losing its grounds; and d) the traditional rational foundation of ethics and science seems to have been lost. An environment of scepticism and relativism is apparently present everywhere. This sort of situation is not very conducive for the nurturing of the idea of responsibility.

Are Scientists *value-driven or interest-driven?* – If they are value-oriented they need to make sacrifices, dialogue, not concerned about the immediate benefits in terms of economy or politics or self-recognition etc; they will be really concerned about the total welfare of humanity, not only of the present but also that of the future generations, they will allow their researches to be guided by the global perspectives, not regional or sectarian interests; if they are interest-driven they may tend to be self-centered and narrow minded. For the total welfare of humanity and environment what today's world needs very badly is science and scientists who are value-driven. That is the reason why socio-ethical issues and humanitarian considerations attract more attention nowadays, more than even before. The need and relevance of ethics and value considerations are felt so deeply that a new discipline, 'Ethics of Science' is fast growing. Scores of books and hundreds of articles see the light of the day every year. Moral considerations of science and the social responsibilities of science and scientists are no more at the periphery of philosophy of science but they take the centre stage.

But one must realize that objective and interest-free science does not imply that it is free from the social and moral responsibilities. The goals of scientific research are not external to science, rather internal. If the goals are made external to the scientific activities the scientists may tend to go in for any kind of research, without having any concern about its consequence on the society at large. Scientific research cannot be removed from the social and moral realms, for "pursuit of empirical knowledge in isolation from their wider social and moral effects, not only provides an easy alibi for scientist to engage in harmful and destructive projects but also encourages the growth of such projects".[27]

The enterprise of science has several *Epistemic goals* (like, advancing human knowledge, aiming at an accurate description of nature; developing explanatory theories, hypotheses; making reliable predictions; eliminating biases and errors; teaching science to next generations; informing the public about the scientific ideas/facts) and many *practical goals* (like, solving problems in engineering, medicines, economics, agriculture etc; increasing technological power, promoting human welfare, happiness, standards of life and so on). Along with these goals of science, *goals of scientists* also do matter: gaining more money, power, prestige, employment, wider recognition… etc. Therefore, the ethical considerations at several levels of the scientists assume importance: at their personal level (honesty, carefulness, openness, freedom, credit, education, social responsibility, legality, opportunity, mutual respect, efficiency, respect for subject), at the level of scientific publications (objectivity in publication, intellectual properties, the media publicity), at the level of issues related to the scientific laboratories (mentor-mentee relationship, harassment, reporting misconduct in science, teaching vs. research, hiring and recruitment, sharing & preserving racecourses) and so on.

Scientists are not just rational people to be concerned only about the reasoning and logic in the scientific enterprise. According to Philip Kitcher points out, as noted earlier down, the real actors in science are not just agents who follow rule-procedures, or socially determined performers but "are those psychologically complex beings, reasonable people".[28] These reasonable people show concern and respect for other, a willingness to participate in dialogues with others. Humans not only reason out, but also love, feel, suffer and desire. All these are not signs of weakness but essential and integrated parts of being human; not that everything that originates apart from reason is bad.[29]

Humans may be economically enriched but impoverished in their *being*. They may 'have' more, but they end up less in their 'being'. Science must not work for knowledge, just for knowledge-sake. Knowledge must contribute to the enrichment of human life and the whole cosmos. The concerns of the environment, the apprehensions involved in biogenetic technologies and many other such domains highlight the moral responsibilities of scientists as social agents, who are supposed to be accountable, not only for the present but also the future generations! As scientists are members of the society, they cannot excuse themselves of their greater responsibilities; they cannot be exonerated from all the negative consequences of the scientific developments. Apart from these social dimensions of the scientists, even scientifically speaking they can't simply ignore the influence of value-considerations in many of their important scientific activities, say, theory choice, as we pointed out in Chapter III.

Untenability of 'Pure' Observation

One of the burning issues in Philosophy of Science is the concept of Observation. Many questions in Philosophy of Science can be either traced back to or derived from the issue of observation. Unlike many other philosophical issues, this one has a very direct relevance to the breakthroughs of contemporary science.

The most intriguing words of Sir Arthur Eddington (1881-1944), I believe, can be a fitting starter for our reflections on the issues of Observation in science and philosophy. He says, "I have settled down to the task of writing these lectures and have drawn up my chairs to my two tables. Two tables! Yes; there are duplicates of every object about me – two tables, two chairs, two pens."[30] The first table is of our ordinary commonsensical experience: a table that is hard and colourful; it supports when something is placed upon it. But the second table, which is beyond our visual perception, is that which is demanded by contemporary physics: it is mostly emptiness, with a number of electric charges moving at an enormous speed in a random manner. He continues, "modern physics has by delicate test and remorseless logic assured me that my second table is the only one which is really there – whatever 'there' may be."[31] These words of Eddington seem to teach us a strong lesson that there is much more in reality than what our naked eyes seem to tell us. Another example would be, though we see on the television screen continuous motions of the pictures it is only a series of instant and motionless pictures as the electrons bombard the screen and it is our eyes which create the sensation of watching a continuous flow of a move. Therefore, we realize that we cannot merely rely on our visual perception and / or other sense perceptions, to learn about the world. The picture of the world given by science seems to be different from what our senses tell us.

At the turn of the 20th century Logical positivists, as staunch empiricists, held on to the following three important theses regarding the notion of observation:

a) In order to decide what is to be considered as a sense-perception, a clear distinction is to be made between purely 'given' in sense perception, which is therefore free of any interpretation; and all 'other ideas' are based on the interpretation of the 'given'. All these beliefs and concepts go beyond the 'given'.
b) All our knowledge-claims are either directly or indirectly traced to observation. When science uses unobservable or theoretical terms they get their meaningfulness only by tracing its origin or relevance to some observational aspects.
c) Observation has two aspects: evidential and perceptual. Both are identical and therefore what is evidential is perceptual.

But, basing on the recent developments in the Contemporary Science, there is a lot of rebellion against such claims of logical positivism. These reactions were mainly due to the contemporary historians of science, like Kuhn and Alexandre Koyrè, have made the philosophers aware of the need of history of science for philosophy of science. Feyerabend[32], N.R. Hanson[33], Robert Palter[34], Stephen Toulmin[35] and many others realized that science would not take place in a 'vacuum' and it would certainly need prior beliefs and theories. So there is nothing like 'pure' observations.

From the recent advancements in the knowledge of history of science and the actual practice of scientists, it can be shown, as discussed earlier, that all the three doctrines mentioned above are extremely problematic: a)* Nc clear-cut distinction can be made between the 'given' and the 'interpretations' of them. There is no pristine or brute 'given' that is ever found. Observation is based on selection. Interpretation of the 'given' is a construction with many presuppositions. The actual situation is: $T_x = f\ [g(O_x)]$.[36] Logical empiricism claims that a theoretical term (T_x) can be shown to be meaningful as a function of an observational term (O_x). But they don't seem to realize that (O_x) itself is not the 'pure given' but coloured with the presuppositions (g). So pure observation is not possible; b)* The second is a natural effect of the denial of the first. Because the 'basis' itself is not purely objective, then all other beliefs might be infected with presuppositions and interpretations. Even if something is taken to be the basis (e.g. unanalysable sense-data), it is not clear how other non-basic terms are based on the 'given'. and c)* Modern science has separated these two, evidential and perceptual, and "the focus of modern science is on observation as evidence, not on observation as perception".[37] It is primarily because sense perception is often unreliable (seeing the half-immersed stick as bent!) and incapable (in the areas of too small or too big in size or with the objects too near or too far!). Since perceptions are often erroneous we bring in instruments to record exactly; evidence is collected by the detectors without any perceptual bias involved. Now these results are to be interpreted.

The arguments meted out against pure observation are:

a. To observe we need principle of selection; we need to decide on what to look for. To make observations one must already have a problem and to tackle that we make observations. So without principle of selection observation is not possible or useless.

b. In science to observe is to describe. To describe we need a language used by a community. Mere observation without description is useless. So there is no observation without theory.

c. Theories are needed to ratify observation. For instance, when Galileo first used telescopes to observe the celestial phenomena, the scientific community was asking for appropriate theories of optics to support his observations. But since he was not able to give it, his observational claims were not accepted.

d. To make observation instruments are used. But these instruments are made on the basis of many theories. When observations are made with those instruments all those theories are taken for granted.

e. Scientific experiments use lots of background theories.

Even if all our knowledge were shown to be ultimately based on sense perception still we would not know or understand knowledge fully. We need something more than that to understand knowledge. For our sense perception covers only a limited section in the whole range of Electro Magnetic (EM) spectrum. For now we learn that the EM spectrum consists of a wide range of rays and with vastly differing wavelengths. It ranges from short wavelength gamma rays through X-rays, Ultra Violet (UV) rays, visible light, Infra Red (IR) rays to long wavelength Radio waves. The normal light, which is visible to our human eyes, is just one portion of the vast spectrum. Gamma rays (with extremely high frequency and wavelength as short as a billionth of wavelength of the visible light) are at the one end and at the other end are the radio waves, with trillion times higher wavelength of the visible light. "The eye" therefore, "comes to be regarded as a particular sort of electromagnetic receptor, capable of 'detecting' electromagnetic waves of the 'blue' to 'red' wavelengths, there being other sorts of receptors capable of detecting other ranges of that spectrum. This *generalized notion of a receptor or detector* thus includes the eye as one type".[38]

From all such revelations many philosophical issues arise, like, a) Do all our observations presuppose some theory or the other? Can any observation be pure? b) Can we draw the dividing line between the observable and the unobservable? Further, there seems to be only a shifting barrier between them. If so, how to define what is meant by unobservable-in-principle and relatively unobservable? c) Is there a need to postulate unobservable in science and the ability of the unobservable to provide explanation and unification? d) What is the ontological status of such unobservable entities?

As our human ability to observe is intrinsically limited in many ways, any strict adherence to the traditional equation of observation with sense perception would not take us very far. It is quite natural for science to give more importance to observable than unobservable. But science must realize that the demarcation of observable and unobservable is a consensus of the scientific community. To quote Fraassen, "It is not irrational to commit oneself only to a search for theories that are empirically adequate, ones whose models fit the observable phenomena, while recognizing that what counts as an observable phenomenon is a function of what the epistemic community is (that *observable* is *observable-to-us*)".[39]

Einstein came up with many revealing factors about the universe which seemed to go against the reason and common sense. He declared that both intangible (like energy) and something substantial (like matter) were basically the same thing. But as the world was just struggling with that baffling idea, he dropped another *bomb-shell*, equally complicated and unexpected marriage between space and time. He brought forth the difficult and different concepts into one intellectually and emotionally overwhelming four-dimensional, continuum. So it seems that only uncommon sense must be taken as commonsensical in the machinery of the universe. There are many more things to be explored and perhaps the future generation, with much more advanced techniques or totally different approaches to reality, may become to learn more about reality; and we and our achievements might appear to them too simplistic to be considered to be anything but knowledge. A time will come when our descendants will be astonished that we had no knowledge of those apparent things.

Thus, we realize that one cannot hang on to the classical understanding of observation, or the whole discipline of empiricism for that matter, in order to have a comprehensive understanding of reality. If one were to insist upon the traditional understanding of observation many insights about reality that we have obtained now, would have been out of question. Further, without these new insights humanity would have been highly handicapped in its efforts to comprehend the universe and the reality around. Humanity certainly needs something more than science and that *more* involves religion, metaphysics, spirituality, mysticism, the extra sensory perception and the like - which have been hitherto regarded as non-scientific, or even non-sense by a few! It would do well if humanity, with this renewed idea of observation, explores the new horizons lying ahead!

Nature of Scientific Knowledge

Scientific knowledge is Relative and Abstract

Reasonableness makes us realize that we need to be aware that science can never give us the absolute knowledge about reality out there, as all the scientific investigations are undertaken from certain perspectives and thus, all the knowledge that we gain from science is relative to those perspectives. The postmodernists, usually criticize science strongly, basing on their mythical understanding of science; often they sound as if the only alternative to the standard account of truth and objectivity is the general skepticism. They exhibit their scientific and historical failings to grasp science.[40] But the notion of reasonableness helps us to see scientific knowledge as essentially relative; it does not mean 'anything goes' but it is relative to the methods and dimensions of inquiry. One can see science as relative without falling into skepticism. It would not be left to the individuals' whims and fancies; there is a check to avoid the danger of relativism. Reasonableness does not demand unconditional or uncritical surrender to the axioms / rules of a theory; one may interpret them differently within the boundaries of the theory. Pluralism need not make theory choice impossible or irrational; for, "There can be reasonable criticism and justification of different positions, and choice can be based on rational, i.e. reasonable grounds".[41]

Feyerabend and Kuhn have gone to the extremes of making science non-objective and showing it incapable of producing any knowledge of reality. It is true, with problems in theory confirmation and theory choice, our beliefs need to be regarded as provisional but that does not mean that science is not reasonable to hold a theory true and objective, however impermanent may be, given the evidence available at a time. Truth is not to be taken as certainty or universal acceptance. Similarly theory leadenness need not make theory choice impossible.[42] Since universal method is denied in science it need not lead to relative skepticism. Science has a privileged epistemological status. Science aims at producing knowledge of the world. "There are historically contingent standards implicit in successful practices"[43] and its success is in coming to gain knowledge of the world.

The fact that science is relative, thus, does not diminish the need and the importance of scientific knowledge in any way. For, though the present knowledge may be questioned or even rejected in future, yet this mere possibility of doubt, is not a reason to doubt them now. Science is "rational" in spite all shortcomings. Science is certainly a product of reason; that reason is of course fallible. This reason is not given a priori to be beyond any shade of mistakes; this reason essentially involves plausibility, subjectivity and controversies. However, it need not lead to a gloomy picture of science; for, as Putnam points out, science "comes up with truths and approximate truths far more than any trial and error procedure could be expected to do".[44] Along with all these elements, the notion of rationality in science has to include *reasonableness*; it has to realize the inevitable role of *induction* and the concrete role of *consensus* of the scientific community.

The human intellect has to pick and choose, compare and contrast, measure and categorize the information that comes from varied sources. The knowledge thus formed will certainly not be able to comprehend all the aspects of the reality presented to it. So human knowledge is intrinsically partial, as it gives only an incomplete picture of reality. A scientific study of rainbow may give us the mechanism of production of the rainbow but never the experience of the thrill, joy and peace that normally accompany the rainbow. Perhaps keeping this insight in mind, Einstein declares, "As far as the laws of Mathematics refer to reality, they are not certain; as far as they are certain, they do not refer to reality".[45]

Some of the Major Axioms of Science

Scientific knowledge is relative for the simple reason science works under several axioms and assumptions. Therefore what is obtained in and through them will be coloured or affected by them. These axioms are generally taken for granted. Reasonableness demands science to be aware of them and the relative nature of science. One must also be cautious in using them, otherwise axioms turn into dogmas and dogmatism is against real science. Some of the axioms are:

a) *Avoiding anything 'spiritual':* Science excludes anything 'spiritual' from its equations and considerations; for it assumes that the universe is made up of atomic and sub atomic particles that obey physical laws. Therefore the question of God or soul lies outside the purview of science. This doctrine of materialism is not proved but taken as a dogma.

b) *Universality of physical laws:* Science believes that all natural laws universally apply to everything and everyone. This claim is never proved to be true.

c) *Measurability of all things:* All things are measurable and those things that can't be exactly measured, like emotions, aesthetic sense, love etc., lie outside the field of science. However a closer look at this claim would reveal that things can never be absolutely measured. For instance, if a table is measured to be 100 cm long with a help of a ruler, a physicist with his better instruments would find it to be 100.124 cm; and if a still more sophisticated instrument it may be 100.12457 cm. But at the quantum level it will be much more challenging to measure as one cannot locate where electrons are exactly. So the exact length of the table will never be known.

d) *Knowability of all things:* We can know everything and there is nothing un-knowable. But the finite speed of light (300, 000 km/Sec) itself is a great limit to our investigations into the universe. We see stars that are billions of light years away but this is only a part of the universe that is visible to us, while the actual and total universe includes also the portion that is beyond our horizon. Though as time passes by, the horizon of our universe expands, astronomy can study only what is within our reach, and nothing can be testably said about what is outside the horizon. Only by 1980s we came to realize that there is a drastic difference between the inside and outside universe. It is to be noted that, "Such limits arise not merely because theories are inadequate, inaccurate, or inappropriate: they tell us something profound about the nature of knowledge and the implications of investigating the Universe from within".[46]

e) *Reliability of Logic and Mathematics*: It is assumed that logic and mathematics work out correctly. Mathematics is said to be the queen of sciences. It even constitutes sciences and without its help many concepts in the contemporary science cannot be understood. However, all is not rosy with mathematics. For instance, there are irrational numbers, which can't be expressed in writing as they have infinite digits, without any pattern of repetition of the digits. E.g. The value of $\pi = 3.14159...$ It shows innate inability.

f) *Reliability of Inductive Reasoning*: Inductive reasoning is the backbone of science. As we have pointed out earlier, there *is no rational or logical justification to use such reasoning in science.* But without induction science will lose its predictive power, and without this predictive power science loses its distinctive character. Science needs to be satisfied with the probabilistic reasoning. Poincare argues that though the Newton's law has been upheld by numerous observations there is an element of chance also; for, there is at least a remote probability that in future it may fail. Due to this essential probabilistic nature of science, "in a multitude of circumstances *the physicist is often in the same position as the gambler who reckons up his chances.* Every time that he reasons by induction, he more or less consciously requires the calculus of probabilities".[47]

g) *Reliability of Scientific Experiments and Pure Observation:* Science is rational and objective as it bases itself on *experiments and pure observation.* But we have problems in both these aspects, as discussed earlier. Any experiment answers a specific question. What is that specific question is decided by the practical and theoretical situation. The relevance and interpretation of experimental results always depend on the theoretical context. The experimental results are constantly revised, or sometimes even rejected, for various reasons: may be, sufficient precaution had not been taken to avoid possible sources of interference, or the measurements may demand new methods of detection, or the problem at hand needs different experiment or the questions they intend to answer become irrelevant in the given context.[48] Therefore experiments don't seem to provide us with the secure foundation.

Total Comprehension of Reality – Ever Elusive!

It is time that science realized that the total comprehension of reality, whatever it might mean, is ever elusive; one can always move towards it but can never possess it. Objective truth, if any, is never attainable in science but we need to strive towards it. A genuine scientist cannot claim that she *possesses* irrefutable truth; rather she has an *irresistible search* for truth and that thirst for truth is what makes her a scientist. For Popper, it is the process, not the product that is more important in the enterprise of science. The fact that we are not the possessor of truth makes our search for truth more interesting and worthwhile. Humanity can never be complacent to think that it has arrived at the final and sufficient knowledge, as at every stage of history, there is the possibility of being falsified. We can have only corroboration of theories and corroboration validates our preference of one theory over another. Rejecting the orthodox attitude of scientific method, Popper held that knowledge is forever provisional and it is the beauty of science to reveal the unimaginable picture of reality.

No Monopoly over Knowledge and Truth

Multidisciplinary approaches seem to gain more and more importance in gaining knowledge of the world. We need to realize that truth cannot be privatized or hijacked by any one particular discipline or so. Truth is never a private property. As Dinis points out, "...truth ought never to be privatized; a multidisciplinary dialogue is the only road along which scientists will avoid engaging in activities that will harm more than help mankind and nature".[49] Questions are complex and complicated and therefore they demand complex solutions. We have certainly not found the solutions. That need not discourage us, rather must make life worth living. We all must make a joint effort in finding the solutions. No one can be excused from it, as Dinis puts it, "But while nobody is to be blamed for not having found these solutions, nobody is to be exempted from the search for finding them".[50]

Removing the elements of non-rationality altogether from science does a great harm to the growth of science. Strict adherence to the fixed methodological rules and elimination of freedom to defy the well-worked frame of thoughts will stifle the developments of science. History of science is full of such examples, determinedly argues Feyerabend, to show that great ventures took place only when the traditional methodological rules were defied: "One of the most striking features of recent discussion in the history and philosophy of science is the realization that events and developments (such as invention of atomism in antiquity, the Copernican Revolution, the rise of modern atomism, ...quantum theory),... occurred only because some thinkers either *decided* not to be bound by certain 'obvious' methodological rules, or because they *unwittingly* broke them".[51]

Need for an Integrated Approach to Knowledge

When we reflect on all points given above, "We can begin to grasp how mystifying is science's claim to be objective, dispassionate, value-neutral, and therefore inherently socially progressive".[52] Therefore science needs to learn to work with other disciplines. Science needs something more than mere observation and experiments. If it confines itself to the domains of empiricism it cannot justify its very existence. For the very principle of induction, which is supposed to be the backbone of the scientific enterprise, cannot be justified purely by experience. As Bertrand Russell stresses, of bringing this fact to light: "(The principle of induction) must therefore be, or be deduced from, an independent principle not based upon experience. To this extent, Hume has proved that pure empiricism is not a sufficient basis for science".[53]

Today science uses lots of probabilistic and statistical theories. Once in a while some drastic failures, like the collapse of space shuttle Challenger II, do take place. By this we cannot undermine science; we undermine only the mythical picture of science. Therefore, today the reasonable position as regards science is that it *gives the best* knowledge possible but *does not give the perfect knowledge*. Science is not able to give full explanations for many natural phenomena that we come across. For instance, more researches in Chrmotherapy are said to reveal that the principal source of energy for life on this earth is Sunlight. The different colours of this light have different powers. Every physical, mental, emotional and spiritual function of ours seems to be light-dependent. For instance, blue colour decreases arterial diseases, purifies the blood and hearing problems and affects throat, thyroid and parathyroid glands; and yellow colour treats lung disorders, kidney problems, regulates the menstrual and other hormonal problems and so on. Now one can't ask why a particular colour has only such qualities and not others. It is the way that they are. Likewise, there are issues where we need to be at ease even though our reason is unable to obtain satisfactory explanation or justification. Finally we need to realize, in the words of John Barrow, "there is more to impossibility than first meets the eye. Its role in our understanding of things is far from negative. Indeed, we will gradually come to appreciate that the things that cannot be known, that cannot be done, and cannot be seen, define our Universe more clearly, more completely, and more sharply than those that can".[54] For instance to learn about values science has to depend on disciplines other than science. As Toulmin puts it, "No science can decide which of its technologies are of real human value".[55] So it has to depend on some other disciplines for its guidance or critical approach.

Learning to Learn

Strict Empiricism Might Strangulate *the Growth of Science*

If we go by strict empiricism to base all our knowledge ultimately upon sense perception then we would not know or understand the nature of knowledge fully. We need something more than that to understand the nature of knowledge. Sense perception is *neither sufficient* (because Sense perception plays only a minor role in this process of getting information. Moreover in many cases sense perception has been shown to be untrustworthy), *nor necessary* (because it has been pushed to the periphery by sophisticated instruments) for knowledge seeking process. May be it was alright during the period from the 16th to 18th centuries to interpret scientific concepts as being exhaustively interpretable in terms of sense perception. But as science developed many features and elements were encountered which were not directly perceivable (e.g. force, inertia, space, electromagnetic radiation etc.). By the 19th and 20th centuries, what is selected to be observed, how to be observed, how it is described, how it is to be interpreted – all such exercises departed from the strict adherence of sensory perceptibility.

There is a rational evolution in the scientific beliefs and principles. For instance, a) Till 18th century it was held that if certain parts were separable from a body, those parts were considered to be the constituents of it. But in the 1930's it was found out that the electrons emitted from an atom were not its constituents; and similarly the radioactive nucleus released in beta decay was not a constituent of nucleus. Further, by 1970's it was proved that quarks could never be separated. Hence, the well-cherished equation 'if separable then constituent; and if constituent then separable' is thoroughly questioned. b) Observability was held to be the criterion to be scientific; if something is not observable it is not scientific. But in the realm of quantum physics, we learn that it is not possible to observe a free quark. If one tries to remove one from its companions in a hardon, one has to introduce a certain amount of energy and that would produce a new quark-antiquark pair, which would get combined with the already existing quarks and produce new hardons. Thus we will never succeed in isolating and observing a single free quark. Thus the primacy of observability with regard to being scientific has been brought under revision. Similarly all the criteria of verifiability and falsifiability too have become questionable. We not only learn about nature, we also learn how to learn about it, and how to think and talk about it, in the light of what we have already learnt. So we are forced *to learn how to learn from nature.*

The Scientific Status of the Unobservables

With a renewed way of learning, we can now look at the role and the scientific status of unobservable in the contemporary science afresh. Some examples for the unobservable: viruses, radio waves, electrons, quarks, genes, atoms and sub-atomic particles, blackholes, etc. We need to revise the nature and the ontological status of the unobservable. We are forced to revise traditional understanding of the (un)observables. Due to the limitedness in our observ'ability', we realize that whether something is observable or not, is a contingent matter of fact, resting on what we have best reason to think we have learned. "What is observable (or unobservable) depends on the current state of accepted scientific belief, including not only the instrumentation involved, but also the theoretical knowledge which tells us the nature of interactions, the theoretical possibilities of detection of a particular interaction, and how the results of a particular interaction give information about its sources".[56]

A caution is required here: One might be led to think that any bizarre and strange items can be accepted as legitimate entities in domain of contemporary science. To avoid any such misgiving Shapere proposes some elementary *guidelines to distinguish between legitimate and wild (or loose) speculation in accepting something unobservable.* Though something is unobservable, yet it is accepted as legitimate in science[57]: a) If that entity is logically and mathematically implied by something that is already observable or has observable consequences; b) If it is needed for consistency considerations, even though it is not implied by the observable parts of the theory; and c) If it provides answers to problems concerning the observable parts of the theory with which no other solution deals successfully.

With the renewal in the understanding of what observation is, we need to revise the nature and the ontological status of the unobservable too. Many ideas in particle physics and contemporary cosmology deal with unobservables, unobservable not only in practice, but even in principle. For instance, today's cosmology speaks about *dark energy* and *dart matter*. Dark matter is an unseen substance whose presence can be felt by the gravitational pull that it exerts upon ordinary matter. The existence of such dark matter is also shown by the bending of the light waves, coming from farther galaxies. As the sun has its own family of planets going around, stars get organized into galaxies, which would be roughly 100 million times larger than the solar family. The recent investigations by NASA reveal that most of the matter in these galaxies is invisible; in some cases, the measurement of the mass of the galaxies greatly differs by different methods. The huge discrepancies are explained by the hidden, dark matter. Now scientists are interested in this dark matter, as it might explain the gravitational forces that operate in the galaxies and also substantiate some of the present theories regarding the origin and the destiny of the universe.

Scientists had claimed that the universe had been constantly expanding for the past about 14 billion years, since its expansion. But after some time, it is at least theoretically predicted, when the gravitation force gets stronger then the universe will begin to collapse. However, in 1998 the Hubble Space Telescope (HST) has revealed a startling fact that the universe was expanding much slowly than what is today. So it is not right to say that the gravitational force is slowly increasing to slow down the rate of the expansion. No satisfactory explanation is given to explain the increase in the expansion. Different assumptions are entertained: a) It is postulated that some sort of strange new form of energy is responsible, known as the 'dark energy', for the acceleration in the expansion. It is calculated that the 70% of the Universe is dark energy, 25%

makes up the dark matter, and only the remaining 5% makes up for all the known matter in the universe; a special kind of energy-fluid fills the space; or a new theory of gravity may include a kind of filed which causes the rapid expansion; so some mysterious force is given as the responsible factor; b) Einstein[58] argued that the space is not empty as it was thought in the classical physics. Space has several properties, including the capacity to create more space by itself. As his gravity theory proposes, 'empty space' has its own energy; when more space is created this energy can increase, which causes the universe to expand; c) Quantum theory suggests that the empty space is actually full of 'vritual' particles which constantly appear or disappear shortly; but when it was calculated how much energy this would add to empty space, it indicated a universe 10^{120} times bigger than what it is now. So the mystery deepens.

All these attempts, (as a property of space, a new active fluid, or a new understanding of gravity) have not proved satisfactory solutions. Our cosmology seems to be getting more and more unexplained and inexplicable. Even after one hundred years of intense investigations and startling revelations in astrophysics, more questions are raised rather than answered. So Haldane, a famous biologist, is right in saying, "It is my supposition that the Universe is not only queerer than we imagine, is queerer than we can imagine".[59]

The theories about dark matter and many other such theories cannot be directly proved in cosmology. At the same time, they are not just speculative. In the present cosmology, new regions which were previously unobservable are now entering our horizon. Similarly, the present unobservable domains may one day enter our horizon, may be 10^{100} years. From this one can't argue that this is observable in principle, which would be a mockery of it. In any case, according to our present understanding the universe is infinite, and therefore, as long as it is infinite, there will always be regions unobservable.

As one may rightly expect classical empiricism would deny / ignore all such theories as unscientific as they turn out to be untestable and they seem to be entering the region of metaphysics, exceeding the bounds of good scientific reasoning. For, they have always claimed that all our knowledge is based on sense experience, or observation or experiment. Thus contemporary science is forced to realize that empiricism alone will be insufficient to tackle with reality.

Reasonable Realism

My focus in this journey towards reasonableness, as I have explained in the introduction, has been mainly on the rational and methodological aspects of science, and not the ontological aspects of science, as to explore whether science reveals reality as it is out there or to investigate the ontological status of the unobservables etc. However, I like to briefly highlight how the construal of rationality in science in terms of reasonableness helps to approach the realist and anti-realist debates in philosophy of science. When science comes out with theories, especially about the unobservables, a crucial question can be raised: Do they portray the world as it is or are they only a convenient tool to understand the natural phenomena? Science not only describes, but also explains how and why a phenonomenon takes place. By explaining the fundamental nature of things, about its existence, about its causation etc, science enters the domain of metaphysics.

The question of realism and anti-realism arises in the context of the unobservable entities. Though they are not observable science is able to acquire knowledge about their nature, behaviour, constituents etc. 20th century physics gives a picture of reality that is more abstract and far-removed from common sense. At this juncture, the question of justifying the ontological status of these unobservable entities arises very strongly. There are two opposing trends in philosophy of science with regard to the existence of unobservable entities: *Scientific Realism* and *Scientific Anti-realism*, where the former claims that the unobservables really exist, and the latter denies it.

The Basic Claims of Scientific Realism

a) Unobservable entities exist; b) They exist independent of our minds; c) We do make statements about them, with truth-value;

d) These statements are true or false, depending on how the world is, and e) We must believe in their existence and consider much of them as knowledge. For instance, the kinetic theory of gases speaks about the molecules in random motion, because that is what actually happens in a gas. Thus, they demand three kinds of philosophical commitments:[60] *metaphysical* (mind-independent existence), *semantic* (literal interpretation of the theories and the correspondence theory of truth) and *epistemological* (the unobservable entities do exist and they count as knowledge).

The Basic Claims of Scientific Anti-Realism

a) Only what is empirically observed does exist, and all the other so-called unobservable entities don't. [*Note*: Most of the antirealists, though they deny the real existence of such entities, yet they agree that science is the most rational enterprise, because of its practical success and growth in accumulation of knowledge. So the issues of 'rationality in science' and 'realistic view of science' are different]; b) The unobservable entities are only useful instruments to make calculations simpler or predictions more feasible, and the description about them are not true reflections of the state of affairs in the world.

The Major Groups among the Anti-Realists

There are *different types of anti-realists* in science. To be an anti-realist it is sufficient to disagree with anyone of the above-mentioned *five claims of the realists*. [Sceptics deny (a), reductive empiricists deny (c), Social constructivists, like Kuhn, deny (b), constructive empiricists, like Bas van Fraassen, deny (e) and they are also unsure of (a)].

a) *Instrumentalism*: i) Scientific theories don't describe the world, rather they are tools to approach the world and to relate a set of observable states of affairs with other sets; Theories about the unobservable entities are not to be taken literally as they are only logical constructs used as useful instruments to relate with the external world; Theories are judged in terms of their efficiency and usefulness, and not in terms of truth / falsity; Since theories are only instruments they don't have predictive powers;

b) *Conventionalism*: i) Scientific theories are only conventions; ii) The theoretical entities don't exist as nothing beyond observational reality exists; iii) The concepts used in a theory, say, 'force' and 'mass' in the Newtonian theories, are mere conventions; however they are empirically significant and useful to explain the world of facts; and

c) *Descriptionism*: The theoretical claims of the unobservable entities are only convenient descriptions of reality, as they don't refer to anything that actually exists

Debates over Realism and Anti-realism

The debates between realists and antirealists seem to go for ever; For, antirealists have to explain how scientific theories, if they are just convenient tools to deal with reality, produce practical successes; on the other hand, realists have to deal with the problems with the theory-ladennes of observation and the fact that the whole enterprise of science is increasingly seen as a product of psycho-social and cultural milieu of a given time. All observational terms are theoretical constructs to a certain extent. Stars, planets, light rays, gases etc. acquire their meaning, to some extent, only from the theoretical contexts in which they occur. For example, instrumentalists argue that only the billiard balls are observable and real, while the force and friction that exist among them are only tools, invented by physicists to understand certain physical phenomena. But one needs to realize that the concept of a billiard ball also involves some theoretical constructs like individuality, rigidity, made of a particular substance etc., without which we cannot speak of those balls. Realists argue that unless we assume the unobservables really exist, how can one justify the efforts of the contemporary scientists to unify various kinds of theories and to arrive at one theory of everything (TOE)? However, instrumentalists would reply that the efforts to unify theories are still worth the trouble, because the scientific community will arrive at one 'big' and 'unified' instrument to deal with nature. Some of them may deny the very effort of unifying as baseless, as science is only a assortment of various tools to deal with variety of problems.[61] Further, the predictive feature of science gives science a distinguishing characteristic for science and this feature is possible only when science treats theories as true descriptions of reality. Given these circumstances, there are authors coming up with nuanced understanding of realism, as instrumentalism seems to have more problems than realism. For example, *'Unrepresentative Realism'* of A. F. Chalmers (1992): It is realism, because we need to assume the independent existence of reality and acknowledge the practical and predictive power of scientific theories. It is *unrepresentative*, because it does not involve correspondence theory of truth between theories and the reality out there.

Reasonable Realism

Considering the problems briefly mentioned above, regarding realism and anti-realism, I tend to think that the 'truth' seems to lie in between these two extremes. A 'reasonable approach to rationality in science' would hold on to some sort of realistic view because the problems against anti-realism seem to be stronger than those against realism. For, unless we assume some sort of stable reality actually existing out there, independent of us, not only our scientific enterprises, but also our practical life activities would lose their meaning. To consider only myself and my mental states to be the only reality would land us in solipsism. So we need to assume the existence of reality out there. Nevertheless, it would be rather too much to claim that I get to know the reality as it is. For the simple reason that when an observer observes reality, there is an interaction between them; for instance, there is an interaction between the object and the photon impinging on it; this photon (light) is essential to make the object visible. This interaction makes a change, very minute though, in the observed reality. This impact becomes more crucial and significant at the sub-atomic level. In the quantum world it is generally accepted that 'the observer creates the observed reality'. The very act of seeing involves a chain of actions created by light (photon), entering the eyes, making an inverted image on the retina, and the brain straightening it up so that the observer can see it as it is outside and so on. All these actions take a considerable time (though negligible for our practical consideration). Thus there is a time-lag between the object outside and its image created in the eyes; so one can never claim to see an object as it is now, but only as it 'was', a minute fraction of a second before!

This can be easily shown by this small exercise: with the right hand we can hold a small piece of paper between the thumb and the index finger of the left hand; when we release the paper to slip through the two fingers of the left hand, we can catch the paper immediately with the left hand. However, if we ask another person to hold and release the paper it is almost impossible for us to catch the small slip of paper, between the thumb and the index finger. This is because, when we see the person releasing the paper, our brain takes some 'time' to register the action and to take further 'time' to send signals to the fingers to catch the paper; and this takes some fractions of a second and by the time the paper slips through our fingers! So seeing an image involves some 'duration' of time. Thus I like to call this approach to realism as 'reasonable realism'.

Therefore reality, as it is right now, always escapes our comprehension. Yet we can't easily deny the independent reality out there, existing independent of our perceptions. So it is reasonable to hold onto realism, but that realism needs to be qualified as 'reasonable'.

Transcending Rationality (Science) for Human Fullness

We need to clarify what we mean by 'human' and 'fullness' here. *Humans* are the products of psycho, social and historical milieu. One finds oneself in the web of relationship at various levels. It is the relationship that makes them what they are. One cannot be understood in isolation, for, *to be is to be related.* In and through such relationships one finds one's potentialities are actualized. Humans are the ones who can produce history and they are the products of history as well. Further, humans are not disembodied spirits. They have concrete bodies to reckon with;

they are naturally connected with all other beings, both organic and inorganic, in the universe. The basic minerals, chemical and salt compounds that make up the human body are found in the yonder stars that shine billions of light years away. For instance, there are 30 essential natural elements, namely, arsenic, cadmium, calcium, carbon, chlorine, chromium, cobalt, fluorine, copper, iodine, iron, magnesium, phosphorous, hydrogen, nitrogen and oxygen etc. which are also found in nature. We share a lot with plants and animal kingdom, in terms of growth, reproduction, self-healing, food-intake, movement and basic five senses. Yet we are something more than all these. However it was soon realized that there are undeniable elements that deeply differentiate humans from others. A few among many others are: reasoning ability, the power to will, the capacity to imagine, the existential experiences of various emotions, including sexual impulses and the ability to control or channelize them, the indubitable aesthetical sense and, above all, the ability to go beyond the immediate environment. Similarly, the concept of *fullness* is also not an easy question; human fullness is also equally mysterious as it involves humans in it and complicated, as it can be asked, fullness for whom? From which angle? How is it to be achieved?... and so on. One way to understand human fullness is, to see it as something that which *enriches human dignity in its totality* – not dehumanizing, not reducing humans to objects, and not to have an oppressive outlook.

Amidst the countless ways of defining or understanding human beings, to look at them as rational beings is one of the most ancient construals. From the time cf Thales of the 6th century B.C., rationality has been taken as the defining characteristic of humans. Thales, known as the Father of Greek Philosophy, emphasized that Nature was rational; Humans had the rational capacity to know about it; and the nature was worth-knowing about. Since then rationality was taken to be the defining characteristic of humans. The rational character was reaffirmed in Plato's *Apology*, for whom reason makes human life complete, and an unexamined life is not worth living. In the *Modern times*, starting from Descartes, more stress was given on reason and its abilities. However, as we know, the

19th century's insightful developments in the fields of human psychology and growth in the awareness of human history have taught us that rationality alone cannot capture the totality of humans. Human beings cannot be reduced only to the rational aspect. The multi-faceted personality of humans and the infinitely complex situations in which they express themselves seem to suggest that one needs to go beyond the confines of rationality in order to understand the real and true nature of humans. When we focus on reason alone, many aspects of life will be dangerously missing. Now researches are on to produce super human, with genetic and hormonal manipulation. We may get super-intelligent person, but that person may end up not having emotions, unable to cry or laugh, and lead a normal human life!

As a reaction to modernism, the era of *postmodernism* came up. It is a reaction to modernity. It brands all the previous categories of thinking and approaches to life as not useful and it invites us to look for fresh answers for newer problems with fresh approaches. It strongly holds that all knowledge is ever provisional, as it is constantly changing. It celebrates plurality of views. It denies meta-narratives or meta-history, as no narrative is suitable for all in all occasions, and at all times. All that we can have are only micro-histories of changing perspectives. It argues for plurality of interpretations as it does not agree with 'the' interpretation. Above all, it fights against the supremacy and glorification of reason. The arrogant view of modernism that there is nothing which cannot be understood by Reason has been mercilessly attacked by postmodernism. There is no 'Meaning' with capital 'M', but only 'meanings', with small 'm'.

Being inspired by this, one may look at humans, as Victor Frankl does, as "seekers of meaning". He suffered untold brutality and all sorts of humiliation and tortures at the Nazi concentration camp. He lost all his relatives except one sister. Though he lost everything fortunately he did not lose meaning in life. With his own suffering he was trying to keep the hope of his fellow prisoners alive. They just hung onto his lips, when he spoke to them. He was convinced that "human life, under any circumstance, never ceases to have a meaning, and that this infinite meaning of life included suffering and dying, privation and death"[62] and he goes further to clarify that "Suffering ceases to be suffering in some way at the moment it finds a meaning, such as the meaning of a sacrifice".[63] If one has faith in God or some divine power, it becomes all the more easy to find meaning in one's life, even in the midst of pain and suffering. By giving meaning to their lives Victor Frankl found meaning in his own. This deep experience led him to develop Logotherapy, which helps those who face *'existential frustration'* to find meaning and responsibility; one is forced to answer the questions posed by life and that makes one to be responsible in fulfilling that answer. Therefore the essence of existence is this *responsibleness*, and it is this *responsibleness* is that which gives meaning to one's life. If these are not found, one is led to a kind of neurosis and life becomes all the more burdensome for the individual and the others as well. When one loses meaning in life one is led to a sort of 'existential vacuum'. This affects them very badly, causing void within themselves, even pushing to the edges of suicide. When 'the will to meaning' is not met with one is easily led to alcoholism, drug abuse, juvenile delinquency, which can be seen as the manifestations of the psychological vacuum.

Thus, Logotherapy goes beyond logical reasoning and moral exhortation but to enable them to find meaning in life. Nietzsche's oft-quoted words, he who has a 'why' to live, can bear with any 'how' are the real motivation in the life of Frankl. The inhuman brutality in the Nazi camp could not remove his will to live and the freedom to make it meaningful. He is convinced that humans have the ability and freedom to transcend the immediate conditions, no matter how delightful or distressful they may be, to have high hopes in life. Every moment one is invited to make a choice and obviously this choice cannot be made on reason alone. Meaning in life should be found, but not in general terms; it differs from person to person, depending on the life circumstances, which are unique to the individuals. Even for a person it changes day to day, or even hour to hour. To look for meaning in life in general terms is like, according to Frankl, asking 'what is the best move in chess?' As the ingenuity of any move in chess depends on the opponent's move, the meaning in one's life depends on one's own challenges and opportunities. As our deeds change and experiences change, naturally "Meaning in life is always changing, but never ceases to be".[64]

What Frankl means by meaning is, I believe, actually meaningfulness; for he says one must not look for an abstract meaning of life. It changes according to one's "own specific vocation or mission in life to carry out a concrete assignment which demands fulfilment".[65] I would like to take it to a higher level saying that one has to look for 'meaningfulness' in order to face the challenges of life. Therefore, it is, in my opinion, more appropriate to see humans as seekers of *meaningfulness*. Though meaning and meaningfulness are used interchangeably in common parlance, there is a subtle difference, I think, between these two. While meaning is fairly available to all, meaningfulness is personal and available to individuals; while the former is fairly objective, mutable and explicable; the latter is rather subjective, immutable and inexplicable. For instance, when one reads a letter, its *meaning* will be available to any reader, provided he/she knows the language; however, the letter will become meaningful only to the one for whom it is written by the sender; because of the love and affection with which it has been written is accessible only to the latter reader the *meaningfulness* is available to him/her, and not to any reader. Thus we can understand that meaning is, in a way, objective and it is available to anyone, while meaningfulness is subjective and not available to all.

More than meaning, it is the meaningfulness which empowers one to face difficulties in life. If one is not convinced of this meaningfulness, life becomes all the more arduous. Therefore, even when one is pushed to the extreme edge of helplessness and hopelessness, it is our attitude that matters. Without meaningfulness it becomes an unbearable pain. One may even make a distinction between suffering and pain, where the former is explainable with adequate reasons, while the latter lacks them. For instance, say, two boys fail in their exams. One knows well that the preparation was not good, and the performance in the exam was also quite bad, and so it was quite understandable that he has failed. Whereas, the other one is fully convinced that the preparation for the exam was very good and the performance too was excellent, but still he failed, and now wonders why and how he failed. Though objectively speaking the phenomenon of failure in the exam is the same for both the students, it is tolerable for the former, and unbearable for the latter, precisely because the latter does not know the reasons for his failure.

This meaningfulness (or meaning for that matter) will be not obtained only with the logical and rational analysis. Any genuine relationship, which makes human lives possible and worth-living, involves commitment. As pointed out earlier there are no rational parameters and logical measurements to measure this commitment exactly. Since science is very often identified with rational approach we need to go beyond science to realize this human fullness. Science may *help us* to achieve human fullness, by way of creating more life-comforts, by producing more food by newer technologies and so on, but *more science does not automatically mean* more human fullness.

Section 2
Implications of Reasonableness for the Contemporary Society

Unlike the sciences of the modern times, which did enhance the life-comforts of the people with more technologies, the developments in the contemporary sciences have greater and deeper consequences and implications in the social and moral lives of the people. Contemporary science seems to provide the world-view for the people; it brings out a new social order, with its own values and principles. It affects the way how one looks at one's self, others, nature and even God. It creates new strata in the society as have's and have-not's, those who enjoy better and safer environment and those who can't, those who can afford to have all advanced treatments, including longevity of life, and those who can't. In short, the breakthroughs of the recent science, especially in the fields of Genetic Engineering and Bio-Sciences, seem to drastically affect the very identity and the destiny of humans, and the cosmos.

In this scenario, looking at science either purely in terms of rationality, devoid of value-considerations, or in terms of reasonableness, which gives 'reasonable' importance to the value-considerations, will certainly have repercussions for the society at large. If science, which has been traditionally treated as the custodian of rationality, is forced to go in for a new understanding of rationality in terms of reasonableness, then the society at large must also take this notion of reasonableness seriously and be prepared for relevant changes in its approaches and principles. Here I like to present a few of them.

The Need for a Holistic Thrust in Education

The young minds of a society must have an equal access to scientific and non-scientific disciplines at their school and graduate levels of education. They must have access to both the intuitive and non-intuitive realms of knowledge. An over-glorified picture of science unconsciously leads to an over-emphasis upon science in educating the children. There is a tendency commonly prevalent among the students that hardcore sciences are meant for the creamy layer of the intellectuals, while the subjects of Humanities, Music, Arts and so on are reserved for the others. Recently, the Chief Minister of a State in India was highly criticized for his advice to the students to go in for technical and professional studies rather than Humanities!

There is no one like Paul Feyerabend who literally fought to deliver the society from the jaws of science. He strongly pleads that any society must be protected from the harmful influence of science, as people demand protection of the State from religion. Science does not have any right to be kept at the pinnacle of all the processes of human knowledge. He is clear that if one attributes any kind of superiority to the scientific enterprise it is not because there is any deserving substance in science, rather because one has arbitrarily decided earlier to give such an importance. In his own words, as mentioned earlier, "(Science) is one of the many forms of thought that have been developed by man, and not necessarily the best. It is conspicuous, noisy, and impudent, but it is inherently superior only for those who have already decided in favour of a certain ideology, or who have accepted it without having ever examined its advantages and its limits".[66] As no special position to science is justified, different disciplines, like astrology, witchcraft, traditional medicines, voodoo etc., must be given equal access in the society. People must be left free to choose any discipline they want. The children must have equal access to various ideologies. Science needs to be dethroned from the monopoly of power and influence. Science also, so pleads Feyerabend, must be brought under the democratic control: "…not only should the subjects that are investigated by scientists be determined by popular election, scientific assumptions and conclusions should also be supervised by committees of lay people".[67] Our education must look for more enlightenment and enrichment. Education must not aim at just imparting more and more knowledge in the young minds; it should not be like dumping the 'intellectual garbage', in an 'intellectual dustbin'. Knowledge must make us better humans, who are ready to work towards the betterment of the society and humanity at large. Such enlightenment and enrichment is what we must look for.

By *enlightenment*, I mean the state of being with better realization and wisdom about the deeper realities of lives. Accumulating more and more knowledge must lead us to the level of wisdom, which alone will teach us how to use the knowledge acquired. Knowledge, without the wisdom that guides how to make use of that knowledge, would be more of a danger than progress, more of curse than a blessing. Today, we have an ocean of information at our finger-tips, but wisdom still seems to be a rare commodity. That is why, T.S. Eliot rightly wonders: "Where is the knowledge that we have lost in information? Where is the wisdom that we have lost in knowledge? And where is the life that we have lost in living?" Precisely because we lack wisdom we have put science into wrong use, whereby it distorts human dignity instead of enhancing it. It is good to remember Gandhiji who speaks about seven deadly sins:[68] "Wealth without work; Pleasure without conscience; *Knowledge without character*; Commerce (business) without morality (ethics); *Science without humanity*; Religion without sacrifice and Politics without principle". Perhaps mere knowledge will not help one to be convinced of the need of character, for which we need wisdom; and science without humanity is not going to be very detrimental to us; he treats all these not just as something undesirable, but as 'deadly sins'!

By *enrichment* I don't just mean better amenities for life, which are also very important. I also include all basic human rights to be respected more dearly and implemented more diligently. All sections of humanity may enjoy the freedom of speech and the freedom to follow any religion. Human and humane care and concern for the whole of suffering humanity must be given. Equal opportunities must be diligently exercised for quality education to all strata of the society. Unfortunately, In India, we have many educational systems like the State Board, the Central Board, the Matriculation Board, the Anglo-Indian Board, the CBSE system and so on. This creates a fundamental difference in the quality of education and the opportunities for higher education and jobs in future. We must not allow such a cruel discrimination, which is systematically designed to suit the vested interests only of some sections of the people in our country. Since the vast majority of the population is deprived of quality education they are not able to compete with the so-called 'privileged' class for higher positions in life. All sections of the society must have easy access to quality education and only that will ensure the holistic growth of society. One sided growth is not desirable, even unhealthy, as we need a symmetric and total growth of the organs in a human body.

The Need for 'Inteliution'

Intellution is the amalgamation of 'intellect' and 'intuition'. No doubt, intellect plays a very important role in science. But often intuition is not given its due importance in science, though no one would deny its role in our practical lives. Taking the role of intuition in science into very serious consideration enables us to have a wider understanding of science. Humans as we are, we need to give equal importance to both, so that we may grow holistically. A thoroughly rational view of science will not have sufficient room meaningful role for intuition, but if one is reasonable, one will certainly have room for intuition. Science needs both intellect and intuition, what William Pollak calls as 'intellution', in an equal measure. First let me highlight some aspects of intuition.

Intuition seems to be needed at every step of the scientific enterprise, not only in framing a hypothesis, but also in constructing experiments to check it, in interpreting the results and in listing out the possible consequences of the test. Intuitive and intellectual faculties of the human consciousness are needed to enrich and strengthen mutually. One-sided stress will not lead to a holistic growth and overall development of humanity. Intuition in science plays an important role at all possible levels of the scientific progress. Carnap reasons out: "Both in the history of science and in the psychological history of a creative scientist, a theory has often first appeared as a kind of visualization, a vision that comes as an inspiration to a scientist long before he has discovered correspondence rules that may help in confirming his theory. When Democritus said that everything consists of atoms, he certainly had not the slightest confirmation for his theory. Nevertheless, it was a stroke of genius, a profound insight, because two thousand years later his vision was confirmed".[69] Rationality, no doubt, is essential for science but that needs to be augmented by intuitive elements. If the intuitive elements are shunned away the growth of science seems to be stunted. Kelly Harris in her write-up, "Reason and Intuition: Science Needs Both", compares the intuitive and intellectual aspects with the Romans' concepts of animus and anima of human consciousness. As any society needs both these for its total welfare science too is in need of them. She insists upon the wedding of reason and intuition: "A marriage of animus and anima, in which reason represses intuition, just as husbands traditionally repressed their wives, cannot achieve scientific greatness because it asphyxiates a thinker's greatest asset. A scientist's mind must be a more progressive union in which reason guides and strengthens intuition as a colleague and friend".[70]

Intuition seems to tell us that there is a purpose behind the universe. With all the fine tunings of the universe down the billions of years, it seems that the universe was expecting the arrival of humans on the earth some day in the future. For instance, the *equal charge of an electron and a proton* in an atom can be given. The mass ratio of a proton and an electron could vary anything between 2 to 2 million; but the actual ratio that we now find is 1:1836, that is, the mass of a proton is 1836 times more than that of an electron. If this ratio had been slightly more or less, molecules would not form and life would have been impossible in this universe. Another interesting factor is that though the mass between them varies so much, their electric charge (1.6×10^{-19} Coulombs) is the same, though proton is positively charged, while an electron is negatively charged. An object has equal number of protons and electrons and that is why they are electrically neutral and the net charge becomes zero. This balance is inherited from nature. If the electrons were to carry more charge than the protons, even if the number of them is the same, the object would be negatively charged and the objects will repel one another. Even a very small difference between the charges, as small as 1 % difference, would result in a catastrophic repulsion; as a result, the book that you read will not stay in your hands, the pages of the book will be torn into pieces and your arms would rip your body, blast away from the earth and fly off into space! Actually the 1% is a too big discrepancy. If the two charges differed by as little as one part in 100 billions, objects like stones, pen, people etc. would fly apart. For massive objects like the Earth and the sun require for their existence a yet more perfect balance of one part in a billion billion. Now all these can be scoffed at as mere coincidences. Be that as it may, but in hundreds of similar cases when one sees the perfect balance even to the level of very very minute fractions seem to suggest something more than a mere coincidence.[71]

Now does this kind of *intuition for purpose* play any role in science? Many think not. But many researches in the contemporary sciences seem to disclose a sort of *purpose* being at a systematic work behind the whole process of evolution. Several biotic and anthropomorphic coincidences that are now revealed seem to strongly suggest that the universe was aware even at the very beginning that once the human race would appear on the face of the earth! Cosmologists speak of at least twenty cosmological constants; if there were even minute changes had occurred in them the universe would have been much different and strange from what it is now and that type of universe would not have allowed life to emerge, much less a conscious life like that of ours to evolve.

It is often claimed that *On the Origin of Species* was the very document that removed teleology from biology once and for all. But its author, Charles Darwin did not think so. In a review of Darwin's work in *Nature*, the distinguished American naturalist Asa Gray claimed that Darwin had brought back teleology to natural science. Darwin immediately sent a note to Gray, saying, "What you say about Teleology pleases me especially, and I don't think anyone else has ever noticed the point. I have always said you were the man to hit the nail on the head".[72] Darwin was an avowed enemy of *anthropomorphic* teleology in biology, but he recognized that explanations by means of natural selection were teleological.

Intuition indeed has a better and bigger role, both in science and in daily lives, than what it is thought to have so far. William Pollak views intuition as the bridge between the inside and outside of humans; it is that which creates the connection between the observer and the observed. Without this intuition it is impossible to see the hidden connections in the received data. Scientific method is capable of creating only various models of interaction with nature and it is the intuition that helps us understand interactions properly. Without intuition, indispensable as it is, science is dead. It is the combination of intellect and intuition, called *intelluition*, that which "enables life to be flexible and stable simultaneously, giving breath and movement to inert matter".[73] The scientific community, and the society at large, will certainly benefit a lot if they pay a sympathetic attention to the call of *intuition*.

Dudley Shapere also sees a definite role for intuition in scientific methods. He sees a similarity in both Inductive and H-D methods in terms of *intuition*: Inductive method *intuitively* holds that observation (or interaction with the world) is the source of our beliefs, especially well-founded ones, about the world, while H-D method's *intuitive* belief is that creative imagination plays a role in building hypotheses. Both methods have weaknesses as well: the weakness of the

Inductive Method is that it needs hypotheses to begin with, but it does not believe in them, whereas the H-D Method has those hypotheses but it is not able to explain the where-from of them. Similarly, both methods have strengths, namely the Inductive Method knows from where the hypotheses come (where H-D Method fails) and the H-D Method gives importance to the antecedent hypotheses (which is neglected by the Inductive Method). Since both the Methods involve some element of intuition we can aim at a method, which combines the strengths of both the methods, while avoiding their weaknesses. In that unitary method the weakness of the one can be the strength of the other.[74] Considering all these, I do think that a holistic account of rationality (reasonableness) in science cannot afford to ignore this great force of intuition.

We need, of course, to be cautious regarding intuitions too. Intuitions have sometimes misled humanity in the past. Galileo had to fight a strong battle against the prevalent intuition of the people of his times in proving that the earth was circling the sun. Similarly, it is normal to think that an unexpected great catastrophe in our lives will affect us deeper and for a longer duration. We also think that some unexpected very good news will keep us joyful and sad news will keep us sorrowful for a longer duration. But recent studies have shown that this intuition is wrong, as they have revealed that our emotional recovery system works faster and so "we accommodate illness, disabilities, romantic breakups and defeats more readily than we intuitively expect".[75] Therefore one can't take it for granted that listening to intuition will always keep us on the path of progress. There reason (intellect) has to come to our rescue. We need both *head* and *heart*, so to say, for a holistic living, of course both in their right places! But because they have misled us often we cannot rule out the very role of intuition in science at all. Intuition may not be always right, but it is not always wrong, or useless, either!

By asserting the essential role of intuition in science, in no way we undermine the need and the role of rationality in science. For, "The contemporary criticism of science does not propagate anti-scientism, it does not intercede in favour of obscurantism and ignorance, but contemporary criticism offers us a criticism of *science's religion* and questions it," and one needs to realize that "The religion of science identifies the technological prosperity with the prosperity of humanity".[76] Among many knowledge-seeking enterprises, science, no doubt, is by far superior rational enterprise. No one denies the fact. However, as many recent philosophers, like E. Husserl, J. Habremas, W. Dilthey, M. Heidegger, H. Marcuse, P. Feyerabend, H. Gadamer, R. Rorty and so on, though they vary in their epistemological and metaphysical commitments, seem to agree on one factor: "that science is not the only knowledge, and that next to it and together with it there are other forms of knowledge – although less powerful and with weaker social authority and influence".[77] The same point is stressed by Fritjof Capra in *The Tao of Physics:* "The rational part of (scientific) research would, in fact, be useless if it were not complemented by the intuition that gives scientists new insights and makes them creative. These insights tend to come suddenly and, characteristically, not when sitting at a desk working out the equations, but when relaxing, in the bath, during a walk in the woods, on the beach, etc".[78]

Finally, As David G. Myers, a well known social psychologist affirms, "Intuition is bigger than we realize. It needs our expertise, creativity, love and spirituality. It is a wonder. But it's also perilous. Today's cognitive science aims not to destroy intuition but to fortify it, to sharpen our thinking and deepen our wisdom".[79] Can one afford to ignore altogether such a big force of intuition in science? *Better Not.* [For more discussion on this, please see my article: Jayard, 2006].

Freedom from the Clutches of Scientism

Modern science has lots of advantages and paves the way for progress. However there are actual and possible threats in the realms of environment and morality as well. Be that as it may, I like to focus on another possible danger: it might revive and reinforce the spirit of scientism, which claims that science has all the answers, if not today, at least in future, for all the questions that humanity can come up with. Scientism is the approach to science, glorifying all that is related to science. It takes scientific knowledge to be the ideal for and superior to any knowledge enterprise. As science reigns supreme there is, and cannot be any other further appeal. In short, it is an inclination to surrender to science totally. Those who are imbibed with the spirit of scientism don't want to believe that science too, like any other human enterprise, has its own ups and downs, failures and successes. They take science to be ever successful and progressive. But investigations into the actual history of science reveal more and more convincingly that science is full of momentous errors which lead to very unpleasant and disastrous consequences. Blunders have occurred out of *arrogance or ignorance of the scientists*, but in both the cases, *humanity has been the losers*. Blunders do take place due to various reasons. Robert Youngson (1998) explains how a wrong hypothesis at the beginning will lead to a disappointing end, and how political and religious ideologies hijack science to end in void, wasting lots of time, energy and money invested in certain research projects. He has elucidated in a lucid manner all the blunders of science down the centuries. No field of science is an exception to this. This trend is rooted in hard-core empiricism, insisting that there is nothing in reality which cannot be brought under some sort of empirical domain. Imbibing the spirit of empiricism logical positivism, a dominant school of philosophy of science at the turn of the 20th century, pitched itself on the spirit of scientism. After a couple of decades when other disciplines like sociology of science, history of science, politics of science, feminist philosophy of science, psychology of science and the like emerged very strongly, logical positivism was made to realize that all were not that glorious in science.

As we are going to see how science also has its own limits and limitations, it is better that we realize that scientists are also limited in many ways. They are not of a special breed that is insulated from errors or emotions. Humans as they are, they have many non-scientific goals in their scientific enterprise. They do behave, as Lars Bergstrom points out, with many extra-scientific goals: "to become famous... to be successful, some want to be invited to congresses and to travel around the world, some want to get rich,... Some want to refute ideas which are proposed by people they dislike and to silence their own critics, some want to contribute to the well-being of others, some want to support certain political ideals, some want their work to have aesthetic qualities, and so on".[80] Scientists too, human as they are, often exhibit the tendency of exaggerating the importance of their discoveries. They like to project themselves and their theories to be of a great substance and supernatural powers. As Phil Ball of *Nature News Service* points out, "It was quite common for nineteenth century scientists to embellish and mythologize their discoveries".[81] Therefore by giving the more or less actual picture of science the society in general, and the students of science in particular, will get to know what to expect, and what not to expect, from science. Thereby they can be saved from the dangers of scientism.

John Dupre (2001) narrates how scientific imperialism distorts human nature by faulty and harmful ways of pseudo scientific thinking. He calls for resisting scientism, a conscious or unconscious tendency to expect science to deliver much more than what it can. Humans, being free agents must not be forced to inquire the universe from just one perspective but they must be allowed to have pluralistic approaches to scientific inquiries. Imbibing the spirit of scientism in an uncritical manner is equally irrational, as one might call anti-science. The hard-core scientism will have serious consequences for humanity; "Believing that only the 'scientific' way of thinking exists can eliminate humanism, morality, altruism from the world. Many human values are not understandable by science. More philosophy and less 'scientism' in today's society of gadgets will prove more than useful...".[82]

Enlightened to Live with Mysteries and Limits of Science

The Limitedness of (Human) Observ'ability'

Rationality looks for concise and precise, final and absolute comprehension of reality. But contemporary science seems to suggest that it is impossible, not only in practice but also in principle. Certain amount of ambiguity in our understanding of reality will ever be there, because of the inadequacy of our very sense organs, which are the primary windows to the reality. In our human relationships we experience many dimensions that are inaccessible to logical proofs or sense perception. We need to rely on our intuition and trust our common sense and human nature. Now it has become normal and essential for science too, to speak about the existences of entities, states and events, which are not accessible to human senses (eyes). There are some unobservables in principle - unobservable because of the limited human sensory capacity. In some cases, other creatures have far advanced abilities in the usage of their senses, compared to the abilities of humans. For instance, migrating birds can observe the polarized sunlight but not we; and sharks, it seems, can sense even small electric fields around the body of their prey, even as far as 1000 miles away! A dog is said to be able to differentiate 40 types of smell and can detect even a mile away. Human ears cannot hear the sound beyond (20,000 decibel) and below certain limit (20 decibel). Our eyes cannot see something too small or too big. Our sense of touch cannot differentiate minute changes in temperature or pressure.

The very concept of colour evades our full understanding. There are only three primary colours, Red, Green and Blue, and all the colours are only mixture of these three colours. Even those three colours are primary because the way human eyes are constructed with cone cells and rod cells in them. In the electromagnetic spectrum of light, as mentioned earlier only the radiation of the wavelength from 0.4 to 0.7 micrometer is visible to our human eyes, while birds can see the Ultra Violet radiation and some animals can see the Infra-red radiation as well. Therefore their world of reality will be entirely different from ours. If we also have sense organs other than these five, the reality will appear different. Similarly scientists are now speaking of 12 dimensions of reality, though we are aware of the familiar three dimensions of space and one dimension of time. If we are able to perceive the other dimensions too obviously we will have an entirely different picture of reality out there. It is reasonableness that enables us to be comfortable with uncertainty and ambiguity of our views of nature, which is "ever approachable and never attainable".[83] As Fraassen pithily puts it, "The human organism is, from the point of view of physics, a certain kind of measuring apparatus. As such it has certain inherent limitations... It is these limitations to which the 'able' in 'observable' refers – our limitations, *qua* human beings".[84]

Limits of Science

No deep-reflecting scientist will ever deny the fact that there are serious limits in science. There are *epistemological limits*, like those arising due to the limitedness of human reasoning; *ontological limits*, like those that are necessitated by the very nature of light; *cognitive limits*, like the restrictions laid down by the human ability to explore nature within the parameters of three plus one dimensions; and finally, needless to say about the *technological limits* that every epoch faces in the process of scientific developments. Though one might say that these technological limits may be overcome by the succeeding generations, yet one does not have much to hope to overcome the limits of the other kinds.

Further, it is believed that the universe is infinite. Our horizons of possible investigation are limited. The part of the universe will not be observable to us, unless it enters our horizon: "Cosmology and physics dictate that there are regions beyond our own horizon that cannot be observed until they enter that horizon, and if the universe is infinite, that, at any given time, there will always be regions which are unobservable".[85] There are some theories in contemporary cosmology for which observational or experimental tests appear impossible. For instance, *Theory of Superstrings*[86] (which aim at unified picture of the universe, to unify the gravitational forces and the standard model of elementary particles and forces) – they deal with energies on a scale of 10^{19} GeV. There is no imaginable technology to test such a high voltage. Science is said to have reached a stage where only the universe itself can be the testing laboratory. The most fundamental limit in the cosmos is that the universe became visible only when photons were emitted from it; that happened only after 300,000 years after the big bang explosion. Till then the universe was opaque and even light did not come out of it. To study the universe before that we need to study the neutrinos, if at all it is possible. Even then we can go back to the time of one second after the expansion. That time the universe was ten billion times smaller than what we observe now. At the most we can go back, if technology allows it, and see when the universe was 10^{32} times smaller than what it is now and never as it was at the *point of its origin.*

The basic argument for the limits of science is, as we saw earlier, grounded on the fact science uses many axioms, which are taken for granted by science. It is further shown that "When you use an axiom, then you start on the wrong foot. You cannot prove axioms and if you base everything on them, then you guarantee that your whole theory cannot ultimately be proved".[87] John Barrow[88] has done an extensive exploration into the limits of science, starting from *practical limits to cosmological limits.* While the former includes cost, technology, mathematical computability and the extreme complexity of the universe, the latter involves the restrictions in learning about the origins, end, nature and structure of the universe. The very position of ours in the universe, the evolutionary growth of our brain that we have achieved over millions of years laid their own restrictions. Our brain, obviously, did not evolve having science in mind. There is no way to go outside of the universe to observe and all that we learn about it has to be done from within. There is no God's point of view to see the universe as it is. An analogy to explain this point: Baby fish while swimming in the ocean asked the mother fish, "what is the ocean?" – the mother fish replied, "This is the ocean, you are swimming in it", to which the baby fish replied, "Don't tell me a lie. This is only water, where is the ocean?" How on earth (rather on water!) can the mother fish explain the ocean to the baby fish!... It can never take its baby up into the air and show the vast ocean spread underneath.

Hempel on the Limits of Science

While discussing the limits of science, Hempel raises another very important question: Do the limits of science tell upon its rational character? Do they make science totally irrational or less rational? Hempel seems to exonerate science from all the usual accusation of making science limited. He looks into the three areas of limits that are usually attributed to science.

Limits in the Reliability of Science: Hempel explains that usually science is expected to do two important things: first, to give us the knowledge about *what* is going on around us and second, to give us the understanding as to *why* it happens. Limits are shown to be strong at both these levels. It is obvious that there are practical limits in answering the infinite number of questions with the finite intellectual capacities, availability of funds and so on of humanity. But Hempel wants to know whether there are limits of philosophical importance, so as to point out some "limits in-principle (which) would have to stem from certain general logical and methodological characteristics of scientific knowledge and understanding".[89] He analyses the classical problem of 'inductive uncertainty'; for instance, Kepler's Laws of planetary motion give us no guarantee for the motions of the hitherto unknown planets would also follow the laws in future too. The laws give us only a strong inductive support but not deductive certainty. He feels that this would be a serious limitation for science if its goal is a search for certainty. But certainty cannot be the goal for empirical sciences. Certainty may be available in Mathematics or other non-empirical sciences. In empirical sciences, a hypothesis always goes far beyond what the content of empirical data gives; moreover, predictions of the unknown future and retrodictions of the unknown past are given utmost importance. So in empirical sciences we can logically never expect total certainty, "Precisely because science seeks knowledge that reaches far beyond the supporting evidence, its empirical claims cannot consistently be required to be certain relative to that evidence; the ideal of empirical knowledge with certainty is logically self-contradictory"[90] and hence the inability to meet such a logically inconsistent requirement of certainty can never be a limitation for science.

Limits in the Potentiality of Science: As the previous limit is about the reliability of knowledge the next one is with regard to the scope of knowledge. Are there questions which are in principle beyond the reach of scientific inquiry? Hempel thinks there are a few: for instance, the question of the existence of ghosts or God, where no empirical assertion can be made to confirm or disprove. But Hempel argues that this incapability is not a limit to the scope of science, as science is expected to prove/disprove only empirical claims.

Limits of the 'Incompleteness of Explanation': Some charge science of incomplete explanation, as science never actually explains everything. For instance, it explains that the rainbow is due to the diffraction of the sun light as it penetrates the water drop, but it does not tell us *why* it happens so: "A scientific explanation is thus always incomplete in the sense that the explanatory facts it adduces are left unexplained and thus not understood. It may even seem that, as a consequence, an explanation in science never does more than reduce the problem of explaining one fact to the problem of explaining several others".[91] *Ideal completeness of explanation* is not possible. An explanation is a set of statements which explain a phenomenon at hand with the help of some supporting facts. But if one expects further explanation for these supporting facts then it will lead to infinite regress. Any type of explanation, scientific, metaphysical or religious, has to stop at some point or the other. This brutal fact is accepted unexplained. Therefore, "No conceptualized account can explain anything without asserting something that it does not explain"[92] and this in turn shows that the fact of incompleteness of explanation is not a limitation for science.

Emil Du Bois-Reymond[93] distinguishes between what he calls 'riddles of the material world' and 'transcendent riddles', where the former may be explained, if not today, some day in future, but the latter may never be satisfactorily explained. For instance, he gives the mind-body relationship, like the question of how thoughts, feelings or consciousness arise out of certain configuration of brain cells. However Hempel opines that even this issue of mind-body relation may be solved in course of time, with more psycho-physical approach, perhaps by developing totally new concepts which would be more fundamental in linking both physics and psychology. Therefore he does not see any logical impossibility in explaining the mind–body problem. Nevertheless Hempel comes out with another possible candidate for transcendent riddles, namely, the question, *why is there anything at all and not rather nothing?* No science or for that matter any other discipline can answer this question, because any answer or explanation would always be in terms of something that is already existing. Even to answer in terms of primary cause or uncaused cause presupposes the agency of that cause. Therefore to expect science to answer this question is a logically inconsistent requirement, as "no theory, no conceptual scheme, can explain the existence of anything without assuming the existence of something".[94]

Mind-Body Relation – Can it be Solved?: Since Hempel mentions that the question of mind-body may be solved in future, I like to make a few brief remarks on that issue. *The debates over the mind-body relation* do not seem to be anywhere closer to a settlement in the near future. There are various theories and models regarding this relation; to put it simply, one can group them under three major models: *Metaphysical Dualism Model* (Brain and consciousness or mind are fundamentally two different realities, with no relation between them whatsoever!); Reductive *Materialism Model* (Consciousness or mind can be reduced to the material aspects of brain, like the billions of neurons there) and *Emergentist Model* (Consciousness or mind is a property that *emerges* from brain at a certain stage of the evolution. It is related to brain, but it cannot be purely reduced the physical or material aspects of the brain). All these models have their own unanswered questions.

Several philosophers and cognitive and neuro-scientists are of the serious opinion that this mind-body problem is far beyond any final understanding in near future, if not forever! For instance, Chalmer's notion of *Hard Problem of Consciousness* can be mentioned here. According to David Chalmers[95] the highly complicated problem for science is the mental phenomena, and within that, to explain consciousness is the most baffling challenge for the scientists. First of all there is no unambiguous definition or understanding of the term 'consciousness'. Scientists and philosophers vary widely in their understanding, and it is not uncommon to label it as 'mystery' due to the queer intangibility and ineffability of the subjective experience of the consciousness. While dealing with the intricacies of the issue of consciousness he categorizes two sets of problems: the *easy* problem and the *hard* problem; the former includes certain phenomena of consciousness, which can be explained in terms of computational or neural mechanisms, the standard methods of cognitive science. For instance, the ability to discriminate, categorize, and react to environmental stimuli, the integration of information by a cognitive system, the reportability of mental states, the ability of a system to access its own internal states, the focus of attention, the deliberate control of behaviour and the difference between wakefulness and sleep - all these issues may be satisfactorily handled by nuerophysiological approaches. Similarly, Robert Miller[96] has sufficient reasons to show why it is that the very concept of conscious mind lies beyond the realm of the physical description. He shows it in four steps: a) A physical system cannot observe itself; b) Physics is not possible without making observations (or an observer); c) Any observation essentially requires a conscious mind; and d) Thus, it is not possible to decipher the working of the conscious mind in purely physical terms.

All said and done, at the end of the day, one thing seems be very clear: the mysterious dimension of reality of the mind-body relations will ever be there; we use our mind to learn about mind, and we can never see it as an objective reality, external to us, as our 'mind' is also involved in seeing the mind, so much so, the quantum physics instructs us that 'observer creates reality'. No observation is possible without the interaction between the observer and the observed reality. Not only mind/soul/ consciousness is mysterious, even matter is mysterious to the core. Matter does no more seem to be 'material' in the usual sense of the word, as matter is said to be 'energy dancing at the bottom'. Einstein has made a breakthrough in showing how matter (mass) and energy are convertible with the famous $E=mc^2$. Whatever approach one might adopt, one has to encounter the mysterious nature of the reality. As Abraham Thanniyiel puts it: "In the reductionistic understanding *nothing is a mystery*, in the dualistic understanding *God is a mystery* and in the non-dualistic understanding *the whole reality is a mystery*.[97] One way to bring together science and religion is to look at human person as psychosomatic unity. The new developments in science don't seem to remove the element of mind/soul, rather they give a new understanding of it. Many quantum physicists speak about a new sort of wholeness and interconnectedness in the whole of reality. David Bohm (1980) sees 'undivided wholeness' in the creation, while N. Herbert holds that quantum wholeness "is a fundamentally new kind of togetherness, undiminished by spatial and temporal separation".[98]

Finally, it is true, as Hempel points out that all these may not be limits of science as such, but I believe, they point out to some deeper lesson: that is, the powers of human cognition are not limitless. It seems that humans cannot see themselves as the masters of shaping their own destiny. It may be seen as a limit of human existence as a whole, as most of the Existential Philosophers have pointed out. Some of them, like Sartre and Nietzsche, have ended up with a pessimistic outlook towards the world, while some others, like Gabriel Marcel, have taken the fact of limits as something adding interest to the very human existence, opening up avenues of transcendence. After all we cannot even know whether we know everything. That is why, Wittgenstein too rightly wonders, "We feel that even if *all possible* scientific questions be answered, the problems of life have still not been touched at all. Of course there is then no question left, and just this is the answer".[99]

Reasonableness Accommodates Scientific Mysteries

Aiming at absolute and clear-cut solutions, classical view of Rationality would not be very comfortable with mysteries, more so unsolved ones, of science. With its strict aim of explaining every complex matter, answering every question, solving every puzzle of the universe it won't have room for accommodating such mysteries. John Malone (2001) explicates twenty one such mysteries, which are mysterious even at the turn of 21st century: the origin and the end of the universe, the origin of life on earth, mass extinctions, Ice Ages, the learning of a language, the nature of light, the number of dimensions in the universe – are some of them. One may argue that some of these may be convincingly solved in future, but one cannot be sure that those solutions will not create further mysteries in turn. Some of these are mind-boggling from time immemorial; for instance, Aristotle came out with some understanding regarding bird migration; now we know he was not fully correct. However, even after about 2300 years we are still not able to understand and explain everything about the issue. The solution to a mystery, whatever that might mean, tends to lead to another mystery; and thus the solution does not solve the mystery but in effect postpones the mysterious nature of reality from one level to another. The insoluble nature of mysteries need not discourage human or dampen their spirit of seeking; for as Malone has it, the realization of the depth of these and other mysteries "is a good part of the pleasure of being alive".[100] Reasonableness in scientific rationality will be very comfortable with such unsolved and even insoluble mysteries of science. For it knows what and how much to expect of the human rational efforts.

Meaningful Role for Spirituality and Mysticism

In spite of such an alarming growth and successes of science, one has to be sadly aware of the painful side of humanity. In the 21st century, we seem to enter a culture of *LESS*; we seem to become lesser day by day. I remember having read somewhere, which interestingly puts what I want to communicate: As our communication gets *wireless*, we get *careless* towards the needs of the fellow humans; as our telephone gets *cordless*, we get *thoughtless* in exploiting nature; as our youth become *jobless*, our relationships become *loveless* towards our own people; as our politics gets *shameless*, our feelings turn *heartless*; as our education becomes *valueless* our mistakes become *countless*! There may be many reasons for this culture of *less*, but one of the important reasons is that our science is seen as *valueless*. People with scientisitic attitude argue for strict dichotomy between facts and values; science is confined to the area of facts and it is value-neutral; factual judgments have no value considerations and value judgments have no factual contents. We have argued against this stand in the previous section. Science is in crisis, as already mentioned in the Introduction, from rationality perspectives and in human perspectives as well. It encounters moral and social problems. Husserl identifies four factors leading to the crisis of science: Galileo's process of mathematization of nature; Descartes' drive for objective rationalism and his mistaken conclusion of identifying objectivity with reason; natural-cognitive theoretical psychology that claims the role of reason (as proposed by Locke, Berkeley and Hume); and finally Kant's projection of transcendental philosophy as the rational foundation.[101] The constituent parts of this crisis, such as the environmental degradation, ethical and social issues in Genetic Engineering and Nanotechnology, the reduction of human life to a commodity, issues of morality in the world of modern medicines – all have far-reaching consequences in the human life and culture. As science plays an inevitable role in today's world it has to be seen in the right and human perspective, so that it enriches humanity. I tend to think that one of the reasons the crisis of science, is that it has totally isolated itself from religion or mysticism. It went to the extent of ridiculing all the engagements with anything that is spiritual as a waste of time, as searching in a dark room, for a black cat, which is, unfortunately, not there!

But science and mysticism can help humanity in striving for a holistic growth. Science may not be at the mercy of religion or vice versa, but we the humans, limited both in our intellectual and intuitive powers, need to be at the *mercy* of both. Science and religion can be seen as two different enterprises, though with two different methodologies and aims, studying the same reality outside. A closer look at science reveals that it seems to miss many elements that are essential for human relationships. But science as such cannot be penalized for that, rather, as Anthony O'Hear points out, "the attitude of mind which would see scientific discourse as the only valid approach to the world",[102] needs to be blamed. Science without religion, or vice versa, seems to land us in a distorted and biased worldview. He further explains that a religious attitude can be a better motivation to understand nature with a reverential approach and similarly growth in science will save us from idolizing human subjectivity.

Science–Religion Dialogue

Science-Religion[103] dialogue is an academic exercise to pave the way for a better world and humanity, by trying to bring together the recent advancements of the contemporary science and the dignified and innermost insights of religions, with the tools of the latest progress and great lessons obtained in philosophy (especially,

Philosophy of Science). It aims at harmoniously unifying the fundamental elements of matter and spirit, the human and the divine, the values and the technology, the scientific achievements and the deepest longing for meaningfulness in life. Enlightenment and Modernity saw religion as the enemy of science; even ridiculed religion. On the other hand, people of the post-modern times are increasingly becoming aware of the shortcomings of science-technology, which not only not solves problems of humanity, but creates newer ones. People realize that they have become the victims of the one-sided growth of science-technology.

There are *several models of Science-Religion relationship*. For instance, Ian G. Barbour discusses four models: between science and religion:[104] a) *Conflict Model:* It is the outcome of the extreme positions held by science and religion, namely, Biblical Literalism and Scientific Materialism. Since both are at the diametric opposite levels, it is not possible to combine both; so, one cannot be a believer in God and in the Theory of Evolution, at the same time, because the evidence of the theory of evolution goes against every aspect of faith in God. Both are in a constant conflict, as there is nothing common between them. (Major proponents: Carl Sagan, Jacques Monod, Edward O. Wilson, Stephen Hawking, Daniel Dennet, Richard Dawkins and so on); b) *Independence Model:* There is no conflict, and at the same time, there is no any complementary relation either, between science and religion. Since the subject matter, the realms of inquiry and methodologies drastically differ, there is no question of any conflict; the strict compartmentalization of the two disciplines maintains and reaffirms their distinctive and unique nature. (Major proponent: Karl Barth); c) *Dialogue Model:* a more constructive approach; it encourages dialogue and integration between them, avoiding conflation or total segregation. While preserving the differences, it focuses on the points of possible interactions and aspects of similarities. The borders and frontiers of science can lead to the issues of religion. Science alone is not sufficient to meaningfully approach those phenomena at this threshold. In the past, religious and philosophical reflections are said to have contributed to the scientific understanding of the origin and evolution of the universe. (Major Proponents: Karl Rahner, David Tracy, Thomas Torrance, John Polkinghorne, Arthur Peacocke, Pannenberg etc.); and d) *Integration Model:* It looks for more substantial type of relationship between science and religion, than the dialogue model. Both are mutually enriching by more extensive and orderly partnership; Religion can positively help science, and be helped by science, in many ways in their great adventures of exploration into the deeper realities of the world.

A deep quest for transcendence is a common feature of human nature. This desire is not easily fulfilled. Any attempt to realize this desire ends up creating more problems than it solves. If one finds fault with religion that it assumes too many things in the God-talk, it would do well for him/her to realize that science too is not free from many fundamental assumptions. The very existence of an independent world, that it is ruled by uniform laws, that it is knowable etc., as explained earlier, are some of the basic assumptions. Without these assumptions it would be nigh impossible for science to progress in any meaningful sense of the word. It is a phenomenon of the past to see science and religion as combatants. Many in the past have tried to solve the disputes between science and religion with a dichotomised view, which holds that both are separated enterprises, having nothing in common, and they cannot contribute in any way for the mutual enrichment. But now the recent revelations in the realm of quantum physics seem to affirm more and more that both science and religion are not only non-combatants but are even mutually enriching for a better understanding of reality. Even great physicists of our own times support the marriage between science and religion. Not just their academic assertions but even their personal convictions affirm this view. I like to spill a bit more ink on the issue of quantum physicists' inclinations towards mysticism and to show that science and religion can be close friends and never foes.

The Understanding of Mysticism – Really Mysterious!

Mysticism, in general, is associated with a sort of consciousness that enables one to have an encounter or a sort of union with the Ultimate Reality, whatever or whoever it may mean. Mysticism seems to be a basic element in most of the world religions and cultures. Nevertheless we don't have a definite understanding of what its nature and characteristics are. It has undergone a long history of evolution. Hence, till date there is no final and decisive definition for mysticism. In fact, in 1899 Dean W.R. Inge, an English author, Anglican priest, professor of divinity at Cambridge, and Dean of St. Paul's Cathedral,[105] listed twenty-five definitions of mysticism and now after an elapse of one hundred years some more types of mystical cults and traditions might have emerged in the world and the list of the definitions also might have grown longer. Further, debates are on over many aspects of mysticism, including the question whether the mystic experiences are the same everywhere. There seems to be some vague consensus to characterize it as "consciousness without content", "the experience of the absolute oneness", "union with the transcendent", "immediate consciousness of the presence of God" and so on.[106]

However, a useful distinction can be made between 'natural' (or cosmic) and 'supernatural' (or religious) mysticism. *Natural mysticism* denotes the kind of profound experience whereby one feels oneself merging with the cosmic totality. *Supernatural mysticism* consists in the individual's feeling one with the transcendence. A deep feeling of oneness with the cosmos may also lead to a sort of religious experience of transcendence. So both can't be held in watertight compartments. All cosmic-mystical experiences can't be equated to religious mysticism. However, denial of any resemblance between the intense, unifying experience of nature and that of a transcendent presence would not be proper.[107]

Similarly, there are diverse opinions with regard to characteristics of mysticism too. One can trace out a few of the common characteristics of all mystical traditions: *Ineffability* – It is an important characteristic of mystical union. No ordinary language is found to be capable of expressing the nature of that union. The mystics use often metaphoric and poetic language to express the ineffability. Though the mystics have written volumes over their experiences, they themselves accept their inability to capture the totality of their experiences. *Noetic quality* of the experience is also seen to be a common characteristic. Mystical experiences don't lead one to a theoretical knowledge, though its insight permeates one's knowledge with a unique, all encompassing sense of integration, which clearly belongs to a noetic order. *Passivity* is another very important feature. In spite of all the efforts that an individual may have taken in terms of ascetic exercises and rigorous meditation and contemplation, it is a gratuitous gift of the transcendence to reveal itself to the individual. When the higher power overcomes and takes hold of the individual, the individual's efforts lose all their need and importance. *Transiency* is yet another important property of mysticism. It is a controversial aspect, as many yogis are said to have prolonged states of mystical consciousness. Some have this ecstasy at regular intervals. And finally *Integration* may be said to be another quality, as the individual overcomes the previously existing opposition by getting integrated with higher reality.

Generally all mystical religious efforts aim at some sort of mystical union, the crowning point of human destiny in its spiritual endeavours. It is a union between the subject and its divine object. Most mystics claim that this climax is never by the conscious effort but achieved by a total surrender and openness towards the divine. One of the fundamental aims of mystical union is to cross the boundaries of time and get diffused into *eternity*. In that state of consciousness, notions of *before* or *after* makes little sense. Natural mysticism and all pantheistic systems see God diffusely present in all things; body is seen as microcosm and through meditation one can move to macrocosm. The sense of 'unity' makes the subject feel totally engrossed in the natural universe. The very awareness of knowing disappears; the duality of subject and object vanishes; in this state of 'absorptive concentration' (*samadhi*), the yogin transcends opposites, unites emptiness and superabundance, life and death, being and non-being. One reintegrates a primordial modality of being; it is trans-consciousness, not so much of regression into 'prenatal sleep' that enables this genuine step into transcendence.[108] At this state of consciousness, there is insensitivity to pain, no experience of the normal time-flow, having higher noetic insights into the essences of divine truth, inexpressible by logical, discursive means. With all the veils removed the reality is seen in all its nakedness.

Mystical Inclinations of the Quantum Physicists

Quantum Physics or Quantum Mechanics,[109] though often taken to be one single paradigm, is not a homogenous body of theories and laws. There are diverse opinions and approaches among the quantum physicists. Some of the differences are very fundamental. For instance, chronologically speaking, the era of quantum physics can be divided into two, as the *Old Quantum Mechanics* (between 1900 and 1925) and the *New Quantum Mechanics* (1925 onwards). The former era includes great quantum giants, like Max Planck, the Father of quantum physics, Albert Einstein, the man of the millennium, Niels Bohr, Sommerfield and so on. The new era has many path-breakers, like Louis de Broglie, Schrodinger, Werner Hesienberg, Max Born, David Bohm, Paul Dirac and many more. Some would prefer to identify Niels Bohr to be the bridge figure between those two eras. The fundamental difference in the approaches of these scientists belonging to these two different eras is that the *former hold deterministic interpretations, while the latter affirm the statistical and indeterministic interpretations of quantum theories*.

Again, those in the old era, known as the Copenhagen school (their orthodox interpretation was proposed by Bohr and his followers), split the world into quantum entities (the indeterminate) and classical measuring apparatus (the determinates). The new era obviously goes against the dualistic understanding of the old era. They affirm that the dualistic description (quantum world and the measuring apparatus) of the universe, strictly speaking, does not hold water. For the whole reality is a unity, as the measuring instrument itself has been made out of the same quantum constituents as those of the objects under observation. For instance, David Bohm attempts to show that the whole reality as "one undivided unity". The basic division between the observer and the observed becomes devoid of meaning, as an observed particle behaves as much like a wave, dependent on the whole environment, including the observing instrument, for its particular form.[110] Further, Heisenberg reiterates that reality is unknowable in itself and "The conception of objective reality... has thus evaporated... into the transparent clarity of a mathematics that represents no longer the behaviour of particles but rather our knowledge of this behaviour".[111]

In spite of all such drastic differences in their interpretations of quantum theories and the deep-rooted understandings of the reality, most of them, both in the old and the new eras, seem to be united in one aspect – that is, they have high tendencies towards mysticism and religious sentiments. We can easily see in them, not only in their writings but also in their personality as such, some sure signs of mystical inclinations, irrespective of their otherwise opposing viewpoints. For our reflection let us consider one scientist each from the old and new quantum era; and they are *Albert Einstein* and *Werner Heisenberg* respectively.

Albert Einstein – An* Evangelist *of Cosmic Religiosity

Albert Einstein (1879 – 1955) was deeply convinced that only a limited portion of reality could be comprehended in logical coherence and meticulous exactness. The main task of physics is to discover the most fundamental laws of nature. In this process, more than logic, an intuition for the order that lies behind the appearance of the nature is very helpful. As science proceeds further humanity becomes more and more aware of the splendid order in nature: "In every important advance the physicist finds that the fundamental laws are simplified more and more as experimental research advances; he is astonished to notice how sublime order emerges from what appeared to be chaos".[112] From time immemorial humanity has this deep inner longing for the pre-established harmony, as Leibniz puts it. The desire to look for this order is compared to the unexplainable yearning of a lover for the partner; "Its inspiration arises from a hunger of the soul".[113]

Types of Religions: Einstein distinguishes three kinds of religion: *Religion of Fear* - One can trace out some sort of feeling and longing behind every human endeavour. Similarly religion too has a feeling as the cause: for primitive people it is 'the feeling of fear' –

fear of death, sickness, beasts, and nature etc. Their knowledge was not so developed as to unravel the principle of causality behind the happenings of nature. So it was very handy to assume a being like them, but with super powers, to control all such fearful phenomena. This super-being, they thought, must be appeased from time to time to get favours from it. The right and the privilege to perform priestly activities (e.g. sacrifices), along with certain secular authorities too were bestowed upon an individual, a ruler or a particular class. *Religion of Morality* – As centuries rolled by, humans began to have a social concept of God. They postulated a Divine Parent, resembling a human parent, who would comfort in sorrows, fulfil their longings and preserve the dead souls. Though the human parents do give love, protection, guidance etc. they are mortal and fallible. So they were forced to conjecture a divine and omni-potent parent and that is God. We do see a gradual movement from the Religion of fear to the Religion of Morality in the Old Testament, continued in the New Testament. One can't categorically assert that the primitive people had the religion of fear and the civilised had the religion of morality. We do find many elements found common in both of them. For instance, anthropomorphic understanding of God was common to both types of religions.

And finally, *Cosmic Religion* - At the third stage humanity arrives at a Cosmic Religious Feeling. Einstein opines that this can't be put into verbal explanation. Hence he declares, "It is very difficult to explain this feeling to anyone who is entirely without it, especially as there is no anthropomorphic concept of God corresponding to it"[114]. Humans slowly realize the futility of human desires. All their endeavours fail to give the real lasting satisfaction. So they aim at the *Sublime Order*, underlying the whole of creation. This Order is readily revealed in Nature as well as in thoughts. Individual existence is seen as a prison, hindering the feeling of the Order. This cosmic order is being explicated in the Psalms of David (a book in The Bible) and the Buddhist scriptures. Great thinkers of all religions have such cosmic religious feeling. This involves no dogmas and God is not seen in a human image. Though many of such geniuses were considered to be heretics or atheists, yet some of them were recognized as saints. In this viewpoint, Democritius, Francis of Assisi and Spinoza are closely related. Such cosmic feeling can't be communicated. However science and art awaken this feeling within us and helps maintain it.

Einstein described his religious feeling as one of the "rapturous amazements at the harmony of natural law". People who were in close association with him agree that he was the most religious person, not in the sense of strictly following any institutionalised religion. Quite often he has expressed his religious conviction as, "My religion consists of a humble admiration of the illimitable superior spirit, who reveals him in the slight details we are able to perceive with our frail and feeble minds. That deeply emotional conviction of the presence of a superior reasoning power which is revealed in the incomprehensible universe forms my idea of God".[115]

"God is subtle but he is not malicious", Einstein used to say often. On this he parted company with his fellow physicists. About his parting, the distinguished German quantum theorist Max Born, a close friend of Einstein, said, "Many of us regard this as a tragedy, both for him, as he gropes his way in loneliness, and for us, who miss our leader and standard-bearer".[116] Einstein was deeply convinced of fundamental rationality in the universe. He termed this feeling as 'cosmic religious feeling', which he thought was greatly offended by the probabilistic interpretation of nature, propounded by quantum physics. "God does not play dice" was his oft-quoted saying.

Science and Religion: With the new understanding of Religion, we arrive at a new dimension of the relationship between science and religion. The traditional outlook of such relation as irreconcilable antagonists is removed. Science may be roughly defined as an endeavor to bring perceptible phenomena by systematic thought into a posterior reconstruction of the external reality. It is more difficult to give even a rough definition for religion, a suitable one to all the religious minded and of all the times. However we can give a fundamental characteristic of a religious person: he or she would be free from the bondage of selfish desires, always being engrossed in the thoughts, feelings and aspirations of super-personal value. This value gives them the topmost meaningfulness in their lives. So, "a religious person is devout in the sense that he has no doubt of the significance and loftiness of those super personal objects and goals which neither require nor are capable of rational foundation".[117]

Science, according to Einstein, can give us only the facts and the relationships among them. It won't lead us beyond what *is* to the realm of what *should be*, namely the goal of our human aspirations. One can't deduce an *ought* from an *is*. Purely rational conception of our existence has its limits. Knowledge, though a very powerful instrument, is limited. It may help us to reach certain ends, but not the Ultimate End. Intelligence is necessary as it shows the relation between the means and the ends, but it is not sufficient. Mere thinking can't give us the sense of the Ultimate End. This Ultimate End, as well as the longing for it, come from another source. Our existence gets meaningful only with the setting up of higher goals and ends. Religion has an important role in explicating this Ultimate End. Our forefathers seemed to have kept this spirit of the search for the End very alive. We the moderns, with all our national and international politics, legislation and organization, seem to have lost this real spirit. If only we kept up "the longing for the achievement of the goal powerfully alive within us, then shall we not lack the strength to find the means for reaching the goal and for transforming it into deeds?"[118]

Science speaks about facts; it can't set up the ultimate goal or principle. Religion evaluates human actions and thoughts; it can't speak of facts and their relations among themselves. A religion can't claim absolute truthfulness. For instance, taking the Bible as the absolute, the Church condemned Galileo. It was merely an encroachment into the territory of science. Though science and religion are separate enterprises, yet both have reciprocal relationship and dependency. Scientists must also have the yearning for truth and understanding, and that yearning can emerge from religion. Thus, "the situation can be expressed by an image: science without religion is lame, religion without science is blind".[119] A genuine scientist easily attains a humble attitude of mind towards the splendour and magnificence of reason. This reason is "incarnate in existence", profoundly deep and inaccessible to human intelligence.

"His (scientist's) religious feeling takes the form of a rapturous amazement at the harmony of natural law, which reveals an intelligence of such superiority that, compared with it, all the systematic thinking and acting of human beings is an utterly insignificant reflection".[120] This attitude is more religious than any follower of religion would have it. Thus science purges religion of its anthropomorphic tendencies and leads to a religious understanding of life. Fundamental ends need no justification for their existence as they exist in a healthy society as powerful traditions.

An Appraisal of Einstein's Ideas: One could easily see the influence of Pythagoras and Spinoza on Einstein in his idea of cosmic religious Feeling. The fact of a central order in the universe is grasped by soul in a mystical union. *The Mystery of the Sublime* is the driving force behind Science, Art, Religion, Ethics etc., though they are distinct in themselves. It is very revealing to learn that Einstein, one of the greatest geniuses ever lived, is very religious minded, of course not of any particular religion as such. His conviction that 'God does not play dice' was so great that he had to go against the majority of the scientists of his times. His faith in the cosmic aspects underlying the universe gave an invincible impetus to work towards a 'unified field theory' to arrive at a unified cosmic formula that would explain all physical laws. Some would name Einstein as agnostic or even atheist. This seems to be unfounded. For he has never denied the existence of God nor that he has expressed his doubts about his existence. He is certainly not favouring any anthropomorphic God, or a personal God. But he is, beyond any shadow of doubt, ready and willing to accept some being, a cosmic force, far superior to humans. According to Einstein, a religious person in the true sense of the word is always selfless and filled with super-personal values and she will never be a source of division, nor hatred.

Today the idea of personal God seems to lead the followers of a religion to be fanatics in their claims.

We have painfully watching the innumerable conflicts and violent blood sheds in the name of religions! The very religion that speaks of love, tolerance, brotherhood and interconnectedness of all reality has been made to be the source of division and hatred. Religious leaders must enlighten the followers that all their claims about God are not absolute, but only partial glimpses of God-reality; all we have is, as St. Paul puts it, "When I was a child, I spoke like a child, I thought like a child, I reasoned like a child. When I became a man, I gave up childish ways. For now we see in a mirror dimly, but then face to face. Now I know in part; then I shall know fully, even as I have been fully known" (The first book of Corinthians, in the New Testament, Chapter 13: 12-13). Einstein's idea of cosmic religion is really enlightening in the world that is torn apart with religious fundamentalism, dogmatism and intolerance. As he insists for an authentic interaction between science and religion, we realise that is the need of the hour. Science can focus on what *is* and never what *should be*. The question of *'ought to'* can be dealt only by religion. Only religion can be the basis of goals, values and morality. Science alone is insufficient for humans. It can't give us the ultimate goals and ends. It might at the most provide us with the means of attaining those ends. Humanity needs both the goals and the means and hence they have to be related well. The real motivation for the scientist to work with truth and understanding comes from religion. A genuine scientist adheres to cosmic religion. He would be more religious than a person following an organised religion. Such realization would help scientists not to isolate themselves from the society and to be aware of their duties towards the welfare of humanity as a whole.

Werner Heisenberg – The Height of Spirituality

The Need for Metaphysics: Werner Heisenberg (1901–1976) was an excellent philosopher, more particularly, a metaphysician or a mystic in the Pythagorean and Platonic sense. Though he was a rigorous analyst and empiricist, he strongly opposed the Logical positivism, developed by the philosophers of science of Vienna Circle, in the first two decades of the 20th century. In his philosophical writings he took pains to show them that what was revealed in any observation was not an absolute, objective datum, but a theory-laden datum, that is, shaped by some background theories and conditioned by the observational contexts. He also wrote on the age-old philosophical problems of the One and Many, Part and Whole, and so on, believing that it might throw some fresh light on the experiments. He was convinced that physics could make valid claims only within limited spheres. Better than the worldview of Democritus, Plato's worldview would explain modern physics. Metaphysics shouldn't be treated as unscientific. 'Meta-' implies a 'going towards fundamental questions' of any given discipline. While *Meta-logic* and *Meta-mathematics* can be allowed, why look at 'Metaphysics' with a despite, argues vehemently.

God as the Central Order: Heisenberg, like many other physicists, is not interested in the idea of a personal God. Behind the universe one could see the 'consciousness', whose intentions are the manifestations in the universe. Human consciousness is not able to fathom that consciousness clearly but only vaguely. The existence of the central order of things and events is beyond any doubt. One can easily become aware of this central order. This central order is to the universe, what the soul is to a person. Soul refers to the inner core of being. As we can become aware of the soul of another person, we can be aware of the central order too. As the magnetic force is important to a compass, so the central order to our existence. The immediately visible environment can't be the source of guiding principles for human conduct. That source lies behind these visible phenomena. That source, for Plato, is the World of Ideas, and for the Bible, it is God, who is spirit. So, ethical and moral values emerge out of this order. For "when people search for values, they are probably searching for the kind of actions that are in harmony with the central order, and as such are free of the confusions springing from divided, partial orders".[122] This order lights up our way. Without the order, we are in the danger of losing our direction, as it happens without a compass. We do encounter attempts in all religions to express human relatedness to such a central order.

This order is very important for science too. Religious people refer to this order when they speak of values and ideals, which lie in being in harmony with the order. Science too points to this order when it subscribes to the plan by which nature has been made. The advances in the area of quantum physics make the central order more clear and vivid to our understanding. Hence there can never be any real conflict, he is deeply convinced, between science and religion. But the apparent tensions arising in this relationship are because by nature religion is said to be static and that of science is dynamic. This tension is not destructive but constructive for the growth of both the disciplines.

Scientific Truth and Religious Truth: That there is a distinction between the scientific and religious truths is acceptable to Heisenberg. But he does not go along with the distinction made by the Positivists, namely between truths that are clear and distinct and those, which are not. They go a step further to deny the latter as meaningless. He considers such a distinction as ingenuous and unnecessary, because, "if we omitted all that is unclear, we would probably be left with completely uninteresting and trivial tautologies".[123] True religions always speak in parables and use imageries extensively. As there is no full correspondence to the meanings they are attempting to express, it is quite difficult to understand them clearly. Usually both the scientific and religious truths are seen to be in conflict, particularly more so, after the Galileo controversy with the Church. Normally science labels religion as an outmoded thinking. Scientific truths may be unconquerable in the scientific fields, but need not be in conflict with the religious truths. Religion focuses only on the divinity aspect of reality, and therefore it is likely to ignore the awareness of the interconnectedness of the whole reality.

Only humans can transcend the immediate environment. This capacity has enabled them to go *within* too. Thus people develop a 'spiritual pattern', where questions of values are decided. Their conducts, morality, aesthetic judgements and even the whole cultural life are shaped and governed by the spiritual pattern. In short, it is here where an individual derives meaning of his/her life. This can be, in a wider sense, called 'religion', a religion of community. Heisenberg elucidates in his *Across the Frontiers* that this element of religious pattern is found in many cultures belonging to different eras, 'even in places where the very idea of God is absent'. Such a comprehensive understanding of religion in human society is necessary, even for the scientists, to think of any relationship between the scientific and religious truths. When Paul Dirac expresses his deepest displeasure with religion, saying that God is nothing but a creation of human mind and the State and the Church need that illusion to exploit people in their own ways, Heisenberg corrects him that it is not the mistake of religion but it is the political abuse. Further he points out, "Dirac disapproves quite particularly of the dishonesty and self-deception that are far too often coupled to religious thought. But in his abhorrence he has become a fanatic defender of rationalism, and I have the feeling that *rationalism is not enough*".[124]

Scientific language, which aims at accuracy, can't help us realise the interconnectedness, but a poetic language makes such an understanding possible. The whole pattern and the implications of scientific and religious languages are different. *When science speaks about 'heaven', it is entirely different form the notion of 'heaven' used in the religious language.* Science projects humans as the objecitfiers who objectify the outside reality and claim a sharp distinction between the subject and the object, whereas religion can't legitimately speak of such a division between subject and object. That's why, one needs to be careful about the mingling of the scientific and religious languages. Religious language always uses parables and images. These always encounter a limitation because what they express falls short of the meaning of what they really try to express. Now contemporary science realizes that it also needs poetry rather than prosaic language to describe the quantum realities. In the words of Niels Bohr: "When it comes to atoms, language can be used only as in poetry. The poet, too, is not nearly so concerned with describing facts as with creating images".[125] So, as Plato held, contemporary physics too treats atoms not as physical objects in the ordinary sense, but can be spoken of only as forms, structures or in Plato's terminology, 'Ideas'. The ultimate reality of sub atomic particles can't be sufficiently described by ordinary language. *Our ordinary language becomes inadequate to describe what goes on at the realm of elementary particles.* Language of mathematics, as Plato maintained, serves well in this regard. Not a precise and direct language, but a language with imageries and likenesses is more helpful in describing the oneness that lies behind the visible domains. In the words of Heisenberg, "If the harmony in a society rests on a common interpretation of the 'one', the unitary principle behind the phenomena, then the language of poetry may be more important here than the language of science".[126]

The Limits of Science: Classical physicists too were well aware that scientific accuracy in any experiment was limited. But it is quantum physics, which has boldly shown that the limitation is not just due to the inadequate technical know-how, rather it is *in principle*. Newtonian laws can be considered as a closed-off theory. It may be final and absolute, but only within that framework. As time goes by, many closed-off theories, like the theory of heat in the 19th century, came up. These theories are idealisations for restricted fields of experience, though they may be valid for all times within those limited ranges. So when sciences speak of 'final' it is final in the limited sense. It is applicable only within the range where the same concepts are applicable. Quantum theory is final in this limited sense. In the sphere of judiciary, laws are final until a totally different situation arises. Legislation can't be binding if an entirely unexpected situation arises. Same thing applies to science too. Scientific theories and explanations are not and even can't be final or absolute; they can't be the basis for acts of faith. Some modern religionists are proud to say that their creed has a scientific basis. *In fact they are not aware of the fact that if faith has science as basis then there is an inner contradiction and they rest on illusion.*

Heisenberg, going along with Heinrich Hertz, who clearly expresses his ideas in the introduction to the *Principles of Mechanics* (1876), asserts that it is not the duty of physics, nor does it have the capacity to reveal the inherent essences of natural phenomena. Physical laws and calculations are only rough pictures of reality, though very much useful for practical purposes. The picture is said to be a good one if it corresponds with our law of thoughts (admissibility), agrees with the experiments (correctness) and gives maximum knowledge of essential features (relevance). New developments in the realms of atomic physics have taught us that the interaction between the observer and the observed cannot be ignored. The dividing line between the observer and the observed became thinner and thinner. This awareness of modesty, which was prevalent in the beginning of modern science, really disappeared by the 19th century. Physics started making lofty claims to have understood nature as a whole. It even crossed into the boundaries of philosophy and demanded that philosophers should be scientists. But today the contemporary physics has to make an about-turn towards the realisation of the original self-limitation of the enterprise of science. Humanity becomes more and more aware of the fact that the realisation of such limitation is essential for momentous discoveries. Science can't insist upon the claim of its having *a priori* knowledge of nature. This consciousness of limitation enables a scientist to have philosophical insights.

An Appraisal of Heisenberg's Ideas: Heisenberg's earnest conviction that there is a 'central order' in the universe helps us realize the need to take efforts to maintain that order in our ordinary life events too. People intuitively know this order. It doesn't need any great learning, reflections or talents to realize it. If a person is genuinely convinced of this order, he or she will be inspired and interested to work towards the orderliness.

Often religionists are afraid of science and scientists seem to reject religion and spirituality as meaningless. It is mostly because both don't seem to understand the basic differences between the languages of both the realms. Both the scientific as well as the religious languages have their own limitations and we can't expect more than what they can actually communicate, rather try to communicate. Scientific language is also only an approximate picture of the reality. It can't make lofty claims of representing the reality as it is. The normal demands for accuracy and precision in the scientific language don't hold water at the realm of atomic and sub-atomic levels. The language of religion has to necessarily have some elements of poetry and metaphors. This awareness is very crucial on both the sides to avoid unwarranted conflicts between science and religion.

The inherent limits in the possibility of accurate knowledge, the complementarity principle and quantum theory in general have shown that the objective world of space and time don't exist at the sub-atomic spheres. The long-cherished mathematical symbols of theoretical physics are projected to be possibilities and probabilities, and no more as facts. But the determinists, like Einstein, could not digest this approach. Einstein was ready to accept quantum theory to the extent as a passing stage. 'God does not throw dice' was, as stated earlier, his oft-repeated conviction, which he thought could not be challenged by anybody. To such a claim Bohr could only reiterate: "Nor is it our business to prescribe to God how He should run the world".[127]

Implications of Their Inclinations

Of all many significant implications of their inclinations let us highlight one particular aspect; i.e. the relation between science and religion. Up and until the recent times, the relation between science and religion has not been very healthy. Both have been at loggerheads. However the recent progress in both the fields of science and religion has forced them to rethink their own long-cherished positions. More than ever before science and religion have come closer. Gone are the days when it was thought science and religion were two separate watertight compartments. In the post-modern times, we see both of them changing their outlooks. Religion readily accepts that the earth is going round the sun. Religionists, do not seem to have any big difficulty in accommodating the scientific theory of evolution with the notion of religious creation accounts. Similarly, science is also ready to give up many of its long cherished misconceptions about facts and faith. Scientists are becoming more and more aware of the fact that the comprehensive understanding of the reality is possible only through the inter-disciplinary efforts, including human sciences and theology.

In as much as both science and religion deal with the questions of ultimate reality they can't afford to be aloof from each other; they need to collaborate with each other. The demarcation between science and religion become more and more misty and fuzzy, as the contemporary physics takes over the issues that were traditionally treated to be the metaphysical issues. Stephen Hawking, acclaimed as the Second Einstein, demands that the question of the origin of the universe, which had been the concern of religion and metaphysics hitherto, must be left to science. The issues like "when and how the universe began and will end; whether the universe needed any other cause other than itself; what the ultimate prime matter and first principle are; what the relationship between mind and matter is; whether the order of the universe is the result of mere chance; and whether realities beyond our testability are scientifically plausible"[128] - are no more in the purview of strict metaphysics but in the realm of new physics. Fowler proposes that a genuine interdisciplinary dialogue between science and religion is essential, not an optional. In his own words, "As an alternative to the two-spheres approach to science and theology in which each disciplines are insulated from the other, I believe that science and theology must be engaged in a dialogue of rhythmic interplay... The interdisciplinary is not a static structure defining reality as it is. Rather it is a dynamic matrix which develops out of the interpenetration and interplay among the different disciplines".[129] We need to have dialogue between both the disciplines. If science rejects religion outright as useless then there is no possibility for any dialogue. Similarly, religion must never consider science a dangerous enemy. Religion cannot treat science contemptuously, blaming that the latter is said to be immersed in the mundane reality. Nor is there a need for religion to be afraid of science. It is against the basic search for the ultimate mystery of reality. Such an attitude of fear denies the presence and the power of God, as discovered by science. Ralph Waldo Emerson succinctly puts it, "the religion that fears science, insults God and commits suicide".[130]

Many theologians and the followers of a religion are happy that today's sciences engage themselves in the ultimate questions about reality, existence and nature of God, creation and consummation of the universe etc. that are so far set apart for metaphysics and theological discussions. However, whether they would agree or not with the conclusions that the sciences would come up with regard to those ultimate questions is a different issue. Both science and religion nowadays do realise that they can't speak of the origin and the end of the universe without the reference to each other. Science can't speak of creation without any mention of the Creator. Similarly, religion can't discuss creation in its full dimensions, without any reference to the knowledge of science, as the modern physics insists upon the likelihood of the universe with a temporal beginning and end, not so much of a static and eternal universe. So, "if theology has to take seriously its contention that God is the Creator of the physical universe, then we can no longer (if this were ever indeed possible) speak about God apart from the context of the knowledge of the physical universe gained through empirical investigation".[131]

With regard to theology science too is sometimes at the receiving end. For "a primary task of theology in regard to physics is to warn against reducing the totality of reality to what can be seen, measured, explained with the context of specific theories, or otherwise accounted for on the basis of empirical inquiry alone".[132] Normally one tends to think that only religious doctrines take recourse to intuition and non-rational concepts to convey their ideas. But even a cursory look at the reflections of quantum theories would make it clear to one that science too is in need of more than reason to explain itself. In the word of John Polkinghore, "Quantum theory provides a striking instance of the general fact that the exploration of the physical often yields surprises, so that if we are to do justice to the way things are, we need a release from an undue tyranny of common sense".[133]

It is sad to note that both scientists and theologians are not, very often, sufficiently informed of the contemporary developments in their own, as well as in the other's field of inquiry. The human experiences that they enquire into may differ; their respective methodologies and epistemologies may also be considerably different but the common platform that both share is the effort to understand reality in its totality and the growth of knowledge for the betterment of humanity. The more they know about the other's enterprise the better appreciative they will become of each other. Both must be enthusiastic about the other's worldview; they need to be eager to learn from them; broad-minded in appreciating the other; and finally firm and gentle in correcting the other as and when the need arises. As Mooney points out, "A coherent vision of all things may indeed be the ultimate goal of the dialogue, but scientists as well as theologians will have to settle in the short run for a friendly collaborative sharing and scrutiny of their fallible insights into very different aspects of reality".[134] In the recent times Physics becomes aware of a thread of unifying force that runs behind the multiple manifestations of Reality. They seem to dictate to us that something of everything is in everything, which was the insight of Anaxagoras (500 B.C.) of the ancient Atomists' school. The same affirmation echoes in the non-dualistic vision of the *Upanishads*, according to which everything is Brahman. The finite, says *Katha Upanishad*, has the infinite as the background and the smallest contains the greatest. "The Spirit of God is in everything", according to the Holy Bible (the book of Wisdom 12:1). Therefore the idea of convergence among science and various religious traditions is certainly not artificial.

Present day physicists can't afford to be isolated from the metaphysical undertones and moral implications of their researches. Scientists need to must realize, as we have maintained earlier, that just as "irrationalism" is anti- scientific and is to be rejected, so also is "scientism", which tries to solve the most basic problems of reality, only by means of scientific *weltanschauung*. Scientism makes metaphysics and religions superfluous, to be rejected. In like manner philosophers, who have been enjoying the self-assumed supremacy till date, in the areas of metaphysical issues, can no longer do that. They need to listen to science to update and correct themselves in their knowledge claims. The eyeballs of 20th century of physics, relativity theories and quantum mechanics, have drastically shaken the age-old conceptions of time, space, determinism, causality etc. Way back in 12th century, there were pictures at the great west door of Chartres Cathedral (France) of four eminent scientists – Euclid, Pythagoras, Ptolemy and Aristotle – along with Christ and the saints. It is a welcome sign to inculcate science into religion as both strive to the understanding of the same reality.

Humans always seem to have some sort of faith in some kind of spirit or force. They recognize that something lies beyond the limits of reason. Intuition stretches a hand of help when reason seems to miserably fail. In the words of Blaise Pascal, a French Mathematician, Philosopher and Physicist (1623-1662) *"The heart has its own reasons which the reason does not know"*.[135] Some elements of faith, intuition and non-rationality have always played, and still to play, a role in science. Niels Bohr's famous anecdote brings this element out beautifully: "One of our neighbours in Tisvilde once fixed a horseshoe over the door to his house. When a mutual acquaintance asked him, 'But are you really superstitious? Oh, you honestly believe that this horseshoe will bring you luck?' he replied, 'Of course not; but they say it helps even if you don't believe in it".[136] Science and Religion are both gigantic forces in themselves. Both can thrive well without the patronage of the other. Fritjof Capra puts it succinctly: "Science does not need mysticism and mysticism does not need science, but man needs both".[137] *But we the human beings need both.*

Having said all these, we should boldly acknowledge in true humility that there is much more to be explored. No one individual or any one particular group, religious or scientific can never ever claim to have comprehended the whole of truth. At the most they may claim, that too with difficulties, that all what they have achieved is true but never that they have achieved all the truth! It is like this: when I hold a piece of bread in my hand, I can boldly and rightly claim that *all that I have in my hand is bread*, but it is not the same to claim to have *all the bread in my hand*, since it is only a piece from the full *bread*. It is good to remember one again what Newton said we are playing with a few little pebbles on the shore while the immensely vast ocean lies before us unexplored. The quantum physicists with mystical inclinations seem to affirm that reality can never be fully comprehended, nor can it be compartmentalized. All we can have is something of it not the whole of it. The words of William Shakespeare, I believe, express better what I struggle to say, "There are more things in heaven and earth, Horatio, than are dreamt of in your philosophy".[138] Our quantum physicists, I am sure, will not have any problem with this Shakespearean claim.

Hence mysticism and physicists can be, and they are, friends and never foes; not just friends at that, but they are *friends, guides and philosophers to each other*! A reasonable approach to science will not have any problem either, in acknowledging this Shakespearean claim. When society takes reasonableness earnestly, I am confident, it will be highly enriched and enlightened.

Notes & References

1. Stephen Nathanson, 1994, 156.
2. For more details on the classification of values, see: L. Reese, William, *Dictionary of Philosophy and Religion*, 1999, 805-807.
3. David B. Resnik, 1998, 41, italics mine.
4. For instance, see: Bernard D'Espagnat, 1990.
5. As reported in "Mumbai or New Orleans, there's only one villain," (no author), *Deccan Chronicle* (4 September 2005), of Hyderabad edition: We have varied dangerous impacts due to global warming: In Jan 2005, in Boston, record-setting 42.2 inches of snow fall and a two-foot snowfall in Los Angeles; Bombay received 37 inches rainfall in one day, which left 1000 dead and disrupted 20 million people's lives; A hurricane (nick named as Katrina) hit Louisiana and Mississippi, in August 2005 and more than 25,000 people were killed; In July 2005, worst drought took place and wildfires broke out in Spain and Portugal and this left water levels in France at their lowest in 30 years and recently lethal heat killed 20 in Arizona. To save humanity from all such ill effects of global warming we need to cut oil and coal use by 70%. But unfortunately the first world countries, while discussing global warming, they concentrate only upon political and diplomatic aspects, not on what the warming is doing to agriculture, water supplies, plant and animal sciences, public health and weather.
6. See: Anthony O'Hear, Anthony, *An Introduction to the Philosophy of Science*, 1989; and Susan Haack, *Manifesto of a Passionate Moderate*, 1998.

7. Ariyapala Perera *Science and Human Values,* 2005.
8. Elizabeth Anderson, "Feminist Epistemology: An Interpretation and a Defense", in *Hypatia* 10(3), 50-84, 1995; and "Knowledge, Human Interests, and Objectivity in Feminist Epistemology", *Philosophical Topics* 23 (2), 27-57; Mary Tiles, "A Science of Mars or Venus?", *Philosophy* 62: 293-306, 1987; Kristen Intemann, "Science and Values: Are Value Judgments Always Irrelevant to the Justification of Scientific Claims?", *Philosophy of Science*, 68 (Proceedings), S506-S518, 2001.
9. Hilary Putnam (1982) "Beyond Fact / Value Dichotomy", *Critica* 14, 3 -12. Reprinted in *Realism with a Human Face* 1990.
10. Popper, *In Search of a Better World*, 1984, 72.
11. van Frolov and Boris Yudin, *The Ethics of Science: Issues and Controversies*, 352-3.
12. Clifford, 1992, 77.
13. Carolyn Merchant and Marilyn Strathern (eds.), *Nature, Culture and Gender*, 1980.
14. Sandra Harding, "Feminist Science Criticism", in Robert Klee, 1999, 278.
15. Ibid., 279
16. David G. Myers, "The Powers and Perils of Intuition". (See: http://cms.psychologytoday.com/articles/ pto-20021209-000001.html)
17. "Not Tonight, Darling", *Times of India* (Sunday Times), Pune Edition, 12 June, 2011, 17.
18. Evelyn Fox Keller, "Feminism and Science", in Richard Boyd et al. (eds.), 1995, 280.
19. Sandra Harding, "Feminist Science Criticism", 290.
20. Ayay Ghosh, "Indian Women Join March for Change in New York", *Indian Currents*, 2000, 32.
21. Ajay Ghosh, "Women Face Torture Across the Globe", *Indian Currents,* 2001, 27.

22. Collins and Trevor Pinch, 1998, 101.
23. Longino, 2002, 8.
24. Restivo, 1990, 120-143.
25. E. D. Klemeke, Robert Hollinger, David Wuss Rudge and David Kline (eds.), 1998, 491.
26. Alfred Dinis, "Moral Responsibility of Scientists". (Online: http:/ /www.facfil.ucp.pt/morals.htm)
27. Augustine Perumalil, *Critical Issues in the Philosophy of Science and Religion,* 2006, 33.
28. Kitcher, 2001, 261.
29. Plazowski, 1995, 183.
30. Arthur Eddington, *The Nature of the Physical World*, 1929, ix
31. Ibid., xii.
32. For instance, his "Explanation, Reduction, and Empiricism," in *Minnesota Studies in the Philosophy of Science*, H. Feigl and G. Maxwell (eds.), 1962, III, 28-97.
33. N. R. Hanson, *Patterns of Discovery,* 1958.
34. Robert Palter, "Philosophic Principles and Scientific Theory", 1956, 111-35.
35. Stephen Toulmin, *The Philosophy of Science*, 1953.
36. For Logical Positivists it would be Tx = f (Ox). They stop with this and don't seem to realize that observation is not a 'pure' observation but a function of some presupposition (s).
37. Shapere, 2000.
38. Shapere, "The Concept of Observation in Science and Philosophy", 1982, 505.
39. Bas van Fraassen, 44.
40. Gary Gutting, 2000, 191-195; Review of *A house Built on Sand: Exposing Postmodernist Myths about Science*, ed. by Noretta Koertge (NY: Oxford University Press, 1998).
41. Deepti Gangavane, 1995, 88.

42. Thomas Nagel, "The Sleep of Reason," in *The New Republic*, Oct 12, 1998, 32-38. (See: http://www.physics.nyu.edu/faculty/sokal/nagel.html).
43. Alan Chalmers, "How to Defend Science against Scepticism: *A Reply to Barry Gower*", 1989, 249-253, 251.
44. Hilary Putnam, "The Place of Facts in a World of Values", 1979, 138. (See: http://www.quotationspage.com/quote/3.html)
46. John Barrow, 1999, viii.
47. Henri Poincare, *Science and Hypothesis* (1905). See: http://en.wikisource.org/wiki/Science_and_Hypothesis
48. A. F. Chalmers, 1990, 64.
49. See: http://www.facfil.ucp.pt/morals.htm
50. Ibid.
51. Paul Feyerabend, *Against Method*, 1975, 179
52. Sandra Harding, 290
53. Bertrand Russell, *History of Western Philosophy*, 1946, 699.
54. John Barrow, 1999, ix.
55. Toulmin, 2001, 80.
56. Dudley Shapere, in Evandro Agazzi and Massimo Pauri (eds.), 2000.
57. Ibid.
58. Einstein included the cosmological constant as a term in his field equations for general relativity because he was dissatisfied that otherwise his equations did not allow, apparently, for a static universe: gravity would cause a universe which was initially at dynamic equilibrium to contract. To counteract this possibility, Einstein added the cosmological constant. However, soon after Einstein developed his static theory, observations by Edwin Hubble indicated that the universe appears to be expanding; this was consistent with a cosmological solution to the original general-relativity equations that had been found by the mathematician Friedman. Einstein later referred to his failure

to predict the expansion of the universe from theory, before it was proven by observation of the cosmological red shift, as the "biggest blunder" of his life. (See; http://en.wikipedia.org/wiki/Cosmological_constant#cite_note-Yale-0).\

59. See: http://thinkexist.com/quotes/j._b._s._haldane/
60. James Ladyman, *Understanding Philosophy of Science*, 2002, 158-9. For more details on the Scientific Realism and Anti-Realism debates one may consult any book on 'Introduction to Philosophy'; [For instance: Edwin H. C. Hung, *The Nature of Science – Problems and Perspectives*, 1997, 210-252.
61. Nancy Cartwright, 1983.
62. Viktor E. Frankl, *Man's Search for Meaning,* 1993, 77-78.
63. Ibid, 102.
64. Ibid, 101.
65. Ibid., 99.
66. Paul Feyerabend, *Against Method*, 1975, 15.
67. See: http://paul-feyerabend.biography.ms/
68. See: http://www.mkgandhi.org/mgmnt.htm#wealth
69. Rudolf Carnap, "The Nature of Theories," in ed. Klemke et al., 330.
70. See: http://www.businessweek.com/reprints.htm.
71. Kuruvilla Pandikattu discusses more than 35 of such coincidences in his interesting work: *The Bliss of Being Human – Science and Religion for Self-Realization,* 2004, 100-198.
72. For this quote and further discussion on these arguments see: Michael T. Ghiselin, T*he Triumph of the Darwinian Method*, 1969; and David L. Hull, *Philosophy of Biological Science*, 1974.
73. William Pollaak D.V.M., "Scientific Method: The Marriage of Intellect and Intuition", in: http://www.astv.hpg.ig.com.br/art-6.html).
74. Shapere, "On the Methods of Science", in Jarret Leplin (ed.), 1995, 14.

75. David G. Myers, "The Powers and Perils of Intuition". (Online: http://cms.psychologytoday.com/articles/ pto-20021209-000001.html)

76. Enver Halilvoiæ, "Feyerabend's Critique of Scientism", in *Enrahonar* 28 (1998), 145-160, p. 149. (Online: www.bib.uab.es/pub/enrahonar/0211402xn28p145.pdf).

77. Ibid.

78. See: http://www.intuition-in-service.org/main/intuition/quotations/science.html#sard

79. David G. Myers, "The Powers and Perils of Intuition". See: http://cms.psychologytoday.com/ articles/pto-20021209-000001.html.

80. Lars Bergstrom, "Some Remarks Concerning Rationality in Science", Risto Hilpinen(ed.), in *Rationality in Science: Studies in the Foundation of Science and Ethics* (Dordrecht: Reidel, 1980). (Online: http://www.philosophy.su.se/texter/remarksonrationality.doc).

81. Phill Ball, "Science History: Setting the Record Straight", *The Hindu* (30 June, 2005), Hyderabad edition.

82. See: http://knol.google.com/k/the-limits-of-science#III(2E)_Limitations_of_exact_science

83. Kuruvilla Pandikattu, 2009, 62 – 79.

84. Bas van Fraassen, "A Defense of the Observational / Theoretical Distinction", 43.

85. Ibid.

86. Superstring theory is a sort of theory which gained popularity in the 1980s. Presently it is the most prominent candidate for *Theory of Everything*. According to this theory, the fundamental units of everything in the world – all particles, forces, even space-time itself, are made up of extremely microscopical strings (about 10-35 meters long), which under high tension vibrate and spin a ten-dimensional super space. Six of the ten dimensions are said to be curled into tiny circles and thus rendered

119. Ibid., 109
120. Albert Einstein, *Ideas and Opinion,* 1954, 29.
122. Werner Heisenberg, *Physics and Beyond,* 1971, 214.
123. Ibid., 213.
124. Ibid., 92. Italics mine.
125. See: http://jcdverha.home.xs4all.nl/scijokes/2_2.html
126. As quoted by Ken Wilber (ed.), 1984, 54.
127. Werner Heisenberg, 1971, 81.
128. Hyung Sup Choi, "Dimension as a Way of Describing the Idea of Ultimate Reality in Contemporary Physics and Theology", 1996, 50.
129. Dean Fowler, "Quantum Physics and Christian Anthropology", 1980, 209.
130. As quoted by Mark William Worthing, *God, Creation, and Contemporary Physics*, 1996, 4.
131. Ibid., 202.
132. Ibid., 210.
133. John Polkinghore, *Science and Creation,* 1988, 334.
134. Christopher F. Mooney S.J. "Theology and Science: A New Commitment to Dialogue", 1991, 329.
135. See: http://thinkexist.com/quotation/the_heart_has_reasons_that_reason_cannot_know/14408.html
136. As quoted by Werner Heisenberg, 1971, 92.
137. Fritjof Capra, *The Tao of Physics* (New York: Bantam, 1977), 297.
138. See: http://www.enotes.com/shakespeare-quotes/there-more-things-heaven-earth-horatio.

Conclusion

By way of concluding the journey towards reasonableness, first I like to sum up the main arguments that run through this work and then I will make a few parting remarks.

Recap of the Journey towards Reasonableness

I have tried to show that the picture of science given by the proponents of both the rational and the non-rational models of rationality in science wanting. The rationalists' glorious picture of science on the one hand, and the non-rationalists' projection of science as relative, and even irrational, on the other hand are at the two extremes of the spectrum of rationality. Philosophers of science belonging to the schools of Logical Positivism and Historicism generally pitch their tents at these extremes respectively. But we have seen that both these views of science are far from the actual picture of science. Science is in fact neither *the* paradigm of absolute knowledge, nor completely relative, where 'anything can go'. I have tried to show that the reasonable and realistic picture of science that emerges from the projects of the Historical Realists, in the past few decades, captures the social, and at the same time, rational nature of science.

In the **First Chapter** I have made a brief presentation of some of the 20th century theories of scientific rationality. Under the category of the proponents of rational model of science, in *the first section*, I have discussed the views of Hempel and Popper. Hempel insists upon, among many other things, the goal-means relation to explicate the rationality in science. As in any other human activity the rationality of scientific activity is also determined by the goal-means relation. An action is rational which maximizes the possibility of achieving the goal. Popper reasons out that rationality in science depends on method and that method is falsification; falsification, therefore, ensures the rationality in science. In *Section two*, I have focused on the views of two important proponents of non-rational models Kuhn and Feyerabend. Kuhn, argues that science is basically a social enterprise. However the rational character of theory choice in science is characterized by five important characters of theories: *Accuracy, Consistency, Broad Scope, Simplicity* and *Fruitfulness*. Feyerabend argues that there is no unique method in science. The highly overestimated view of science needs to be corrected and even the society needs to be liberated from the clutches of science as once it needed liberation from religions.

In the **Second Chapter** I have made a quick evaluation of the views discussed in the previous Chapter. *Section 1* critiques the proponents of rational model. I have attempted to show, with the help of four prominent critiques, Philip Kitcher, Thomas Kuhn, Robert Nozick and Wesley Salmon, that Hempel's construal of rationality in terms of goal-means relationship is inadequate, both in normal life situations and in science as well. The complex nature of human goals and the multi-faceted nature of human actions tell us that the rich notion of rationality will only be insufficiently captured if rationality is analyzed only in terms of goal-means relationship. In assessing Popper's contributions I have pointed out, among other things, the shortcomings in his theory of Falsificationism, which plays a very crucial role in his construal of the notion of rationality in science. In *Section 2*, I have evaluated the views of the proponents of non-rational model of science. I recognize the valuable role of Kuhn in bringing out the social dimensions of science, which was in fact an eye-opener, not only to philosophers and historians of science, but to most of the scientists themselves. Nevertheless, I have tried to show how certain precautionary steps need to be taken to save the Kuhnian position from slipping into relativism. His ideas on the role of the community in construing rationality help me in arriving at an integrated notion of rationality. Moving on to appraise the views of Feyerabend, I have explained how he demands that society needs to be liberated from the iron clutches of science, as science is wrongly considered to be the ideal of human rationality and progress. While I agree with him that science is not *the* rational enterprise, I disagree with him and maintain that it is not the case that 'anything goes' in science either.

The whole of the **Third Chapter** is devoted to outline my account of rationality in science. In *Section one,* I focus on some of the traditional accounts of rationality. I have briefly analyzed them to show their limitations. Often rationality is taken to be the fundamental existential experience, which may be treated as the gift of the long evolutionary process. But this, in my opinion, would simplify the concept of rationality and one may not be enthused to explore the great potentialities of humans. Often rationality is identified with "logical consistency" or "justification", but the shortcomings of these approaches have been pointed out. Since rationality is fundamentally related with the agent (human being) it is not proper to reduce rationality merely to logical consistency or justification. Moreover, since there is a deep-rooted interaction among justification, rationality and beliefs, justification is a complex phenomenon. Seeing rationality in terms of "true / false beliefs" has been shown to be faulty because all false beliefs need not lead us to irrational actions. Rationality is sometimes equated with "scientificity". But this rational model of science, however, has been severely criticized during the second half of 20th century. It is also claimed by the proponents of the non-rational model that the scientific changes and the theory choice can't be rationally explained as there are many social, psychological and non-rational elements are at work. Rationality is also seen in terms of "autonomous (a priori) principles", but this need not be an absolute view as there is a possibility of a naturalistic account of rationality, as provided by

Churchland (1979), Hooker (1988), Rouse (1986) and others. Confining the notion of rationality to "goal-orientedness" is also not without problems. As we noted earlier in critiquing Hempel's ideas, the very idea of 'goal' is a multi-dimensional concept, as it involves the agent's desires and aspirations; further, as Kitcher (2000) points out these desires and aspirations don't approve of a notion of a single and all-binding rationality, as human life-situations are infinite in themselves. With the help of all these critical analysis I have tried to argue that the traditional ways of treating rationality are inadequate in capturing the notion of rationality in life (and in science), given the multi-dimensional and complex nature of human experiences (and scientific activities). An integrated view of rationality may not go *against* all these, but needs certainly to go *beyond* them.

I have made it a point to show that it is not that all the aspects of traditional accounts of rationality are useless. We find *inductive reasoning*, a legacy of the past, to be indispensable in life and in science. In spite of the problem with the inductive reasoning we can keep the rational character of science intact. One way to vouch for its rational justification is the "Bayesian theory of reasoning", according to which confirmation of a hypothesis in science depends basically on the ratio between the prior probability and the posterior probability of the hypothesis. A piece of evidence can be confirming or undermining or neutral to the hypothesis under the test. Thus Bayesian approach tries to capture the rational character of inductive reasoning. Of course, this approach is also not without difficulties. Induction may also be seen as a part of our neural wiring. No logical or deductive evidence is needed to support its working. All our knowledge is to a large extent tacit. No logical or cognitive procedures are available that "can unambiguously grasp the meaning dependent upon empathic indwelling of the human subject".[1] The recognition of all such tacit and personal elements does not make knowledge relative but helps us to become more critical and realistic in understanding the rationality in science.

In *Section 2* I have briefly discussed the ideas of some of the philosophers, who bring out the need to revise the traditional understanding of rationality. The alternative accounts to the traditional accounts of rationality proposed by Hilary Putnam ('Integrated View of Rationality'), Stephen Toulmin (denial of over-reliance of rationality), Christopher McMahon ('Collective Rationality'), Stephen Nathanson ('Reasonable form of Rationalism') and Habermas ('Consensus as a Form of Rationality'), capture the notion of rationality in a much more comprehensive manner. Modern science seems to see the need for dialogue with other disciplines. Dialoguing with others is one of the best ways of clarifying and correcting, acquiring and transmitting one's knowledge. As the modern world becomes more interactive and inter-disciplinary this dialogue becomes an essential dimension for science and scientists as well. Science also needs the 'communicative virtues', like tolerance, patience, respect for pluralism, readiness to listen to others[2] and further 'virtues of reasonableness' like being objective, fallible, pragmatic and judicious.[3]

I present, in *Section 3,* my understanding of rationality in science in terms of ***Reasonableness***. Like the notion of rationality, the notion of reasonableness is also not an easy concept to define without any residue. I have briefly mentioned how Confucius has seen this notion and how he has understood its nature and functions. It has helped me to clarify my own understanding of reasonableness. Still an all-embracing definition of reasonableness is not feasible and so I have tried to explicate it in terms of various features. For the sake of clarity and convenience, I classify these features into three flexible categories: *Reasonableness at the Personal Level of the Scientists* – Imagination, the Agent, Judgment and Intuition; *Reasonableness in Scientific Methodology* - Rejection of Zero-tolerance, An Enriched Notion of Objectivity, The Need for Skeptical Attitude, and The Need for Common Sense; and *Reasonableness in Scientific Practice* – Embracing Pragmatism and the Consensus of the Scientific Community. All these features are, though overlapping sometimes, I have argued, so *important and constitutive of science* that it is more meaningful to speak of *'Reasonableness in Science'* rather than 'Rationality in Science'.

It is not only that **Imagination** plays an important role in the understanding of rationality but also, as many authors like Ruth Burne[4] consider, imagination itself to be rational. I have shown that a revised account of rationality (reasonableness) needs to include imagination as one of the important components. The **role of the agent** in science is clearly brought out by Henry Harris. A scientist explores a hypothesis in her head first whether to proceed further or not. [5] When we give due importance to the agent and the context in which she acts, hard and fast rules take the back burner. The ability to make a **judgment** is central to our notion of rationality. Our human capacity and potentialities are so very rich that they can't be limited by rigid rules and regulations. Harold Brown holds an account of rationality which necessarily acknowledges the critical role of judgment.[6] The role of judgment in science is involved at various levels: in choosing a hypothesis for further investigation, as all the hypotheses may not be taken up for the test; in deciding upon the exact pursuit of the chosen hypothesis; in devising testable consequences; and above all in attributing the falsity to any of the hypotheses or the auxiliary hypotheses, as there is no logical and error-free way of doing it. **Intuition** is yet another important element, which has so far been neglected in the domain of science. But more realistic reflections on the historical episodes of science seem to suggest that intuition too has an important role in the scientific enterprise at various levels. Its role becomes apparent when we see that scientists do often have true, but not justified, beliefs! For instance, John Brockman (2006) has recorded the open and sincere confession of more than one hundred scientists and experts from diverse fields of natural and social sciences. Each of them comes out, quite boldly and realistically, even humbly, I would say, to declare what one believes in, even though one cannot prove it. They sometimes arrive at the knowledge of the truth of a proposition without any inference or mediation. They also seem to have some immediate knowledge of some concepts but are not able to define it exactly.[7] Many of the contemporary scientists, especially quantum physicists, seem to be convinced of the role of intuition and mysticism in science.[8] They seem to realize very strongly that science needs to transcend the traditional epistemologies, rationalism or empiricism, or even the combination of both, to make sense reality. It needs some other valuable modes of knowing, like intuition, aesthetic perspectives and so on.[9]

Rejection of Zero-tolerance seems to be a demand as human situations are infinitely complex. The rule-makers could not have, humanly speaking, anticipated all types of possible life-happenings. A situation might arise where human concerns demand deliberate breaking of those rules. A rational mind may hang on to zero-tolerance, but a reasonable one, I think, will certainly appreciate the rejection of zero-tolerance as and when the situations at hand demand. **An Enriched view of Objectivity** is needed to understand the actual practice of science. For Popper objectivity is possible even at the level of community, for "Objectivity of science is not a matter for the individual scientist but rather the social result of mutual criticism".[10] **Skeptical Attitude** is an essential dimension of rationality in general. Rationality in science is no exception. Skeptical attitude is essential in science because of the tentativeness of beliefs. It demands openness to check whether one is deceived by self or by others. **Common Sense** is too important to be overlooked in science. Rationality in science, in my opinion, can't be incompatible with common sense. Science starts with such common sense and goes beyond that. Science never eliminates it, but illuminates it. It is understandable that **Pragmatism** in a general sense and not a particular school of thought (of Dewy, James, or Peirce) is needed in science. It refers to an attitude to tolerate uncertainty, imperfection, and incompleteness at the existential conditions of human thoughts, values, and actions; and yet it demands for persistence and flexibility in confronting such difficulties.

In *Section Four,* I have argued how the feature of ***Consensus in Scientific Community*** plays a very important role in the contemporary science, especially in the domains where the experiments or empirical observations often don't seem to arrive at unambiguous conclusions. Certain experiments need the consensus of the community to reach some sort of meaningful and useful conclusions. With the analysis of gravitational wave experiments, I have tried to explicate how *the role of scientific community begins to have a very prominent role* in construing important concepts in science, for instance, observation. With the help of the neutrino experiments Shapere arrived at three conditions of something to be observed or observable. They are: "a) information is received by appropriate receptor; b) that information is transmitted directly, without any interference, to the receptor from the entity"[11], and c) "The information is transformed by appropriate devices into humanly-accessible information which is (eventually) perceived (and used appropriately as information) by a human being".[12] I have explored the notion of observation with the Gravitational Wave experiments to show that along with those Shapere's three conditions, the contemporary science seems to need *the fourth condition*: d) The consensus of the scientific community is also required to consider something to be observed (observable) or not. Without this consensus very often the researches in modern science are inconclusive and ineffective.

In ***the Fourth and the Final Chapter*** I have shared my thoughts on some of the possible implications of the notion of reasonableness (its nature and functions) for the contemporary science and society. The implications that I have considered here are not exhaustive; they cannot be given in water-tight compartments either. I assign *Section one* for highlighting some of the implications for the contemporary science. Taking reasonableness in science earnestly we get a newer and more comprehensive picture of science that seems to be more realistic and moderate. We are made to feel comfortable with the view of science as relative and mutable. Science is no more seen to be value-neutral, but responsible for all its actions. The choice of the problems to investigate, the methods of such investigations, the possible beneficiaries of these activities – all need to be thoroughly scrutinized. Science is taught how to learn from reality, how to make meaning of the data revealed by nature and how to transcend the traditional outlook of empiricism and rationalism in order to make sense of what is going on behind the natural phenomena. Science is forced to be 'reasonable' in its claims, because, as we have argued, among many other factors, no 'pure' (i.e. 'theory-free') observations are possible; and scientific experiments are no more shown to be fool-proof. The hard-core rationalistic picture of science has been built upon the edifice of such pure observations and infallible experiments and when both of them are shown to be untenable, science is shaken to the core. The need and the relevance of inter-disciplinary approaches, the essential and even existential, need for openness to dialogue with other disciplines, including Philosophy, Religion, and Humanities, are much more easily realized, when science is seen to be 'reasonable' rather than rational. *In Section two,* I have turned my attention to the implications of reasonableness for society. If science has been taught to take reasonableness seriously then society and all the dimensions of human lives must also take reasonableness very seriously. It helps the society to be relieved from the clutches of 'the spirit of scientism' and to give a holistic thrust to education. Society learns to give equal importance to *intellution,* the equal combination of 'intellect' and 'intuition'. Intellectual quotient alone does not determine the quality of human life, but also emotional quotient and emotional intelligence. The parameters to measure 'quality of life' are obviously unambiguous; and that, in my opinion, forces us to be more cautious and credible in our approaches to enhance the quality and dignity of life. With the reasonable approaches, society can more easily see the meaning and the need of integrating science with mysticism and spirituality. Since fundamental questions about the origin of the universe, the evolution of life and consciousness, the destiny of the universe and humanity and many other similar issues always evade the full comprehension of science and intellect, we learn to rely on other disciplines for insights to satisfy our quest.

Parting Remarks

Reasonableness – Only a 'Reasonable' Claim, not an Absolute One

At the end of the investigation into the understanding of rationality in science, I tend to think that the objectives of my journey, laid out in the Introduction, have been fulfilled, at least to an extent. As I begin to end this journey, I realize that *it is only the end of the beginning*. At the culmination of journey I have not arrived at a final and absolute understanding of rationality in science in terms of reasonableness. Though for time being the journey is halted it is not that I have fully grasped the notion of rationality in science. I am not under any illusion that I have arrived at a very clear and final notion of rationality in science. The notion of reasonableness is also not the final answer or explanation to one and for all settle the issue of rationality in science. As it has been pointed out more than once, it is not easy to define reasonableness conclusively. Max Black gives five reasons to show why "the concept of rationality is very elusive".[13] Our Universe is such a multifaceted one that it appears that we need different types of rationality to comprehend it. Michael Polanyi (1983) categorizes various kinds of rationality which include: the purely deductive systems in mathematics, the deductive systems concerned with elementary entities in physics, geometric systems such as crystal physics and organic chemistry, structural-functional systems in biology…, ordering and operational principles, … many complex systems of thought, relation and action in the social sciences, and quite generally, the filling of gaps in the ever-growing edge of the web of science. The debates on the construals of rationality, it appears, will not be (never?) settled in the near future. Perhaps, as Mario Bunge advocates regarding the construction of Philosophy, "let us hang a sign saying 'Building under permanent reconstruction'",[14] one can hang a similar sign for the exercise of reconstruction of rationality as well. I have only tried to show that there is a need to revise the understanding of rationality in science. I have not claimed that I have come up with the final solution to the issue of rationality; if I claim so, that will reveal my *ignorance* more rather than my *arrogance*!

Rationality Itself Never Denied

Generally any theory of rationality has to have *logical consistency* (as the necessary condition and for some as also the sufficient condition), *epistemological truthfulness* and *practicality* as the characteristics of rationality.[15] I have not denied all these; but the problem arises only when one tries to confine the whole notion of rationality, in science or otherwise, only to this straight jacket. Rationality has to go beyond these elements to include a sense of reasonableness and this reasonableness, as Collins has argued, is a social category and not an *apriori* notion derived from, say, a hard core physics.[16] I want to reiterate in unambiguous terms that only the classical ideals of rationality are questioned and not rationality itself. Reason, perhaps, needs to be put into use in a reasonable way, which would be in harmony with the ethical and aesthetic, social and cultural, psychological and emotional, rational and religious, human and humane aspects of the larger spectrum of human life. The rational aspects, therefore, need to be complemented by emotional, intuitive and psychological aspects, without which humans would be far from complete. Ruben Thomas argues for a deliberate undermining of Consistency and Rationality, and this sometimes leads to an unexpected solution of a problem, which was not solved by pure logic and rationality: "Most importantly, man is an emotional creature, and it is ridiculous to ignore his capacities and capabilities in that direction… Emotion is not like the appendix, a mere encumbrance, but manifestly directs much of our lives".[17] The rationality in science, therefore, can also be enriched if it is complemented by emotional, intuitive, imaginative, common-sensical, psychological and non-rational aspects, without which science would be far from complete.

Lessons from Philosophy of Science

Philosophy of Science has made us realize that to get a clear picture of science we need reasonableness, rather than the notion of rationality. Similarly, it is time that we realized that humans, who claim to be rational, need to go beyond the confines of reason to see themselves as reasonable. Otherwise the picture that emerges of human beings will be one-sided. Philosophy of science has made us realize that the total comprehension of reality is an ever-receding horizon for science. The debate over realistic and anti-realistic approach towards scientific theories is the topic of the day in the circles of philosophy of science. Whether scientific theories reveal reality as it is, or they are only instruments to give a rough picture of the reality outside, is hotly discussed. Karl Popper would say scientific theories are like fishing nets; the size of the fish that is caught is determined by the size of the net and the nodes (holes) in it; similarly the understanding of the reality depends on the kind of experiments we use and the type of theories that we use in them. Nowadays authors, like Bipin Indurkya (1990), speak of scientific theories as metaphors, though even this concept itself has to be taken metaphorically.

At times being rational might even demand inconsistency; one may be rational to act in a particular way, even on the basis of unjustified belief. Similarly rationality can't be equated to truth; and this allows one to act in a perfectly rational way even on the basis of false beliefs. Scientists have many assumptions that affect not only their beliefs and behaviours but also their scientific activities. Of course, Rationality does play a role but that is not everything. The traditional view of rationality of science does really help in science but it is neither sufficient nor necessary for understanding science. According to Harris, it is beneficial to recall the requirements to identify rational scientists are:[18] a) A rational scientist, as a radical empiricist, does not worry about logic but only concerned whether his science yields useful information about the real world; b) he is aware that he is fallible but he trusts the ability of scientific investigation to verify or falsify propositions; c) he chooses his best fruitful hypotheses and takes pains to test them himself; d) he presents the well-tested hypotheses to others, and he is prepared to correct himself if needs be; and e) Thus rationality helps, but not as a prescription, in making discoveries. He concludes that rationality is very helpful in discoveries but it is not mandatory. Similarly, science largely bases itself upon 'observation and experiment'. But it is not that science has always been so. As A. F. Chalmers shows it, by studying the history of the law of refraction, "The antics of the early investigators serve to show that the experimental method that modern science takes for granted did not always exist".[19] So the rational methods that we value very highly today, have not been so indispensable in the history of science.

Human Response Preferable to 'Computer' Response!

As I have been arguing, being rational alone is not enough for a happy, healthy and holistic human life, nor for science, as science is meant for humans, not the other way about! *Being just rational can be compared to a computer-response, whereas being reasonable can be equated to a human-response.* An analogy may elucidate this point. A man and his small boy accompany their guest to the bus stop after their dinner at home. After walking a while the guest suddenly remembers that he has left his pair of spectacles back at home; and searching his pockets, he tells the boy, "I am afraid, I have forgotten my spectacles at home on the dining table. Could you just go and see?" The boy runs home and finds it on the table and comes back fast and reports to them, "Yes uncle, it is there". The guest asks in a wonder, "Why have not you brought it then?" The boy gaspingly replies, "Because you told me just to see whether the spectacles was there or not". This boy's behaviour may be *very rational but not reasonable;* he is exactly answering the guest's question and fulfilling the purpose of the action of his running back home to find out whether the spectacles is there or not; he is behaving like a computer! But a reasonable boy would have certainly carried the spectacles along with him, because he knows the purpose

wrong place, or vice versa, but in both the cases we will be the losers. *Why should we, the most rational (reasonable!) creatures on the face of the earth, be the losers?*

Notes & References

1. Zycinski, 1995, 253.
2. Burbules and Suzane Rice, 1992, 411.
3. Burbules, 1993.
4. Burne, 2005.
5. Harris, 1981.
6. Brown, 1988, 139.
7. Rorty, 1967.
8. For an elaborate exposition on the role of intuition in science, see my paper: Jayard, 2004.
9. Tarnas, "The Great Initiation," in http://www.intuition-in-service.org/main/intuition/quotations/creativity.html#tag.
10. Popper, 1998, 17; as quoted by L. B. Bazhenov, 2002, 22.
11. Shapere, "Rationalism and Empiricism: A New Perspective", 1988, 308.
12. Shapere, "The Concept of Observation in Science and Philosophy", 1982, 509.
13. Black, 1986, 27; for the explanation of those five reasons, see pages 29-31.
14. Bunge, 2001, 225.
15. Sayeed has an elaborate discussion on these characteristics to show how these characteristics are not adequate for a theory of rationality. For rather a comprehensive discussion on the characteristics of rationality, see: Sayeed, 2007.
16. Collins, 1994, 503.
17. Ruben Thomas, "The Problems of Philosophical Consistency and Rationality", 1991. Source: http://rrt.sc3d.org/The_Problems_of_Philosophical_Consistency_and_Rationality

18. Henry Harris "Rationality in Science", in *Scientific Explanation,* ed. A. F. Heath (Oxford: Clarendon Press, 1981), 51.
19. Chalmers, A.F., *Science and Its Fabrication,* 1990, 132.
20. Toulmin, 2001, 80.
21. Dilthey, *Gesammelte Schriften* VII, p. 359, As quoted by Herbert Schnädelbach, "Against Feyerabend", in Gonzalo Mune¢var (ed.), 1991, 433-448, 440.
22. John Preston, *Feyerabend: Philosophy, Science and Society,* 1997, 210.
23. The book of Genesis (Chapters: 1-3) in the Bible narrates about how the first parents, Adam and Eve, disobeyed God and got punished for the same. With this *First Fall,* they, and in and through them, the whole humanity lost the blissful and graceful existence that they enjoyed in the Garden of Eden. Francis Bacon argues that modern science must take us back to that blissful state of life enjoyed by the first parents.
24. Edmund Husserl, 1954, 20-104, as quoted by Doan, 2001, 17.

Bibliography

Agassi, J. (1988). Sir Karl Popper in Retrospect: The Positive Power of Negative Thinking. In *The Philosophical Art of Philosophical Polemics. Selected Reviews and Comments,* 479-501, 497-498.

Anderson, Elizabeth (1995). Feminist Epistemology: An Interpretation and a Defense. In *Hypatia,* 10(3), 50-84; and Knowledge, Human Interests, and Objectivity in Feminist Epistemology. *Philosophical Topics,* 23(2), 27-57.

Ball, Phill (2005). Science History: Setting the Record Straight. *The Hindu,* 30 June,, Hyderabad edition.

Barrow, John D. (1999). *Impossibility – The Limits of Science and the Science of Limits.*

Bergstrom, Lars (1980). Some Remarks Concerning Rationality in Science. Risto Hilpinen (ed.), in *Rationality in Science: Studies in the Foundation of Science and Ethics,* Dordrecht: Reidel. (Online: http://www.philosophy.su.se/texter/remarksonrationality.doc).

Bernays, Paul (1974). Concerning Rationality. In Paul Arthur Schilpp, *The Philosophy of Karl Popper*, 600.

Bohm, David (1980). *Wholeness and the Implicate Order.*

Burbules, Nicholas C. (1993). Rethinking Rationality: The Importance of Being Reasonable. *Philosophy of Education*, Audery Thompson, ed.

Capra, Fritjof (1977). *The Tao of Physics,* New York: Bantam.

Chalmers, A. F. (1992). *What is This Thing Called Science?.*

— (1990). *Science and Its Fabrication.*

Chalmers, David J. (1995). Facing Up to the Problem of Consciousness. *Journal of Consciousness Studies,* 2(3), 200-19. (see: http://consc.net/papers/facing.html).

Conant, James and John Haugeland (ed.) (2000). *The Road Since Structure.*

Dupre, Louis (1987). Mysticism. *Encyclopaedia of Religion*, X, 247.

Eddington, Arthur (1929). *The Nature of the Physical World.*

Einstein, Albert (1949). *The World As I See It,* New York: Philosophical Library.

Feigl, H. and G. Maxwell (eds.) (1962). Explanation, Reduction, and Empiricism. In *Minnesota Studies in the Philosophy of Science*, III, 28-97.

Feyerabend, Paul (1987). Putnam on Incommensurability. In *Farewell to Reason*, 265-72, 266.

— (1998). How to Be a Good Empiricist – A Plea for Tolerance in Matters Epistemological. *Philosophy of Science – The Central Issues*, Marci Curd and J.A. Cover ed., 923.

— (2002). Explanation, Reduction and Empiricism. In *Philosophy of Science – Contemporary Readings*, Yuri Balashov and Alex Rosenberg, eds., 143.

Fleming, Marie (1992). Women's Place in Communicative Reason. In *Women and Reason*, Elizabeth Harvey and Kathleen Okruhlik, ed, 245-62.

Gutting, Gary (2000), Review of *A house Built on Sand: Exposing Postmodernist Myths about Science*, ed. by Noretta Koertge, NY: Oxford University Press, 1998, 191-195.

Hacking, Ian (1983). *Representing and Intervening – Introductory Topics in the Philosophy of Natural Science.*

Halilvoiæ, Enver (1998). Feyerabend's Critique of Scientism. In *Enrahonar* 28, 145-160, p. 149. (Online: www.bib.uab.es/pub/enrahonar/0211402xn28p145.pdf).

Harris, Henry (1981). Rationality in Science. In *Scientific Explanation,* ed. A. F. Heath, Oxford: Clarendon Press, 51.

Hempel, C.G. (1981). Turns in the Evolution of the Problem of Induction. *Synthese*, 46, 389-404.

— (1995). A Brief Guide to the Work of Carl Gustav Hempel. In *Erkenntnis,* 42, 3-7.

— (2000). The Irrelevance of Concept of Truth for the Critical Appraisal of Scientific Claims. *Unpublished manuscript*, as quoted by Robert Nozick, 295.

Holton, Gerald (1978). Subelectrons, Presuppositions, and the Millikan-Ehrenhaft Dispute. In G. Holton, *The Scientific Imagination: Case Studies,* 25-83.

Husserl, Edmund Die (1954). *Krisis der Europaischen Wissenschanften und die transcendentale Phanomenologie*, hrsg. Walter Biemel, Del Hass: Martinus Nijihoff, 20–104; as quoted in Tran van Doan, 2001, 16.

Inge, William Ralph (1899). *Christian Mysticism,* Bampton Lectures.

Intemann, Kristen (2001). Science and Values: Are Value Judgments Always Irrelevant to the Justification of Scientific Claims? *Philcsophy of Science*, 68 (Proceedings), S506-S518.

Kuhn, Thomas (1977). Logic of Discovery or Psychology of Research. In *Essential Tension*, 267-8.

— (1977). Objectivity, Value Judgment, and Theory Choice. In *The Essential Tension.*

— (1983). Commensurability, Comparability, Communicability. In P.D. Asquith and T. Nickles, (ed.), 669-88, 679-80.

— (2000). Commensurability, Comparability, Communicability. In *The Road Since Structure*, 57.

— (2000). Rationality and Theory Choice. In *The Road Since Structure: Philosophical Essays,* 208.

— (2000). The Trouble with the Historical Philosophy of Science. In *The Road Since Structure: Philosophical Essays,* 12.

Laudan, Larry (1989). For Method: Or, Against Feyerabend. In J.R. Brown and J. Mittlestrass (eds.), *An Intimate Relation,* 299-317, 302.

Michael, A.A. and E.W. Morley (1887). The Original Michelson Morley Interferometer Description. In *American Journal of Science,* 34.

Nola, Robert (1987). The Status of Popper's Theory of Scientific Method. *British Journal of Philosophy of Science*, 38, 441-480, 455.

Popper, K. (1963). On the Status of Science and Metaphysics. In *Conjectures and Refutations: The Growth of Scientific Knowledge*, 192.

Ravindra, Ravi (1987). Albert Einstein. *The Encyclopedia of Religion,* 72.

Reese, L. and William (1999). *Dictionary of Philosophy and Religion,* 805-807.

Rouse, Joseph (1987). *Knowledge and Power – Toward a Political Philosophy of Science*, 33.

Taylor, Charles (1982). Rationality. In *Rationality and Relativism,* Martin Hollis and Steven Lukes, eds., 87-105, 105.

Weber, J. (1994). *Observation of the Thermal Fluctuations of Gravitational Wave Detector*, 1966; P. Saulson, *Fundamentals of Interferometric Gravitational Wave Detector,* and *Physics of Gravitational Wave Detection: Resonant and Interferometric Detectors*, 1998.

Index

A

Account 3, 9, 82, 92-93, 145, 149, 152, 156, 174, 181, 191-92, 198, 240, 245-46, 309, 355, 357-58,

Agassi, Joseph 103, 139

Agent 1-2, 9, 12, 81, 88-89, 145, 147, 149-50, 176, 190-91, 193, 196, 204, 276-77, 312, 355-58

Anarchy 54

Anomaly 31, 99

Anthropomorphic teleology 308

Anti-realism 256, 293-96, 348

Anything goes 13, 28, 56-57, 93, 108, 124, 130-31, 241, 282, 355

Astrology 22, 66, 304

Attitudinal change 265, 268-69

B

Balashov, Yuri 74, 77

Barbour, Ian G. 323

Bartley, W.W. 95

Bayes, Thomas 162

Bayesian theory 157, 162-63, 165, 356

Bazhenov, L. B. 97, 106, 138-40, 367

Belief perseverance 198

Bernard, Claude 95

Biased assimilation 197

Big bang 207

Biotechnology 202

Black hole 209-10, 224-25

Brockman, John 358

Brown, Harold 192, 206, 233, 358

C

Capra, Fritjof 310, 343, 352

Carnap, Rudolf 167, 348

Causal theory 119, 134

Causality in Observation 227

Chalmers, Alan 237, 347

Charity 151, 389

Choice theory 171, 199, 277, 282-83, 354-55

Circle, Vienna 333, 393
Communicative virtues 185, 357
Competence reasoning 151
Computer response 365
Concept schema 120-21
Conditions for Rationality 10
Confirmation bias 197
Confucian notion of reasonableness 189
Consistency condition 28, 59-60, 63, 130, 132
Context of discovery 16, 50, 111, 159, 395
  justification 16, 50, 395
Contextual theory of meaning 65
Conventionalism 295
Copernican revolution 53, 100, 112, 287
Cosmic inflation 210
Cosmological constant 207, 307, 347
Crisis 30-31, 108-10, 155, 268-69, 321-22, 367
Critical thinking 96, 146, 197, 204, 248

D

Daniel Dennett 126
Dark energy 291
  matter 291-92
Darwin, Charles 81, 308
Deductive reasoning 152-54, 156-57, 183
  support 160, 165
Denial of induction 17
Dennett, Daniel 151
Descriptionism 295
Dhurandhar, S. V. 250, 251
Disciplinary matrix 36-37, 112-13
Doan, Tran van 245, 350

E

Ecological concern 201, 275
Einstein, Albert 327-28, 351-52
Electromagnetic radiation 210, 289
Empirical cognitive psychology 122
Enlightenment 131, 194, 304, 323
Enriched notion of objectivity 146, 190, 195, 357
Epistemic value 12, 171, 256, 262
Epistemological limits 314
Essential tension 29, 35-36, 70-72, 75, 140-42
Existential frustration 300
  vacuum 300
Experimenters' Regress 146, 235-36, 238
Explanatory-normative methodology 12

F

Factors in science 146
Fact-value dichotomy 172, 260-61

Faraday's electrolysis experiment 121
Feminist epistemology 263-64, 345
Fetzer, James H. 137
Feyerabend, Paul 2, 27, 50, 75-77, 123, 144, 303, 347, 348
Ford, Kenneth 238, 253
Frankl, Victor 299-00
Franklin, Alan 238
Functional model 80, 84-86

G

Gangavane, Deepti 183, 245, 346
Gattei, Stefano 69, 96, 103, 105-06, 138-40
Gender issues in science 255, 263
Genetically modified food 202
Goal-institution rationality 91
Ground-based detectors 211, 215, 219-20, 223
Grunbaum, Adolf 104, 139

H

Haack, Susan 344
Habermas, Jurgen 145, 177
Hacking, Ian 75, 134, 140, 144, 148, 239, 252
Hanson, N. R. 346
Hard problem of consciousness 318
Harding, Sandra 265-66, 269, 345, 347
Harre, Rom 101, 114, 139, 141
Harris, Henry 181, 191, 198, 245-46, 357, 368
Heisenberg, Werner 328, 333, 352
Hempel, Carl G. 67, 137
Hidden-hand of reality 181
Historicism 13-14, 29, 353
Holy Bible 157, 243, 246, 342
Hoyningen-Huene, Paul 44, 74, 107, 118, 140-42
Hubble, Edwin 207, 347
Human fullness 255-56, 297-98, 302
response 364-65
Husserl, Edmund 350, 368
Hypothetico-deductive method 3, 19, 159

I

Ideal completeness of explanation 317
Imagination 16, 24, 26, 67, 129, 145, 150, 190-91, 194, 308, 357
Imperialism 52, 312
Incommensurability 28, 30, 33-35, 38-41, 58, 63-65, 71-73, 77, 80, 107-09, 114-20, 122, 133-35, 141-42, 144, 172

Incompleteness of explanation 317

Inductive inference 19-21, 157, 160-61, 165, 167, 170

metaphysics 106

reasoning 1, 3, 17, 19, 21, 79, 81, 87, 96, 105-06, 145-47, 156-60, 162, 164, 166, 167-71, 208, 285, 356

risk 168

Indurkhya, Bipin 69, 159, 168

Instrumentalism 294, 296

Intellution 255, 305-06, 361

Intuition 12, 16, 24, 159, 190, 193-95, 247, 259, 267, 305-10, 313, 328, 341, 343, 345, 348-49, 358, 361, 367

J

Jayard, Stephen 367

Judgment 12, 58, 74, 113, 142, 146, 168, 182, 190, 192-93, 206, 246, 256-57, 261, 357-58

K

Kitcher, Philip 4, 80-81, 137, 150, 182, 276, 354

Klee, Robert 28, 70, 75-76, 140, 345

Klemeke, E. D. 346

Knowability 284

Krauz, Michael 114

Kuhn, Thomas S. 71, 73, 142

Kulkarni, S.G. 142

L

Ladyman, James 31, 33, 71, 348, 403

Lakatos, Imre 67, 70, 72, 74

Laudan, Larry 4, 143

Learning to Learn 289

Lloyd, Elizabeth 196, 248

Logical falsificationism 100

positivism 4, 13, 14, 15, 19, 29, 48, 84, 92-93, 111, 153, 278, 311, 333, 353

Logotherapy 300

Luther, Martin 397

M

Malone, Michael E. 72, 141

Masculinity 203, 265, 266

Maxwell, Grover 123, 139, 252

McMahon, Christopher 176, 244, 356

McMullin, Ernan 76

Meaningful Life 365

Meaningfulness in life 323

Measurability 284

Methodological monists 53-54, 131

Miller, Robert 319

Mind-body relationship 317

Mob psychology 6, 42, 80

Mooney, Christopher F. 352

Musgrave, Alan 161, 243
Mysteries, Scientific 320
Mystical inclinations 327-28, 343

N

Nagel, Ernest 84
Nathanson, Stephen 88, 145, 174-75, 244, 344, 356, 367
Naturalism 5
Newton-Smith, W. H. 77, 138
Nola, Robert 97, 138
Normal science 30-31, 39, 49, 108-09, 112-13, 206, 252
Novum Organum 157
Nozick, Robert 4, 80, 137-38, 354

O

O'Hear, Anthony 322, 344
Objectivism 132, 141
Observational terms 61, 64, 295

P

Pandikattu, Kuruvilla 348-49
Papineau, David 151
Pascal, Blaise 343
Perera, Ariyapala 345
Performance reasoning 151
Personal probability 164
Philosophy of science 3-4, 14-16, 18, 23, 27, 29-30, 50, 54, 69-72, 76-77, 79-80, 84, 87, 102, 106, 108-11, 120, 122, 137-38, 140, 142, 193, 208, 275, 277-78, 287, 293, 311, 323, 344-46, 348, 363
Phlogiston 31, 42, 71, 143
Plausibility consideration 164
Poincare, Henri 347
Polkinghorne, John 324
Pollak, William 306, 308
Popper, Karl 1, 3, 13-14, 50, 69, 75, 94, 98, 138-40, 243, 363
Posterior probability 163, 356
Postmodernism 299
Practical rationality 398
Pragmatism 146, 190, 199-200, 357, 359
Preston, John 51, 76, 131, 144, 368
Prior probability 163-165, 356
Putnam, Hilary 134, 139, 145, 260, 345, 347, 356
Puzzle-solving activity 30, 49

Q

Quantum mechanics 121, 327, 342, 351
  physicists 193, 319, 324, 327, 343, 358
  physics 23, 32, 45, 261, 289, 319, 324, 327, 330, 335, 337, 351-52

R

Rational reconstruction 3, 12-13, 93

Rationalism 5, 14, 95-97, 111

Rationality 1-16, 26-27, 29-30, 33, 45-46, 51-56. 63, 67, 70, 79-97, 106-09, 113-14, 123-29, 131, 135, 137, 140, 143, 145-58, 169, 171-75, 177-93, 195-200, 202-06, 208-09, 233, 239-41, 245-46, 255-56, 263-65, 273-74, 283, 287, 293-96, 298-99, 303, 309, 321, 330, 343, 353-59, 361-64, 367-68

Reason 7, 18-19, 55, 58-59, 82, 88, 95-96, 107, 110, 115, 127, 129-30, 133, 150, 152-53, 157-58, 161, 167, 169-70, 172-73, 176-80, 182-83, 185-86, 190-91, 193, 202-03, 229, 232, 234, 241, 261, 265, 269, 275-76, 281, 283

Reasonable 145, 174-75, 182, 356, 367

Reasonable realism 255, 293, 296-97

Reasonableness 82, 123, 145-46, 150, 172-74, 178, 180-91, 193, 195, 199-01, 233, 239, 241, 245-46, 255-56, 282-83, 293, 303, 309, 313, 344, 353, 357, 360-63, 367

Reductionism 61, 63, 395

Relativism 33-34, 52, 58-59, 72, 80, 107, 114-18, 132, 139, 141, 172, 181, 239, 241, 275, 282, 354

Reliability of inductive reasoning 285

logic and mathematics 285

scientific experiments 286

Religions 114, 322, 325, 328-29, 332, 334-35, 342, 354

Resnik, David 344

Revolutionary science 108, 109, 112, 123

Rorty, Richard 194

Rosenberg, Alex 74, 77

Rouse, Joseph 71

Russell, Bertrand 248, 288, 347

S

Sabbath 246

Salazar, Boris 96, 138

Salmon, Wesley 80, 354

Sankey, Howard 39, 199,

Schilpp, Paul Arthur 69-70, 139, 243

Science 1-19, 22-33, 35, 38-40, 42, 45-63, 65-67, 76, 79-80, 82-87, 89, 91-97, 99-03, 105

Scientific knowledge 21, 48, 57, 69, 85, 97, 106, 109, 113, 139, 157, 237-40, 273, 282-83, 311, 316,

Scientific realism 47, 75, 293, 348

Scientism 255, 273, 309-12, 342, 349
Sense-Perceptual 229
Shapere, Dudley 34, 118, 252, 308, 347
Skeptical attitude 146, 190, 197, 357, 359
Skepticism 21, 26, 114, 170, 182, 197, 198, 241, 282, 283
Space-based detectors 211
Spirituality 256, 282, 310, 321, 333, 338, 350-51, 361, 366
Stone, Mark, A. 112
Strawson, P. F. 160, 242
Sublime order 328-29
Substantial arguments 155-56, 183
Swann, Andrew 104

T

Tarnas, Richard 194
Taylor, Charles 147
Testability 5, 12, 20, 23, 26, 52, 93, 97, 100, 102, 124, 164, 206, 209, 340
Theoretical pluralism 53, 55, 59
    rationality 151
Theory superstrings 314
    ladenness of observation 86, 98, 113
Toulmin, stephen 34, 118, 145, 173, 182, 244, 278, 346, 356, 367
Transcendent riddles 317

U

Underdetermination theories 86
Unrepresentative Realism 296
Urbach, Peter 244

V

Value-neutrality 259-60
van Fraassen, Bas 294, 346, 349
Verification 19, 22-23, 26-27, 83, 101, 129, 153, 163, 172, 247
Verisimilitude 19, 79, 95, 102-03, 106, 262

W

Wave detectors 249-251
    experiments 200, 205, 210, 359-60,
Waves 209-10, 214-15, 222, 233, 250
Weber, Max 92, 138
Wilber, Ken 351-352
Wittgenstein, L. 141

Y

Youngson, Robert 311

Z

Zero-tolerance 146, 190, 195, 357, 358, 359